Phycology: Principles, Processes and Applications

PHYCOLOGY
Principles, Processes and Applications

Editor
Amrik Singh Ahluwalia
Ph.D, F.B.S.
Professor of Botany
Panjab University, Chandigarh - 160 014
(INDIA)

2003

DAYA PUBLISHING HOUSE
Delhi - 110 035

ISBN 81-7035-296-7

Published by : **Daya Publishing House**
1123/74, Deva Ram Park,
Tri Nagar, Delhi - 110 035
Phone: 7103999
Fax: (011) 7199029
e-mail: dayabooks@vsnl.com
website: www.dayabooks.com

Showroom : 4762-63/23, Ansari Road, Darya Ganj,
New Delhi - 110 002
Phone: 3245578, 3244987

Laser Typesetting : **Classic Computer Services**
Delhi - 110 035

Printed at : **Chawla Offset Printers**
Delhi - 110 052

PRINTED IN INDIA

PREFACE

Phycology is the study of algae, a ubiquitous and an important group ecologically because of the dependence of other species on its primary production. Algae is an important tool and wonderful system for the scientists working in the fields of morphogenesis and differentiation, ecology, agronomy, physiology, biochemistry, microbiology, geology, toxicology, aquaculture, molecular biology, oceanography and developmental biology because of their short life span, broadly autotrophic mode of nutrition and fine examples of functional specialization.

A recent spurt in the use of algae as experimental organisms in a variety of disciplines to decipher the secrets of biology has brought the *Phycology* in the center stage. It includes the oldest organisms which appeared on this earth few billion years before, showing a variety of evolutionary lineages.

The aim of this volume is to provide insight and recent progress on different aspects of phycological research. Research articles – reporting results of substantially completed work, and review articles, presenting critical appraisals of specific topics of interest, are included. Yet it may not be a comprehensive treatise on this vast subject. The book starts with a brief life sketch of Dr. Mohinder Singh Randhawa, D.Sc., F.N.I., I.C.S. to whom this volume has been dedicated. A total of 24 chapters contributed by eminent scientists from various parts of the country have been included.

Section I: Floristics, Phylogeny and Ultrastructure (Chapters 1-4), discusses important aspects of phylogeny and principles of classification of green algae, diversity of desmids and blue-green algae and current taxonomic status of *Azolla*. Section II: Physiology, Bio-chemistry and Ecology (Chapters 5-14) presents various processes operative in algae, nitrogen fixation and metabolism in free-living and symbiotic ones, interactions of heavy metals and petroleum oils with various algae and includes other important topics like UV absorbing pigments, algicide production, functional properties of algal proteins, biochemical studies of *Hypnea* and biology of akinetes. Section III: Molecular biology and biotechnology (Chapters 15-16) illustrates the significance of molecular studies in algal biotechnology. Section IV: Applications of algae (Chapters 17-21) discusses several aspects of production and use of various macromolecules for health food, industrial purposes, fish farming, biofertilizers *etc*. Important aspects of methodology in the cultivation and study of algae is discussed in Section V (Chapters 22-24).

Thus in totality, the volume illustrates the principles, processes and applications of *Phycology*. Chapters are presented in a manner that can be understood by readers with basic and applied knowledge. Although efforts have been made to keep the technical contents of the contributions intact but minor alterations have been made to present the matter in a consistent and uniform pattern. However, the onus of technical contents rests with the contributors.

I am indebted to authors for submitting their valuable work for this volume. I am grateful to Professor H.D. Kumar FNA for constant encouragement and guidance during preparation of this volume. I express my gratitude to Drs Inderjit, Manjit Kaur, and Anjuli Pabby for their help in various ways. I owe special thanks to my wife Dr (Mrs.) Kanwaljit Kaur Ahluwalia, and sons, Preetpaul Singh and Navpreet Singh for their emotional support during painstaking work of the volume. I also extend my appreciation to Mr Avtar Singh of N.R. Book Distributors, Mr Anil Mittal of Daya Publishing House and Mrs Daya Goyal for their cooperation during the preparation of this volume.

I trust the information presented in this volume will provide recent insights into *Phycological Research and Teaching*.

June 2002 **Amrik Singh Ahluwalia**

CONTENTS

SECTION IV: APPLICATIONS OF ALGAE

SECTION V: CULTURE METHODS AND TECHNIQUES

LIST OF CONTRIBUTORS

Adhikary, S.P.
P.G. Department of Botany, Utkal University,
Bhubaneshwar- 751 044 (Orissa)

Ahluwalia, Amrik Singh
Phycology Laboratory, Department of Botany,
Panjab University, Chandigarh -160 014

Anand, N.
CAS in Botany, University of Madras,
Guindy Campus, Chennai-600 025 (T.N.)

Angadi, S.B.
Phycology Laboratory, Department of Botany,
Gulbarga University, Gulbarga –585 106 (Karnataka)

Bagchi, Suvendra N.
Department of Biological Sciences, R.D. University,
Jabalpur – 482 001(M.P.)

Behera, B.K.
Division of Biotechnology, Department of Biosciences,
M.D. University, Rohtak-124 001 (Haryana)

Bhagyalakashmi, N.
Department of Plant Cell Biotechnology,
Central Food Technological Research Institute,
Mysore – 570 013 (Karnataka)

Bisen, Parkash S.
Director, Madhav Institute of Technology & Science,
Gwalior – 475 005 (M.P.)

Chauhan, Vinay S.
Institute of Microbiology and Biotechnology, Berkatullah University, Bhopal – 462 026 (M.P.)

Devi, G.A.
Department of Life Sciences, Manipur University, Canchipur, Imphal – 795 003 (Manipur)

Dorycanta, H.
Department of Life Sciences, Manipur University, Canchipur, Imphal – 795 003 (Manipur)

Dua, Saroj
Bio-chemistry Laboratory, Department of Botany, Panjab University, Chandigarh -160 014

Gaur, J.P.
Laboratory of Algal Biology, CAS in Botany, Banaras Hindu University, Varanasi – 221 005 (U.P.)

Gnanasekaran, Selvi
CAS in Botany,University of Madras, Guindy Campus, Chennai-600 025 (T.N.)

Habib, Iqbal
Head, Department of Botany, Govt. P.G. College, Mahoba – 210 427 (U.P.)

Jeeji Bai, N.
Parry Agro Industries Ltd., R & D Centre, Bungalow 6, Carborundum Universal, Chennai – 600 019 (T.N.)

Kannaiyan, S.
Vice-Chancellor, Tamil Nadu Agricultural University, Coimbatore – 641 003 (T.N.)

Kaur, Manjeet
Division of Biotechnology, Department of Biosciences, M.D. University, Rohtak – 124 001 (Haryana)

Kaur, Manjit
Phycology Laboratory, Department of Botany, Panjab University, Chandigarh -160 014

Kaushik, B.D.
Division of Microbiology, Indian Agricultural Research Institute, New Delhi – 110 012

Khosa, Gagandeep Kaur
Phycology Laboratory, Department of Botany,
Panjab University, Chandigarh -160 014

Krishnamurthy, V.
Director, Krishnamurthy Institute of Algology, 19,
Saraswathi Street, Mahalingapuram, Chennai –600 034(T.N.)

Kumar, K.
Azolla Laboratory, Department of Agricultural Microbiology,
Tamil Nadu Agricultural University, Coimbatore – 641 003 (T.N.)

Maheshkumar, D.G.
Phycology Laboratory, Department of Botany,
Gulbarga University, Gulbarga – 585 106 (Karnataka)

Narayan, M.S.
Department of Plant Cell Biotechnology,
Central Food Technological Research Institute,
Mysore – 570 013 (Karnataka)

Pabby, Anjuli
National Centre for Conservation and Utilization of Cyanobacteria,
Indian Agricultural Research Institute, New Delhi – 110 012

Pandey, Pramod K.
Institute of Microbiology and Biotechnology,
Berkatullah University, Bhopal – 462 026 (M.P.)

Pradhan, Subhashree
Laboratory of Algal Biology, CAS in Botany,
Banaras Hindu University, Varanasi – 221 005 (U.P.)

Prasanna, Radha
National Centre for Conservation and Utilization of Cyanobacteria,
Indian Agricultural Research Institute, New Delhi – 110 012

Rai, L.C.
Laboratory of Algal Biology, CAS in Botany,
Banaras Hindu University, Varanasi – 221 005 (U.P.)

Ray, Samit
Department of Botany, Visva Bharati University,
Santiniketan – 731 235 (W.B.)

Ray, Soma
Department of Biological Sciences, R.D. University,
Jabalpur – 482 001(M.P.)

Rengasamy, R.
CAS in Botany, University of Madras, Guindy Campus,
Chennai – 600 025 (T.N.)

Roy, Amarpalli
P.G. Department of Botany, Utkal University,
Bhubaneshwar- 751 044 (Orissa)

Santosh, M.K.
Phycology Laboratory, Department of Botany,
Gulbarga University, Gulbarga – 585 106 (Karnataka)

Saxena, Rishi K.
Institute of Microbiology and Biotechnology,
Berkatullah University, Bhopal – 462 026 (M.P.)

Shrivastava, Rahul
Department of Biological Sciences, R.D. University,
Jabalpur – 482 001(M.P.)

Singh, A.K.
Laboratory of Algal Biology, CAS in Botany,
Banaras Hindu University, Varanasi – 221 005 (U.P.)

Singh, Bhanumati
Institute of Microbiology and Biotechnology,
Berkatullah University, Bhopal – 462 026 (M.P.)

Singh, N. Irabanta
Department of Life Sciences, Manipur University, Canchipur,
Imphal – 795 003 (Manipur)

Singh, P.K.
Director,
National Centre for Conservation and Utilization of Cyanobacteria,
Indian Agricultural Research Institute, New Delhi – 110 012

Singh, Surendra
School of Studies in Botany, Jiwaji University,
Gwalior – 474 011 (M.P.)

Sivakumar, K.
Department of Botany, (DDE Wing), Annamalai University,
Annamalai Nagar – 608 002 (T.N.)

Tyagi, Rashmi
Division of Microbiology, Indian Agricultural Research Institute,
New Delhi – 110 012

Uttam, V.G.
Phycology Laboratory, Department of Botany, Gulbarga University, Gulbarga – 585 106 (Karnataka)

Venkataraman, L.V.
Sudarshana, 236, 8th Cross, Gokulam, 3rd Stage, V.V. Mohalla, Mysore – 570 002 (Karnataka)

MOHINDER SINGH RANDHAWA
A Multifaceted Personality

Mohinder Singh Randhawa (born February 2, 1909 at Zila, Punjab), had his early education at Khalsa High School, Muktsar and got his M.Sc. in Botany from the Government College, Lahore. He was selected for Indian Civil Service (ICS) in 1934 and was allotted to Uttar Pradesh cadre. Sardar Patel during the partition posted him as Deputy Commissioner, Delhi. As Deputy Commissioner, he made all arrangements for the historic event of August 15, 1947, when at the destined hour, Pt Jawaharlal Nehru unfurled the Tricolour from the ramparts of the Red Fort in Delhi and delivered his famous *"Tryst with Destiny"* speech. One can well imagine the kind of responsibility of being the Deputy Commissioner of Delhi during those turbulent days when historical events of global significance were unfolding in the capital of the Republic of India.

Later Dr. Randhawa was appointed Director General, Rehabilitation at Jalandhar (Punjab) where he performed the most daunting task of rehabilitating millions of migrants from Pakistan allotting them land and residences left behind by the evacuees in Punjab.

Dr Randhawa was the founder Chief Commissioner of Chandigarh (Union Territory). During his tenure (1966-1968) he established the much famous Rose Garden and named it as Dr. Zakir Husain Rose Garden. He guided the plantation of flowering trees on various roadsides in the city. Dr. Randhawa has been synonymous with the beautification of Chandigarh. It is only due to his efforts that Chandigarh is popularly known as the *City Beautiful*. In addition, credit for all the artistic activities in the city since 1966 goes to him, as he struggled hard to develop this town into an important art centre and in the process founded *Punjab Lalit Kala Akademi*, *Sangeet Natak Akademi* and *Sahitya Akademi*. Because of his untiring efforts, Chandigarh Museum, Punjab Arts Council and Museum of Cultural Heritage of Punjab at Ludhiana could be built. Even Nek Chand, the creator of Rock Garden had Dr Randhawa's unstinted support and patronage in the early years of this great creation.

Dr Randhawa also served as the Vice-Chancellor of Punjab Agricultural University, Ludhiana. He was considered the soul of the University and is remembered as a great scientist, administrator, educationist, art critic and sport lover. As the Vice-President of Indian Council for Agricultural Research, he guided all the research that later on bloomed as the Green Revolution in Punjab.

As an administrator, Dr Randhawa's acumen was universally recognised though no less important was his contribution as a phycologist or an agricultural scientist or as a connoisseur of art and painting or as a culture scientist and folklorist and as a builder and promoter of institutions.

As a writer some of his books like *Kangra Paintings of the Gita Govinda*, *Basholi Paintings of the Rasmanjari*, *Flowering Trees*, *Beautifying India*, *Beautiful Trees and Gardens*, *A History of Indian Agriculture* (4 volumes totalling about 2157 pages), *Developing Village*

India, Farmers of India (4 volumes in collaboration), four collections of folk songs of Punjab, Kangra, Kullu and Haryana are everlasting.

Dr. M. S. Randhawa, D.Sc., F.N.I., I.C.S. was much more than an administrator. He continued his love for 'Science' in general and algae in particular; all this with a high degree of enthusiasm, employing modest means and facilities arranged with his own private resources. He made some important discoveries particularly in relation to terrestrial algae of the Himalayas and the Indo-Gangetic plains. He collected numerous algae while travelling for his official tours. He normally carried a microscope (now protected in M.S. Randhawa Library at Punjab Agricultural University, Ludhiana) and a box of tubes and empty ink bottles. In the evenings, he used to examine those collections and preserving the interesting ones. He had been utilizing his free time or vacations for making sketches, and clicking some others. He published more than three dozens original papers in leading national and international journals of plant sciences (Annexure 1). He authored a monograph on Zygnemaceae (which was the first of the series of monographs on algae). This monograph is still recognized as a masterpiece and rated very high all over the world. As Vice-President of Indian Council for Agricultural Research during sixties, he encouraged preparation and publication of monographs on various groups of algae – A great service to the field of Phycology, indeed. He was instrumental in organizing an International Seminar on Algology in December 1959 at New Delhi under the joint auspices of ICAR and UNESCO.

Since early days of my research at PAU, Ludhiana, the fascinating personality of Dr. Randhawa inspired me to take up Phycology as my research career. It is my pride to dedicate this volume to this great mentor, whose passion for Phycology will continue to inspire students for several generations to come.

June 2002 **Amrik Singh Ahluwalia**

ANNEXURE 1: PUBLICATIONS OF DR. M.S. RANDHAWA IN PHYCOLOGY

1. 1934a. Current Science. **2**: 55.
2. 1934b. J. Indian Botanical Soc. **13**(1): 11.
3. 1936a. Proc. Indian Acad. Sci. **3**(5): 401.
4. 1936b. Proc. Indian Acad. Sci. **3**(5): 407.
5. 1936c. Proc. Indian Acad. Sci. **4**(1): 36.
6. 1936d. Proc. Indian Acad. Sci. **4**(2) :97.
7. 1936e. Proc. Indian Acad. Sci. **4**(3): 239.
8. 1936f. Proc. Indian Acad. Sci. **4**(3): 246.
9. 1936g. Proc. Indian Acad. Sci. **4**(5): 405.
10. 1936h. Proc. Indian Acad. Sci. **4**(5): 408.
11. 1937a. Proc. Indian Acad. Sci. **5**(6): 297.
12. 1937b. Proc. Indian Acad. Sci. **6**(4): 230.
13. 1938a. Proc. Indian Acad. Sci. **8**(3): 109.
14. 1938b. Proc. Indian Acad. Sci. **8**(4): 336.
15. 1939a. Arch. Protistenk. **92**: 131.
16. 1939b. Hedwigia. **78**: 273.
17. 1939c. Arch. Protistenk. **92**(3): 537.
18. 1940a. J. Indian Bot. Soc. **19**: 247.
19. 1940b. Proc. Indian Acad. Sci. **11**(5):222.
20. 1940c. Current Sci. **9**(1):26.
21. 1940d. Current Sci. **9**(7):326.
22. 1940e. Current Sci. **9**(8):373.
23. 1941a. Current Sci. **9**(10):459.
24. 1941b. Proc. Indian Acad. Sci. **12**(4):129.
25. 1941c. Bot. Gaz. **103**(1):192.
26. 1941d. Trans. Amer. Micros. Soc. **60**(4): 417.
27. 1941e. Current Sci. **10**(6): 292.
28. 1942a. Bot. Gaz. **103** (4): 810.

29. 1942b. J. Indian Bot. Soc. **21**(5-6): 263.
30. 1943a. Proc. Indian Acad. Sci. **18**: 73
31. 1943b. Current Sci. **12** :78
32. 1944. Bot. Gaz. **105** (4): 484.
33. 1946. New Phytol. **45**(2): 278.
34. 1948. Proc. Nat. Inst. Sci. India **14**(8): 367.
35. 1958a. Bot. Gaz. **119**(3): 201.
36. 1958b. Bot. Gaz. **120**(1): 25.
37. 1958c. J. Indian Bot. Soc. **37**(3): 380.
38. 1959a. Current Sci. **28**: 24.
39. 1959b. Zygnemaceae. I.C.A.R., New Delhi. 478 p.

Section I

FLORISTICS, PHYLOGENY AND ULTRASTRUCTURE

Chapter 1

CLASSIFICATION AND PHYLOGENY OF CHLOROPHYTA

Samit Ray

Abstract

Division Chlorophyta has been established as a natural group of green algae and is characterized by the presence of chlorophyll *a* and *b*, starch as storage food formed within the chloroplast, chloroplast envelope of two membranes, lack of chloroplast endoplasmic reticulum and girdle lamellae, cellulose production and in the motile cells a flagellar transition region with unique stellate structure and H-piece. Based on the ultrastructural features, chloroplasts of green algae belong to the simple plastid (chloroplast) category. The chloroplast genome is the only type of plastid genome that has transferred its rbc S gene from the chloroplast to the nucleus. With 500 genera and 8000 species, Chlorophyta is a big division. The members show a wide range of diversification with regard to the thallus organization, structure of vegetative cells, different kinds of mitosis, chemical composition of cell wall and life-cycle patterns. Traditionally, major groups within Chlorophyta were distinguished on the level of thallus organization. With the application of electron microscopy, diversified ultrastructural features, especially of cell wall, flagellar basal body and cell division patterns have been revealed. With these cytological evidences, Chlorophyta was divided into five classes – Micromonadophyceae, Charophyceae. Ulvophyceae, Pleurastrophyceae and Chlorophyceae. Later, combining traditional criteria and the ultrastructural features Chlorophta is divided into eleven classes – Prasinophyceae, Chlorophyceae, Ulvophyceae, Cladophorophyceae, Bryopsidophyceae, Dasycladophyceae, Trentepohliophyceae, Pleurastrophyeae, Klebsormidiophyceae, Zygnematophyceae and Charophyceae. The first radiative evolution of the Chlorophyta is thought to have taken place at the flagellate level resulting in a multitude of ancient lineages of flagellates, some of which then evolved into higher organizational levels.

Key words: Algae, classification, Chlorophyta, phylogeny, ultrastructure.

Introduction

The term "green algae" refers to algal organisms possessing chlorophyll *a* and *b* and often grouped as Chlorophyta, Charophyta and Prasinophyta in a conservative sense and including the Euglenophyta and Prochlorophyta in the broadest sense. However, the division Prochlorophyta has been rejected (Urbach *et. al.*, 1992) and the members are regarded as green cyanobacteria. Euglenoids, with many distinctive features are quite distant from green algae. There is another group Chlorarachniophyta, members of which also have chlorophyll *a* and *b* in the chloroplast but the presence of nucleomorph in the periplastidial compartment indicates their separate origin. Green algae as a natural assemblage, separate from other algae, were first recognized by Harvey (1836) and named Chlorospermeae (the greens). Considering the pigment composition, nature and location of reserve food and the ultrastructural features revealed by electron microscopy (EM), there was general agreement on a natural group of green algae, the Chlorophyta, that shared the following characteristics – presence of chlorophyll *a* and *b*, starch as storage food within the chloroplast, chloroplast envelope of two membranes, lack of CER and girdle lamellae, cellulose production, and in the motile cells a flagellar transition region with unique stellate structure and H-piece. It is very relevant to mention here that structure of chloroplasts is significant in tracing the origin and classifying a particular group of algae.

Structurally, plastids of eukaryotic algae can be grouped into simple plastids i.e. plastids surrounded by two envelope membranes and complex plastids i.e. plastids surrounded by chloroplast endoplasmic reticulum (CER) in addition to the chloroplast membranes. One of the three varieties of simple plastids is chloroplasts (plastids of green algae and embryophytes). They contain chlorophyll *a* and *b* and lack phycobilins. Their thylakoids are stacked and they deposit starch within the stroma. The chloroplast genome is the only type of plastid genome that has transferred its rbc S gene from the chloroplast to the nucleus. Comparisons of genome organization and coding capacity, gene content and arrangement and gene sequences between extant cyanobacteria and plastids provide conclusive evidence that plastids (green chloroplasts) originated from cyanobacteria through a phyletic primary endosymbiosis, of a cyanobacterium into an aplastidial host (Melkonian, 1996). Similarly, ultrastructures of flagella and flagellar basal body have also played a crucial role in dividing the division Chlorophyta into classes. Considering the diversity of this division with regard to the thallus organization, structure of vegetative cells, different kinds of mitosis and cell division, composition of the cell wall and life-cycle patterns, the division could not be divided solely on the basis of ultrastructure of one or two organelles. The primary purpose of the present review is to present a detailed account of the recent findings of morphological, cellular, ultrastructural and molecular features of the green algae. The classification and phylogeny of this group based on these features are discussed, preceded by a historical account of various classifications.

General characters of chlorophyta

Pigments present in the vegetative cells are chlorophyll α and β; ϒ-carotene; neoxanthin, siphonoxanthin, violaxanthin, zeaxanthin, antheraxanthin, siphonein, lutein, lycopene (in most of the green algae), loroxanthin, b-cryptoxanthin and astaxanthin (rarely present). Chlorophyll *c* is present in some members of the class Prasinophyceae. In addition to primary carotenoids that are present during the active growth of the cell, some secondary carotenoids – echinenone, canthaxanthin, and haematochrome (a keto derivative of β-carotene) are found.

Chloroplasts are surrounded only by the double membrane chloroplast envelope, and chloroplast endoplasmic reticulum (CER) is absent. The thylakoids are grouped into bands of two to six thylakoids with grana in some species. Girdle lamellae are absent. Pyrenoids are embedded in the chloroplast and are often penetrated by thylakoids and surrounded by a shell of starch grains. The circular molecules of chloroplast DNA are concentrated in numerous small nucleoids, which are distributed throughout the chloroplast. The DNA is never organized in a single ring like nucleoid.

Starch is formed in the chloroplast in association with a pyrenoid if one is present. In microalgae much of the Rubisco is localized in the pyrenoid. Carbon dioxide concentration is locally elevated within the chloroplastic microenvironment. Bicarbonate from the stroma could diffuse into the pyrenoid and be converted to carbon dioxide by carbonic anhydrase and the starch sheath could act as the diffusion barrier to slow the escape of carbon dioxide (Badger and Price, 1992). The starch is similar to that of higher plants and is composed of amylose and amylopectin.

Most of the green algae have cellulose as the main structural polysaccharide of their cell wall. The siphonaceous algae (caulerpales of Ulvophyceae) have a mannan or xylan. In *Tetraselmis*, a member of Prasinophyceae, organic scales coalesce to form a theca and the main chemical constituent is 3-deoxy-manno-2- octulosonic acid along with D-galacturonic acid. Members of volvocales have walls composed of glycoprotein.

Most of the flagellated cells that show phototactic movement have an eyespot. In the chlorophyta, the eyespot or the stigma is always in the chloroplast, usually in the anterior portion near the flagellar bases. The eyespot consists of one to a number of layers of lipid droplets, usually in the stroma between the chloroplast envelope and the outermost band of thylakoids. The eyespot is usually coloured orange-red from the carotenoids in the lipid droplets. The photoreceptor in *Chlamydomonas* is in the plasmamembrane above the eyespot and consists of a chromophore linked to a protein. The chromophore is 11-cis-retinol, a rhodopsin that functions as a photoreceptor in animals.

Contractile vacuoles are present in vegetative cells of most volvocales. Usually there are two contractile vacuoles at the base of the flagella in biflagellate genera.

Range of thallus structure: The thallus structure in the green algae ranges from unicellular forms to a parenchymatous structure. The main forms of thalli are given below –

Unicellular motile forms – *Chlamydomonas* (ii) Unicellular non-motile forms– *Chlorella*. There are two variations within this category (a) Palmelloid – Loose assemblage of cells in a gelatinous matrix – *Tetraspora* and (b) Coccoid – Small isolated, unicellular spherical cells – *Chlorococcum* (iii) Colonial thallus – This is not a mere assemblage of flagellated cells, but is a coordinated group, behaving as a well organized unit with a definite number of cells arranged in a specific manner known as coenobium. The individual cells are complete in themselves and independent of each other. a) Motile colony – *Volvox*. b) Non-motile colony – *Pediastrum*. (iv) Filamentous forms – A filament results from repeated cell divisions of a non-motile cell in a single plane–*Spirogyra, Ulothrix, Oedogonium, Chaetophora* (filamentous branched form), (v) heterotrichous forms-This is the most advanced type of filamentous construction. Plant body is composed of two portions, a prostrate portion consisting of creeping threads attached to the substratum, and a more or less richly branched erect portion growing from the prostrate portion-*Draparnaldia, Draparnaldiopsis*. (vi) Siphonaceous form – In a number of green algae, considerable enlargement of the plant body may take place without the occurrence of any septation producing a large multinucleate structure known as a coenocyte – *Caulerpa, Protosiphon*. (vii) Siphonocladus form – In this form filaments are with multinucleate long cells – members of Cladophorophyceae (viii) Parenchymatous thalloid forms – It is produced as a result of division of cells in more than one plane – *Ulva*.

Two modes of cytokinesis are: (i) Cell division in which the mitotic spindle disperses after nuclear division and the two daughter nuclei coming close together. A new set of microtubules (phycoplast) arises perpendicular to the former position of the microtubules of the mitotic spindle. The new cell wall is formed along these microtubules (Fig. 1.1c) (ii) Cell division in which the spindle of the nuclear apparatus does not disintegrate but holds the two daughter nuclei apart. New microtubules proliferate so that, by the end of telophase, the spindle microtubules are surounded by many other microtubules running parallel to them. The whole complex of microtubules from the old spindle and the peripheral arrays of new microtubules are termed a phragmoplast (Fig. 1.2b). Numerous vesicles appear within the phragmoplast, derived from the golgi apparatus and form a cell plate of golgi derived vesicles. Vesicles contain cell wall material and coalescence of the vesicles form the septum.

Types of nuclear division: In Chlorophyta, the mitosis may be open or closed. In open mitosis, the nuclear envelope breaks down and disappears (Fig. 1.2). The fully condensed chromosomes have distinct kinetochores. In most of the green algae, except the conjugales, have centric chromosomes with localized centromeres in the median, submedian or terminal position; the conjugales on the other hand has diffused centromere. At metaphase the chromosomes become aligned at the center of the spindle forming a compact metaphase plate. In closed mitosis (Fig. 1.1), the elongated nuclear envelope remains almost intact at metaphase. At the poles, there are polar fenestrae, which are 300-500 nm wide openings in the nuclear envelope, through which the spindle microtubules penetrate into the nucleus.

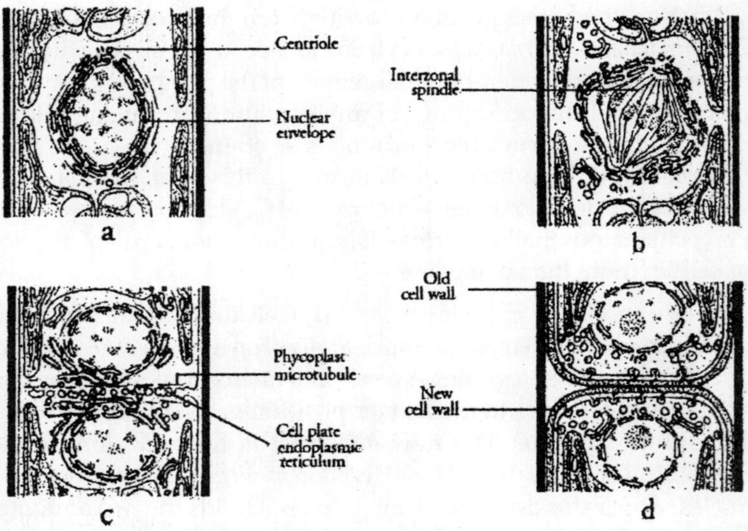

Fig. 1.1: Closed mitosis with non-persistent telophase spindle, cytokinesis by phytoplast.
(a) Early prophase, (b) Metaphase, (c) Late telophase, (d) Early interphase.

There are various combinations of type of mitosis, cytokinesis and spindle types in various members of Chlorophyta. In *Pyramimonas*, it is open mitosis with persistent telophase spindle and cytokinesis is affected by cleavage furrow, while in *Chlamydomonas* mitosis is closed with non-persistent telophase spindle and in cytokinesis, cleavage furrow operates within the phycoplast. In filamentous forms like *Uronema*, mitosis is closed with non-persistent telophase spindle and cell plate is formed by

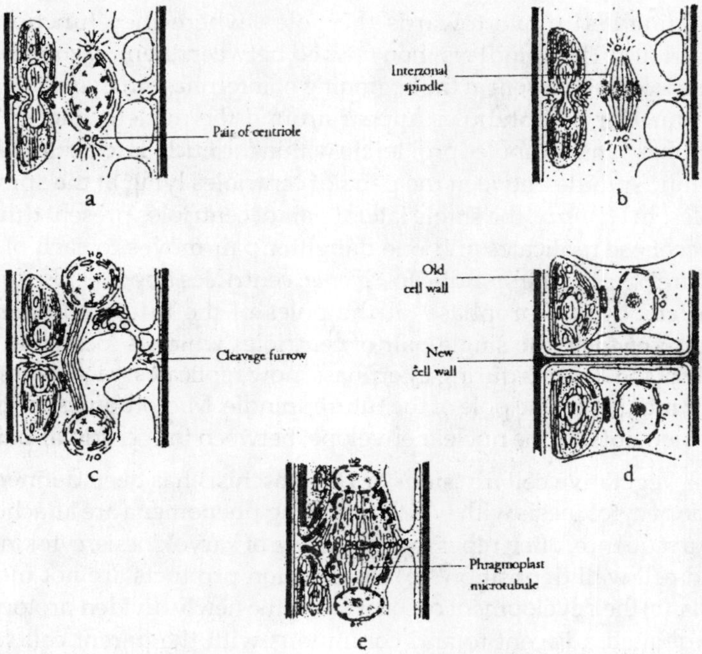

Fig. 1.2: Open mitosis with persistent telophase spindle, cytokinesis by cleavage furrow, cell plate formation with phragmoplast microtubule.
(a) Early prophase, (b) Metaphase, (c) Late telophase, (d) Early interphase, (e) Late telophase.

golgi derived vesicles within a phycoplast during cytokinesis. In *Ulothrix,* mitosis is closed with persistent telophase spindle and cytokinesis is affected by cleavage furrow to which golgi derived vesicles are added. In *Coleochaete,* mitosis is open with persistent spindle and the cell plate is formed of golgi vesicles in association with a phragmoplast.

Centrioles in cell division: In *Pyramimonas,* a unicellular flagellate, the basal bodies function as centrioles. The basal body separates into two sets of four and move towards the future poles. The spindle is formed between these two sets of basal bodies. In *Chlamydomonas* also the basal bodies function as centrioles. During the early prophase, the basal bodies replicate and the two sets separate and move towards the poles, where they function as centrioles. The spindle is then formed between them. Two pairs of centrioles are present at the beginning of interphase in *Cylindrocapsa*. Perinuclear microtubules appear around the nucleus and in late prophase, microtubules proliferate within the nucleus to form a tilted mitotic spindle between the pairs of centrioles lying at the spindle poles. In *Ulothrix*, the single lateral pair of centrioles present during interphase replicates and one daughter pair moves to each of the future poles of the spindle. In *Valonia*, centrioles appear close to the nucleus during prophase, at the poles of the future spindle. In *Klebsormidium*, the single pair of centrioles which is located at the side of the nucleus during interphase, now replicates, each daughter pair moving to one pole of the future spindle. Microtubules are then formed outside the nuclear envelope, between the centriolar pair.

Vegetative cell division- or Desmoschisis has been defined as type of cytokinesis with which following phenomena are attached – (a) a sequence, often repetitive consisting of karyokinesis, cytokinesis and cell wall deposition, (b) the division products are not motile cells, (c) the development of cell walls of the newly divided protoplast is initiated adjacent to and continuous with the parent cell wall, (d) the walls of the young protoplasts remain closely contiguous with the parental cell wall. These characteristics are in marked contrast to eleutheroschisis. Here, the products of cell division are either naked or surrounded by completely new walls that are not intimately related with the parental cell walls. Furthermore, the parental walls are promptly discarded when their enclosed division products (zoospores, aplanospores, hypnospores or gametes) mature and are liberated.

The motile vegetative cells and most zoospores and gametes generally have one, two or four and rarely many anteriorly inserted whiplash flagella of equal length. The flagellate cells are isokont, similar in structure but may differ in length. Flagella lack tubular hairs but may bear delicate hairs and scales. The flagella have the '9+2' structure, characteristic of the flagella of almost all eukaryotic cells. This '9+2' structure contributes in the formation of the flagellar axoneme. Near the base of the flagellum, however, just above where it emerges from the cell, the axoneme passes into the so-called transition region, between the flagellum proper and the basal body. In *Chlamydomonas* the transition region contains structures that appear stellate in cross-section; this stellate pattern is characteristic of Chlorophyta, Bryophyta, Tracheophyta and differs from the structures found in the transition region in other algal divisions. Inside the cell each axoneme ends in basal body. This is composed of nine interconnected triplets of microtubules, the inner two of which are continuous with the doublet microtubules of the flagellar axoneme. In the lowermost part of the basal body, the triplets are connected to a central cylinder by delicate spokes, the whole structure presenting a cart-wheel pattern when sectioned transversely. This pattern is not restricted to Chlorophyta. The flagella are anchored in the cell by four microtubular roots and fibrous root (rhizoplast). The microtubular roots are arranged in a cross if the cell is viewed from the anterior end. Two of these four cruciately arranged microtubular roots contain two microtubules, while the other two contain four; the two stranded roots lie opposite to each other (Fig. 1.3b). The roots diverge from the basal bodies and run beneath the plasmalemma towards the posterior of the cell. The cruciately arranged microtubular roots have thus an X-2-X-2 arrangement. This notion refers to the fact that two of the microtubular roots are usually composed of two microtubules, whereas two other roots can have different numbers of microtubules in different organisms. Thus *Chlamydomonas moewusii* has a 4-2-4-2 arrangement, whereas motile cells of *Ulothrix* sp. have a 5-2-5-2 arrangement. Rhizoplasts (fibrous roots) run from the basal bodies posteriorly toward the nucleus. A rhizoplast is usually a cylinder containing filaments of 5-10 nm diameter, interrupted at approximately 80nm intervals by bands of electron dense material. The cruciate system is neither truly radially symmetrical, nor even bilaterally symmetrical; it can be symmetrically divided in only one plane and this yields identical, not mirror image, halves that will superimpose if one of the halves is rotated by 180°. Floyd *et. al.* (1980) have previously discussed this arrangement and have called it 180° rotational symmetry. A hypothetical "ancestral configuration" has also been included, where the basal bodies are in line with each other (12-6 o'clock configuration, Fig. 1.4a). There are two variations in the green algae with regard to cruciate arrangement and 180° rotational symmetry. There may be clockwise rotation (1-7 o'clock configuration, Fig. 1.4c) and anti-clockwise rotation (11-5 o'clock configuration, Fig. 1.4b) with respect to ancestral arrangement. O'Kelly and Floyd (1983, 1984) have recommended that the terms "counter-clockwise absolute orientation" and "clockwise absolute orientation: replace the 11-5 o'clock and 1-7 o'clock descriptions respectively. In Charophyceae the flagella are anchored in the cell by one broad unilateral band of many microtubules, which runs from the flagellar bases along the side of the cell towards the posterior (Fig. 1.3a). Near the flagellar bases the microtubular band forms part of a multilayered structure (MLS). In addition to the microtubules, the MLS contains two laminate plates, each composed of a stock of differently oriented platelets. The two basal bodies are linked by a transversely striated connective. Melkonian recognizes two chief types of fibrous structures (1) those which connect different

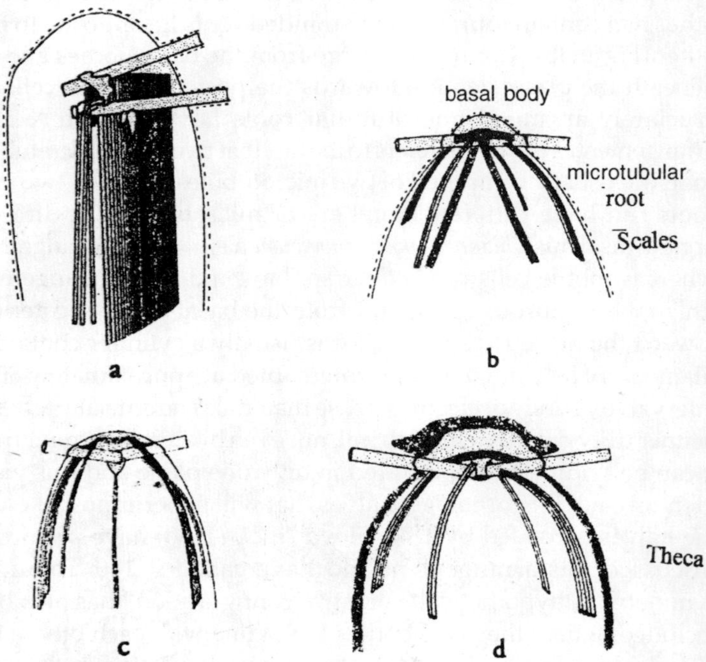

Fig. 1.3: Schematic representation of various arrangements of flagellar roots of different classes of Chlorophyta.
(a) Motile cells covered with scales, a thick band of microtubules, (b) Motile cells covered with scales and flagellar roots arranged in a cruciate pattern, (c) Cruciate roots with no cell covering, (d) Cruciate arrangement of roots and presence of a theca.

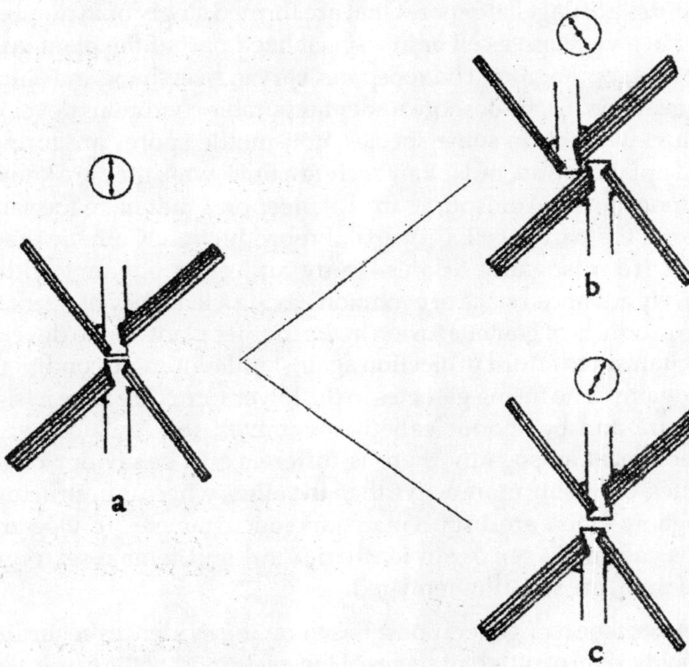

Fig. 1.4: Cruciate arrangements of microtubular roots in the flagellar apparatus of green algae.
(a) Hypothetical ancestral type of flagellar apparatus (12 o' clock- 6 o'clock), (b) Cruciate arrangement of 11 o'clock – 5 o'clock type, (c) Cruciate arrangement of 1 o'clock – 7 o'clock type.

basal bodies – distal connecting fibres and proximal connecting fibres, (2) those which originate at basal bodies and terminate somewhere else in the cell – (a) striated fibres associated with microtubular roots – system I fibre. These are associated with cruciate microtubular roots and run parallel to the root microtubules, (b) striated fibres made up of a bundle of filaments – system II fibres. These fibres run towards the cell nucleus and in this case very often represent the rhizoplasts of classical phycologists.

Reproduction: The methods of reproduction may be considered under the headings-vegetative, asexual, sexual (i) Vegetative propagation: It involves cell division, fragmentation or akinete formation. (ii) Asexual reproduction: most common method of asexual reproduction is by the formation of Zoospores. The zoospores are flagellate spores that are formed singly or in numbers either in a vegetative cell or in a specialized part of the plant body known as sporangium. The zoospores vary in size, shape and nature of flagellation. Each zoospore under favourable conditions develops into a new plant. In some species, non-motile spores are formed called aplanospores, but if they secrete a thick wall, they are known as hypnospores. Autospores are aplanospores that have the same shape as the parent cell. (iii) Sexual reproduction: Gametic union ranges from isogamy to anisogamy and even oogamy with a relatively advanced stage of evolution having a tendency of retention of the products of gametic fusion in the mother plant and to develop a mechanism to afford protection against unfavourable conditions. (a) Isogamy: The fusing gametes in the lower forms are alike in size, structure and behaviour called isogametes, (b) Anisogamy: In physiological anisogamy there is difference in behaviour of the gametes, one being more active than the other, whereas in structural anisogamy – there are distinct macro and microgametes. (c) Oogamy: In this case male sex organ (antheridium) and female sex organ (oogonium) are well differentiated.

Alternation of generations: Based on the evidences regarding the ploidy level of different stages of life-cycle most of the green algal groups have a haplontic life-cycle where only the nucleus of the zygote is diploid (the life-cycle is haplobiontic, as there is only one free living phase), eg. Chlorophceae, Zygnematophyceae, Klebsormidiophyceae and Charophyceae. In the classes Ulvophyceae, Cladophorophyceae and Trentepohliophyceae, the life-cycle is of diplohaplontic type. The life-cycle of *Ulva lactuca* includes a haploid gametophytic generation and a diploid sporophytic generation and the life-cycle is described as isomorphic diplohaplontic type. Members of Bryopsidophyceae and Dasycladophyceae were described to have a diplontic life-cycle. However, present karyological evidences do not support this.The life-cycle of the members of these two classes are interpreted as haplontic.

The division Chlorophyta contains around 500 genera and approximately 8000 species. These are primarily freshwater, about 10% of the algae being marine and 90% freshwater. Some orders (siphonocladales, dasycladales, codiales) are predominantly marine, whereas others are predominantly freshwater (ulotrichales, coleochaetales) or exclusively freshwater (oedogoniales, conjugales). The division includes many unicellular and colonial planktonic algae but there are also many unicellular or multicellular forms that live as benthos, growing attached to rocks (epilithically) or on other plants (as epiphytes). On rocky sea coasts green algae are particularly abundant and dominant in the upper part of the intertidal zone. Here rocks are often completely covered with green algae, including species of *Ulva, Enteromorpha* and *Ulothrix*. Species of *Ulva* and *Enteromorpha* can also form dense growth on sandy shores and mudflats in protected and still environments forming green carpet over the sediment at low tide. The sandy or muddy bottoms of tropical lagoons often bear impressive growth of species belonging to the genera *Caulerpa, Udotea, Penicillus* and *Halimeda*. Some green algae are aerophilic, living on the tree trunk, soil and rocks e.g., *Trebauxia* (unicellular, coccoid) and *Trentepohlia* (filamentous). Species of these genera are also found as the symbiotic algae within lichens. Green algae can also be found in the mountains on permanent snow, colouring it red.

Classification and phylogeny

Kutzing (1845) recorded a very impressive range of green algae which he split into a whole range of families many of which are recongized entities even today e.g., desmideae, ulotricheae, zygnemeae, ulvaceae, dasycladeae, chareae and changed the name of the group as Chlorophyceae. During whole of the nineteenth century the concept of five clusters of green algae based on thallus organization developed, a protococcoid, a palmelloid, a confervoid (Chlorophycean), a conjugate (Zygnematophycean) and a macroscopic multicellular (Charophycean) cluster. Blackman (1900), building on earlier scattered work, crystalised the concept of motile, coccoid, tetrasporine and stephanokont lines of development. His writings enabled workers to understand the phylogenetic relations between different forms and he came to the conclusion that the different forms are all derived from a *Chlamydomonas* like organisms. The major line of algal morphological advance were elaborated by Pascher (1931) who also elevated the classes to divisions. Classes having some basic similarities are put under a division, which is supposed to represent a separate line of evolution. He also commented perceptively that it was not possible to find connections between Chlorophyceae and other algal groups and also that the charophytes could not be drawn into the Chloropohyta. After Pascher, Fritsch (1935) developed apparently logical and simplified nine order system of classification of green algae. The orders are – Volvocales, Chlorococcales, Ulotrichales, Cladophorales, Chaetophorales, Oedogoniales, Conjugales, Siphonales and Charales. From 1950 onwards, studies of the green algae have been approached from three different angles – one is the rebirth of cytology using electron microscopy, second is the stimulus arising from Margulis's writing on endosymbiosis, and third is the intense study of biochemical/molecular genetic system. The principal feature used for distinguishing the major groups (orders and classes) within the Chlorophyta has traditionally been the level of organization exhibited by the thallus. To illustrate this, Fott's classification (1971) is reproduced, where Chloropohyta is divided into three classes-Chlorophyceae, Conjugatophyceae.(coccoid organizational level and reproduction by conjugation) and Charophyceae. This classification reflects the hypothesis that unicellular flagellates are primitive within the Chlorophyta and they evolved into coccoid and sarcinoid green algae and later into filamentous and siphonous forms. Round (1971) split the division into four classes-Chlorophyceae, Zygnemaphyceae, Oedogoniophyceae and Bryopsidophyceae. From 1960's onwards, the power of the EM was quickly put to use by various researchers in the field of Phycology, most notably by Picket-Heaps (1975). The ability to visualize through TEM photomicrographs, the ultrastructure of cells provided a whole suit of characters that have proved to be highly conserved, evolutionarily and thus invaluable for giving evidence concerning ancient divergences. Cavalier-Smith's Chlorophyta (1981) and Silva's (1982) Chlorophycota (both including Chlorophyceae, Prasinophyceae and Charophyceae) are considered monophyletic. Round (1984) believed that the evidences are not conclusive and therefore, preferred to retain a number of divisions – Chlorophyta, Prasinophyta and Charophyta. A fourth division – Gamophyta (= Zygnematophyta or Conjugatophyta) is proposed for the conjugate green algae by Margulis (1981). Silva's (1980) survey lists the basic entities (classes) variously discussed during the last 150 years and crystallized by Ettl and Komarek (1982) into a flagellate series (Chlamydophyceae), a filamentous series (Chlorophyceae, Ulotrichophyceae, Ulvophyceae), a heteromorphic series

(Codiolophyceae), a conjugate series (Zygnematophyceae), and a siphonaceous series (Bryopsidophyceae). In addition, on an overall vegetative morphology and reproductive basis, a stephanokont group (Oedogoniophyceae) was created.

Picket-Heaps (1975) and Stewart and Mattox (1975) divided the green algae into two classes, Chlorophyceae and Charophyceae, mainly on the basis of cytological features. In addition, modes of cell division and ultrastructure of flagellar apparatus were also used. Charophyceae is characterized by (i) persistent interzonal spindle at telophase (Fig. 1.5b) , (ii) cytokinesis is affected by phragmoplast, (iii) motile cells when present are covered by minute organic scales, (iv) motile cells with flagella attached subapically and having a single band of microtubular roots, (v) cells with enzyme glycolate oxidase. Chloropohyceae is characterized by (i) non-persistent interzonal spindle at telephase (Fig. 1.1c) (ii) cytokinesis is affected by phycoplast formation, (iii) motile cells without scales, (iv) flagella attached apically and having cruciately arranged microtubular roots, (v) cells have glycolate dehydrogenase. Melkonian (1980, 1981) classified the green algae into three classes, relying mainly on ultrastructural characteristics of motile cells, mitosis and cytokinesis – Chlorophyceae, Charophyceae and Ulvophyceae. The name Ulvophyceae was first coined by Stewart and Mattox (1978). Six years later, Mattox and Stewart (1984) gave a latin diagnosis of the Ulvophyceae. This new phyletic line was proposed because a number of green algae could not be assigned to any of the two classes that these authors had conceived earlier i.e., Chlorophyceae and Charophyceae. Initially, the class Ulvophyceae was circumscribed only in negative terms: its members had neither the MLS, which characterizes the Charophycean motile cells, nor the phycoplast that characterizes the Chlorophyceae. Subsequent efforts found features unique for Ulvophyceae; cleavage furrow as a mode of cytokinesis, scales present on the flagellated cells, cruciate microtubular roots showing 180° anticlockwise rotational symmetry, presence of a rhizoplast, glycolate degradation by glycolate dehydrogenase and urea degradation by urease.

In 1984, Mattox and Stewart used two decades of comparative cytological evidences to synthesize a classification scheme for the Chlorophyta based on features of flagellated cells, such as type of cell covering, types of cell division and structure of flagellar

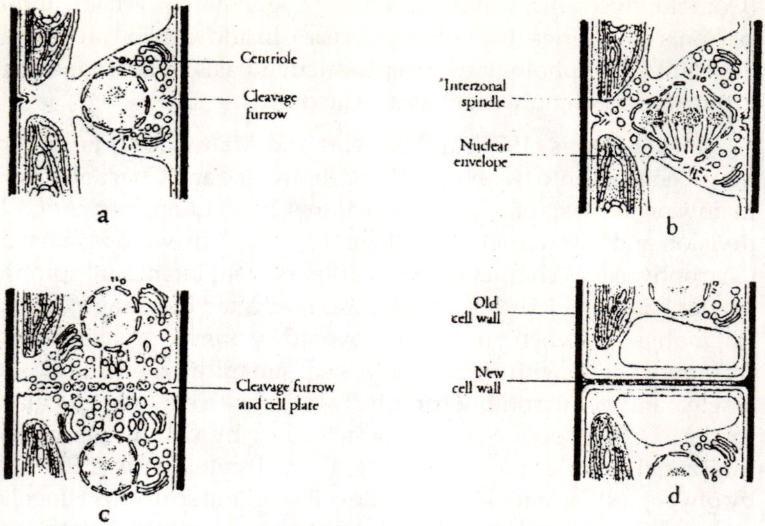

Fig. 1.5: Closed mitosis with persistent telophase spindle, cytokinesis by cleavage furrow. (a) Early prophase, (b) Metaphase, (c) Late telophase, (d) Early interphase

apparatus. They divided Chlorophyta into five classes – Micromonadophyceae, Charophyceae (Chlorokybales, Klebsormidiales, Zygnematales, Coleochaetales, Charales), Ulvophyceae, Pleurastrophyceae (Tetraselmidales, Pleurastrales) and Chlorophyceae (Chlamydomonadales, Volvocales, Chlorococcales, Sphaeropleales, Chlorosarcinales, Chaetophorales, Oedogoniales). Van den Hoek *et. al.* (1995) have evaluated the full range of 'organismal data' (i.e. morphology, ultrastructure, habitat, life-cycles, biochemistry and other non-molecular features) and proposed an altrernative classification of green algae. The evolutionary tree proposed no longer implies a step-wise evolutionary progression through the various organizational levels, in which the flagellated level represents one lineage, the coccoid and sarcinoid levels of organization represent two lineages of intermediate derivation, while the filamentous, siphonocladous and siphonous levels represent three most derived (advanced) lineages. In the new phylogeny, the first radiation of the Chlorophyta is thought to have taken place at the flagellate level, resulting in a multitude of ancient lineages of flagellates, some of which then went on to give rise to non-flagellate coccoid, sarcinoid, filamentous, siphonocladous or siphonous representatives. The more elaborate types of thallus seem to have evolved many times in parallel in the main lineages (Classes). The new system of classification is based on the idea that there are at least four main evolutionary lineages, each characterized by its own type of flagellate cell (zooid). These four types are (i) zooid with organic body scales, (ii) zooids with four microtubular flagellar roots in a cruciate arrangements and a 1 o'clock – 7 o'clock configuration of the basal bodies, (iii) zooids with four microtubular flagellar roots in a cruciate arrangement and an 11 o'clock-5 o'clock configuration of the basal bodies and (iv) zooids with a unilateral arrangement of the flagellar roots. Originally, it was suggested that these four lineages should be recognized as four classes called the Prasinophyceae, Chlorophyceae, Ulvophyceae and Charophyceae respectively. These authors consider that the traits of the flagellated cells are insufficient on their own to characterize classes within the green algae. Other features must also be taken into account, such as the structure of the vegetative cells, mitosis and cell division, the composition of the cell walls and the life-cycle. A comparable spread of different unrelated forms occur among the unbranched filamentous green algae previously classified in the old ulotrichales. *Ulothrix* has zooids with a cruciate 11 o'clock -5 o'clock arrangement of rootlets and mitosis is closed with a persistent telophase spindle and cytokinesis is affected by a cleavage furrow to which golgi derived vesicles as added. *Uronema* has zooids with a 7o'clock configuration and mitosis is closed with non-presistent telophase spindle, cytokinesis is affected by a cell plate of golgi derived vesicle within a phycoplast. *Klebsormidium* on the other hand, has zoospores of the unilateral type, mitosis is open with a prominent persistent telophase spindle and cytokinesis is affected by a cleavage furrow. Under the light microscope the filaments of these three genera look so similar that all have often been identified as *Ulothrix*. Convergent evolution has apparently produced a very similar plant habit with unbranched, parallel sided filaments in three distantly related phylogenetic lineages of the Chlorophyta, namely the class Ulvophyceae (*Ulothrix*). Chlorophyceae (*Uronema*) and Klebsormidiophyceae (*Klebsormidium*). The result of this approach is that the classes in this system are more narrowly circumscribed and probably more 'natural' than in the four class system. Most of the classes Cladophorophyceae, Bryopsidophyceae, Dasycladophyceae, Trentepohliophyceae, Zygnematophyceae, and Charophyceae are so distinctive that they have long been considered natural but they have hitherto been recognized generally as orders rather than classes. The classes under this system are – Prasinophyceae, Chlorophyceae, Ulvophyceae, Cladophorophyceae, Bryopsidophyceae,, Dasycladophyceae, Trentepohliophyceae, Pleurastrophyceae, Klebsormidiophyceae, Zygnematophyceae and Charophyceae. They take the view that classes of green algae should be distinguished by sets of derived characters common to the members of the class.

The structure of the flagellar apparatus and the nature of cell coverings in the swarmers of higher green algae, along with the structural and developmental features of mitosis and cytokinesis in their vegetative cells, provide clear evidence concerning the nature of their ancestral flagellates. A probable explanation for this is that the most fundamental evolutionary changes in general cellular structure occurred among green flagellates rather than after the origin of higher forms. Most of the differences in cell division, the flagellar apparatus, and cell coverings that have recently been used to distinguish between classes of higher green algae also occur in green flagellates. Green flagellates possess further variations not known among higher green algae, so that if one considers only basic cellular structure and development, the flagellates are even more diverse than their descendants. The probable characteristics of ancestral green flagellates (AGF) are as follows – (i) the AGF possessed chlorophyll a *and b*, the H-piece of the flagellar transition region, starch storage in the chloroplast, a chloroplast envelope of two membranes and lack of CER, (ii) a covering of scales, (iii) an interzonal mitotic spindle that persisted during cytokinesis, (iv) AGF was asymmetrical, had only one of each kind of flagellar root (asymmetrical or cruciate pattern) and subapically inserted flagella. The facts, presently known, suggest that one of the flagella of the AGF had two different microtubular roots associated with it. This flagellum and its associated roots become duplicated to give rise to the cruciate root system, (v) one of the microtubular flagellar roots of the AGF was apparently of the type that is called a multilayered structure, (vi) the AGF probably had a pit or at least a groove or depression at the point of flagellar insertion.

The class Prasinophyceae sensu Moestrup et Throndsen (1988) and Moestrup (1991) includes unicellular algae containing chlorophyll *a* and *b* and are separated from other chlorophyll *b* containing algae essentially by the combination of characters such as a unicellular, flagellate life form and the presence of organic scales on the cell body and flagella. Characters observed under light microscope are a flagellar pit, parabasal bodies and muciferous bodies. EM has revealed new characters for this group, such as presence of organic scales of various types on cell body and flagella, flagella (1-8 per cell) arise from a depression, flagellar root system vary from cruciate to unilateral, both open and closed mitosis and cytokinesis by cleavage furrow and cell plate formation within the phycoplast. In some members of Prasinophyceae, there is presence of chlorophyll c like pigment MgDVP. The great diversity in the architecture of the flagellar apparatus and the different modes of mitosis and cytokinesis are interpreted as reflecting the great age and primitiveness of the Prasinophyceae (Nakayama *et. al.*, 1988). Prasinophyceae is a class of green algae that have no derived characters in common with each other, the Prasinophyceae would therefore be an unnatural group, less closely related to each other than they are to members of other classes within the Chlorophyta. They are thought to be the most primitive algae within the Chlorophyta, one argument for this being the great diversity of cellular architecture within the group, which seems to represent a pool from which the other types of green algal zooids have evolved. Another reason is that the presence of body scales is shared with the members of several other classes of green algae, which have scales on the reproductive zooids.

The current results of molecular and ultrastructural analyses recognize two major lineages of green plants. In analyses with multiple representatives of several different outgroups, it is clear that these green plants form a monophyletic group and following Bremer (1985) would be called the subkingdom – Chlorobionta. The two major lineages are represented as two divisions – Streptophyta for the charophycean green algae and land plants, and division – Chlorophyta for the remaining green algae (Bremer 1985).

It is widely held that scaly green flagellates were presumably ancestral to both the Streptophyta and the Chlorophyta. The main rationale for this assumption has been the fact that flagellate reproductive cells (zoospores and gametes) of some taxa in both the phyla are covered by a layer of square shaped scales which also occur as an underlayer in many Prasinophytes but nowhere outside the Viridiplantae. Green plants, the Viridiplantae sensu Cavalier-Smith (1981) and Sluiman (1985) which comprise the green algae and the embryophyte land plants. Phylogenetic analysis involving comparisons of complete nuclear-encoded SSU rRNA sequences of selected Prasinophyte taxa with other green algae have supported this notion, as the scaly green flagellates have invariably occupied basal positions in the phylogenetic trees (Friedl, 1997; Huss and Kranz, 1997).

Characteristics of the classes under chlorophyta

Prasinophyceae

Free living, flagellate green algae; cell body and flagella covered by organic scales produced in golgi apparatus; 1-8 flagella arising from base of a depression; flagellar root system vary from cruciate to unilateral; mitosis open or closed and with or without persistent telophase spindle.

Chlorophyceae

Unicellular or colonial, free living flagellates, non-motile coccoid or palmelloid forms, others are multicellular, a few siphonous; body of the flagellate cells (zooids) surrounded by a glycoprotein envelope; non-flagellate forms have firm polysaccharide wall; flagellate cells have cruciate type of flagellar root system with a 1-7 o'clock configuration; mitosis is closed; telophase spindle non-persistent; cytokinesis by phycoplast, reproduction by the formation of flagellate reproductive cells, mostly haplontic life-cycle.

Ulvophyceae

Contains unicellular, multicellular, siphonocladous non-flagellate green algae; each cell contain one parietal chloroplast, cup shaped with one to several pyrenoids; mitosis is closed and telophase spindle is persistent; cytokinesis by cleavage furrow; reproductive zooids have 2-4 flagella with cruciate 11-5 o'clock configuration; sexual reproduction iso- or oogamous; haplontic life-cycle; mostly marine.

Clapophorophyceae

Thalli with siphonocladous level of organization; cell wall made up of crystalline cellulose; mitosis closed, prominent persistent telophase spindle; each cell multinucleate containing numerous chloroplasts; reproductive zooids have 2-4 flagella with basal bodies having 11-5 o'clock configuration; sexually reproducting species are diplohaplontic and isomorphic.

Bryopsidophyceae

Organization of the thallus is siphonous, a single giant multinucleate cell; main polysaccharides of the cell wall are xylan, mannan and glucan (not in crystalline form); mitosis closed and telophase spindle persistent; cells contain numerous ellipsoidal chloroplasts and amyloplasts; cytoplasm shows streaming; reproductive zooids have 2-4 flagella; flagellar basal bodies have 11-5 o'clock configuratioin; sexually reproducing species have haplontic life-cycle; all forms are marine.

Dasycladophyceae

Thalli are siphonous; principal polysaccharide in the cell wall is a β-1,4 linked mannan; mitosis closed with persistent telophase spindle; numerous ellipsoidal chloroplasts without pyrenoid; streaming of cytoplasm present; reproductive zooids have 2 flagella and cruciate 11-5 o'clock configuration, haplontic life-cycle; all marine.

Pleurastrophyceae

Some are coccoid, some sarcinoid, others branched filamentous; flagellate cells strongly compressed; four microtubular roots in cruciate arrangement; the pair of nuclear centrioles is not involved in the formation of the mitotic spindle; cytokinesis by cleavage furrow.

Klebsormidiophyceae

Three kinds of thallus organization include coccoid, sarcinoid and filamentous; cell wall made up of crystalline cellulose; mitosis is open and telophase spindle persistent; cytokinesis by cleavage furrow and cell plate formation within phragmoplast; cells uninucleate with cup or band shaped chloroplast; flagellate zooids with two flagella, unilateral construction of flagellar apparatus with multilayered structure; haplontic life-cycle; all freshwater.

Zygnematophyceae

Species are either coccoid or filamentous; structural component of cell wall crystalline cellulose; mitosis semiclosed with persistent telophase spindle; cytokinesis by cleavage furrow or cell plate formation with phragmoplast; cells uninucleate; chloroplasts may be spiral or axial plate like; non-flagellate reproductive cells; sexual reproduction by conjugation; life-cycle haplontic; all freshwater.

Trentepohliophyceae

Thalli are filamentous and consist of uninucleate cells, some show heterotrichy; chloroplasts are parietal; cytokinesis is by phragmoplast formation; reproductive zooids produce 2-4 flagella, having four microtubular roots in cruciate arrangement in 11-5 o'clock orientation; life cycle diplohaplontic, isomorphic; grow in subaerial habitat.

Charophyceae

Thalli macroscopic with differentiation into nodes and internodes; there are branches of limited and unlimited growth; young cells uninucleate, older cells may become multinucleate; structural fraction of cell wall is crystalline cellulose, microfibrils arranged in a helicoidal crossed fibrillar pattern; mitosis open with persistent telophase spindle; cytokinesis by cell plate formation within phragmoplast; countless round discoid chloroplasts without pyrenoid, present in the peripheral cytoplasm; flagellate cells are produced which are highly specialized version of the unilateral type of zooids covered with organic scales; all the members are oogamous haplonts; most of the species are freshwater, a few grow in brackish water.

References

Badger, M.R. and Price, G.D. 1992. The CO_2 concentrating mechanism in cyanobacterial and microalgae. *Physiol. Plant.* **84**: 606-615.

Blackmann, F.E. 1900. The primitive algae and the flagellate. An account of modern work bearing on the evolution of algae. *Ann. Bot.* **14**: 647-688.

Bremer, K. 1985. Summary of green plants: phylogeny and classification. *Cladistics* **1**: 369-385.

Cavalier-Smith, T. 1981. Eukaryotic kingdom: Seven or nine. *Biosystems* **14**: 461-481.

Ettl, H. and Komarek, J. 1982. Was besteht man unter dem Begrift "Coccale Grunalgen". *Arch. Hydrobiol.* **60**: 345-374.

Fott, B. 1971. *Algenkunde.* 2nd Ed. VEB Fischer, Jena.

Friedl, T. 1997. The evolution of the green algae. In: *The origin of Algae and Plastids.* Edited by Bhattacharya, D. , Springer Verlag, Wien, pp. 87-101.

Fritsch, F.E. 1935. *The Structure and Reproduction of the Algae.* Vol.I Cambridge University Press, London.

Harvey, W.H. 1836. Algae. In: *Flora Hibernica.* Edited by Mackay, J.T., Willium Curray, Dublin, Vol. 2, Part 3 pp. 157-254.

Huss, V. and Kranz, H.D. 1997. Charophyte evolution and the origin of land plants. In: *Origins of Algae and their Plastids.* Edited by Bhattacharaya, D., Springer Verlag, Wein, pp. 103-114.

Kutzing, F.E. 1845. *Phycologia Germanica,* W.Kohne, Nordhausen.

Margulis, L. 1981. *Symbiosis in cell division. Life and its environment on the Early earth.* Freeman, San Francisco, California.

Mattox, K.R. and Stewrat, K.D. 1984. Classification of the green algae: a concept based on comparative cytology. In: *Systematics of the Green Algae.* Edited by Irvine, D.E.G. and John, D.M., Academic Press, London. pp. 29-72.

Melkonian, M. 1980. Ultrastructural aspects of basal body associated fibrous structures in green algae: A critical review. *Biosystems* **12**: 85-104.

Melkonian, M. 1996. Phylogeny of photosynthetic protists and their plastids. *Verh. Dtsch. Zool. Ges.* **89**: 71-96.

Moestrup, A.E. 1991. Further studies of presumedly primitive green algae, including the description of Pedinophyceae class nov. and Resultor Gen. nov. *J. Phycol.* **27**: 119-133.

Moestrup, A.E. and Throndsen, J. 1988. Light and electron microscopical studies on *Pseudoscourfieldia marina* (Prasinophyceae) with posterior flagella. *Can. J. Bot.* **667**: 1415-1434.

Nakayama, T., Marina, B., Kranz,H.D., Surel, B., Huss, V.A.R., Inouye, I. and Melkonian, M. 1998. The basal position of scaly green flagellates among the green algae (Chlorophyta) is revealed by analyses of nuclear encoded SSU rRNA sequences. *Protists* **144**: 367-380.

Pascher, A. 1931. Systematische ubersicht uber die mit flagellatenin Zusammenhang stehenden Algenreihen und versuch einer Einerihung dieser Algensatamme in die Stamme des Pflenzen eichs, Beith . *Bot. Zentralb.* **248**: 317-332.

Picket-Heaps, J.D. 1975. *Green Algae: Structure, Reproduction and Evolution in Selected Genera*. Sinauer, Sunderland, Massachusetts.

Round, F.E. 1971. The taxonomy of the Chlorophyta II. *Br. Phycol. J.* **6**: 235-264.

Round,F.E. 1984. The systematics of the Chlorophyta: An historical review leading to Modern concepts. In: *Systematics of the Green Algae*. Edited by Irvine, D.E.G. and John, D.M., Academic Press, London, pp. 1-27.

Silva, P.S. 1980. Names of classes and families of living algae. *Regman Veg*. **103**: 1-156.

Silva, P.S. 1982. Chlorophycota. In: *Synopsis and Classification of Living Organisms*. Edited by Parkar, S.P., McGraw-Hill, New York, Vol. I, pp. 133-166.

Sluiman, H.J. 1985. A cladistic evaluation of the lower and higher green plants (Viridiplantae). *Plant Syst. Evol.* **149**: 217-232.

Stewart, K.D. and Mattox, K.R. 1975a . Comparative cytology, evolution and classification of the green algae with some consideration of the origin of other organisms with chlorophylls *a* and *b*. *Bot. Rev.* **41**: 105-135.

Stewart, K.D. and Mattox, K.R. 1975b. Structural evolution in the flagellated cells of green algae and land plants. *Biosystems* **10**: 145-152.

Urbach, E., Robertson, D.L. and Chisholm, S.W. 1992. Multiple evolutionary origin of prochlorophytes within the cyanobacterial radiation. *Nature* 355: 267-269.

Van den Hoek, C., Mann, D.G. and John, H.M. 1995. *Algae- An Introduction to Phycology*. Cambridge University Press, Cambridge. pp. 1-623.

Waters, D.A. and Chapman, R.L. 1996. Molecular phylogenetics and the evolution of green algae and land plants. In: *Cytology, Genetics and Molecular Biology of Algae.* Edited by Chaudhary, B.R. and Agarwal, S.B., Academic Publ., Amsterdam. pp. 337-349.

Chapter 2

CONTRIBUTION TO THE KNOWLEDGE OF DESMIDS OF KUMAUN HIMALAYA

Iqbal Habib

Abstract

The present communication deals with morphotaxonomic enumeration of 40 taxa of desmids collected from different localities situated in and around Ranikhet in Kumaun Himalaya during 1998-99. Notes on forms showing slight variations in their morphological attributes have also been given. All these taxa have been recorded for the first time from this region. The following genera of the desmids were represented- *Closterium* (17), *Pleurotaenium* (1), *Euastrum* (1), *Cosmarium* (19), *Arthrodesmus* (1) and *Staurastrum* (1).

Key words: Desmidiaceae, freshwater algae, Kumaun Himalaya, morphotaxonomy.

Introduction

The region of Kumaun lies between the latitudes of 28°-44′ and 30° -49′ North and longitudes of 78° -45′ and 81° -5′ East along the eastern and southern part of the Central Himalaya. In the north, the natural waterparting separates it from Tibet, while towards the east the river Kali marks its border with Nepal. Towards west lie the Himalayan district of Chamoli and Pauri 'Garhwal within Garhwal division. Favourable tropical climate with mineral rich alluvium affords luxurious growth of diverse algae. The area is interspersed with a network of rivers, rivulets, stagnant water bodies and dam which renders high fertility to the soil. A perusal of the literature shows that this area has not been studied so far and therefore, a survey of this area was undertaken during 1998-99.

Material and methods

The collections were made from different collecting points during 1998-99. Studies were made in the Department of Botany, Govt. P.G. College, Kashipur along with their accession number (DK1 to DK100). In all, taxa of desmids were collected and identified (Fig. 2.1). The algal samples were collected in polythene bottles and preserved in 4% formalin for detailed morpho-taxonomical study.

For the identification of taxa, standard publications (West & West, 1904-12, lyenger & Vimala Bai, 1941, Scott & Prescott, 1961, Croasdale & Gronblad, 1964, Scott *et. al.*, 1965, Croasdale, 1956, 1965, Suxena & Venkateswarlu, 1968a,b, Nurul Islam, 1970, Gronblad & Croasdale, 1971, Agarkar & Agarkar, 1973, 1977, Suxena *et. al.*, 1973, Bharati & Bongale, 1975, Croasdale & Scott, 1976, Prasad & Mehrotra, 1977a,b, Patel and Kumar, 1979, Agarkar *et. al.*, 1979, Prasad *et. al.*, 1982, Pandey & Pandey, 1983, Prasad & Misra, 1985, Isaacs & Hegde, 1986, Pandey & Habib, 1987, Pandey *et. al.*, 1988, Habib *et. al.*, 1989,1998, Habib, 1990a,b, 1993a,b,c, 1994, Habib & Chaturvedi, 1995) were consulted. The numbers in the brackets at the end of description of each taxon, indicate the code numbers of the algal material.

Systematic account of the taxa

***Closterium acerosum* (Schrank) Ehr.**

Patel & Kumar, 1979,p. 112, pl. 1, fig.1.

Cells: 45-58 µm broad, 578-600 µm long, apices 3.5-4.5 µm broad.

Habitat: Planktonic in road side ditches, choubattiya, Ranikhet, October, 1999, (DK-27).

***Cl. acerosum* (Schrank) Ehr. var. *angolense* West et West**

West & West, 1904,p. 149, pl. 18, fig. 6.

Cells: 25-35 µm broad, 500-506 µm long.

Habitat: Planktonic in a small pond near Bus Station, Ranikhet, March, 1998 (DK-28). This is a shorter form than the type in dimensions.

Cl. calosporum **Wittrock (Fig. 13)**

Agarkar *et. al.*, 1979, p. 213, pl. 1, fig. 13.

Cells: 8-11.5 µm broad, 85-96.5 µm long.

Habitat: Planktonic in road side pools, near Govt. P.G. College, Ranikhet, October, 1999 (DK-24). The alga resembles the type but for smaller size.

Cl. corum **Ehr.**

Croasdale & Gronblad, 1964, p. 154, pl. 2, figs. 13 & 14.

Cells: 5-7.8 µm broad, 85-95 µm long.

Habitat: Planktonic in road side pools, Choubattiya road, November, 1998 (DK-32). The present taxon is slightly broader from the type species.

Cl. diane **Ehr.**

West & West, 1904, p. 130, pl. 15, figs. 1-6.

Cells: 30-35 µm broad, 380-385 µm long.

Habitat: Planktonic in a small pond near Kali river,October, 1998 (DK-37). This is a longer form than the type in dimension.

Cl. gracile **Breb.**

Scott & Prescott, 1961, p. 11, pl. 1, fig. 19.

Cells: 4-6 µm broad, 100-125 µm long, apices 1-3µm broad.

Habitat: Planktonic in a small pond near Kali river, May, 1999 (DK-42)

Cl. intermedium **Ralfs.**

Scott & Prescott 1961, p. 11, pl. 1, fig.19.

Cells: 10-14 µm broad, 210-225 µm long, apices 8-10 µm broad.

Habitat: Planktonic in a small pond near Bus Station, May, 1999 (DK-55).

Cl. kuetzingii **Breb. var.** *vittatum* **Nordst. (Fig. 7)**

Isaacs & Hegde, 1986, p. 104

Cells: 10-14 µm broad, 400-418 µm long, apices 14.5 –30.7 µm broad.

Habitat: Planktonic in Kali river, November, 1999 (DK-59). This taxon differs from the type species in having longer dimensions.

Cl. lineatum **Ehr.**

Croasdale & Gronblad 1964, p. 157.

Cells: 30-45 µm broad, 485-546 µm long, apices 3-8 µm broad.

Habitat: Planktonic in a small pond, near Govt. P. G. College, Ranikhet, July, 1998 (DK-3). This species differs from the type species in having larger dimensions.

Cl. lunata **var. *massartii* (Wildem) Krieger**

Scott & Prescott, 1961, p. 12, pl. 1, fig. 29.

Cells: 90-95 µm broad, 680-700 µm long, apices 8-10 µm broad.

Habitat: Planktonic in a small pond near college road, June, 1999 (DK-4). This is a shorter form than the type in dimensions.

Cl. parvulum **var. *cornutum* (Pluyf.) Krieger**

Scott & Prescott, 1961, p. 13, pl. 2,fig.9

Cells: 24-30 µm broad, 120-130 µm long, apices 4-6 µm broad.

Habitat: Planktonic in road side ditch near Cantt. Area, August, 1998 (DK-56).

Cl. porrectum **Nordst. (Fig. 11)**

Scott & Prescott, 1961, p. 12, pl. 2, fig. 6

Cells: 18-25 µm broad, 175-180 µm long, apices 3-6 µm broad.

Habitat: Planktonic in a small pond near Apple Garden, Ranikhet, January, 1998 (DK-72)

Cl. pussillum **Hantzsch**

Suxena & Venkateswarlu, 1968, p. 183, pl. 1, figs. 12 & 13.

Cells: 5-8 µm broad, 25-36.6 µm long.

Habitat: Planktonic in a small pond near T. V. station, March, 1999 (DK-73).

Cl. ralfsii **Breb.**

Patel & Kumar, 1979, p. 118, pl. 2, fig. 6.

Cells: 32-36.5 µm broad, 320-370 µm long, apices 6-8 µm broad.

Habitat: Planktonic in a small pond near Pine forest, March, 1998 (DK-93).

Cl. rectimarginatum **Scott & Prescott**

Scott & Prescott, 1961, pl. 14, figs. 27 & 28.

Cells: 20-25 μm broad, 175-220 μm long, apices 3-4 μm broad.

Habitat: Planktonic in a ditch near Main Bazar, August, 1999 (DK-96).

***Cl. turgidum* Ehr.**

Patel & Kumar, 1979, p. 119, pl. 2, fig. 8.

Cells: 40-55 μm broad, 750-800 μm long, apices 7-9 μm broad.

Habitat: Planktonic in a small pond near Deodar forest, Ranikhet, February, 1998 (DK-9).

***Cl. venus* Kuetz.**

Scott & Prescott, 1961, p. 14, pl. 1, fig. 15.

Cells: 7-9.3 μm broad, 65-83 μm long, apices 1.5-2.5 μm broad.

Habitat: Planktonic in a road side ditches, near Apple Garden, February, 1998 (DK-7).

***Pleurotaenium ehrenbergii* (Breb.) De Bary var. *undulatum* Schaarschm. (Fig. 10)**

Suxena *et. al.*, 1973, p. 323, figs. 25 & 26.

Cells: 20-22 μm broad, 220-245 μm long, apices 12-15 μm broad.

Habitat: Planktonic in river Kali, November, 1999 (DK-17).

***Euastrum evolutum* var. *glaziovii* (Boerg.) West et West (Fig. 5)**

Prasad & Mehrotra, 1977a,p. 344, fig.4.

Cells: 25-27.5 μm broad, 40-45 μm long, isthmus 4-6 μm broad.

Habitat: Planktonic in a small pond near Ranikhet Roadways, March, 1999 (DK-99)

***Cosmarium bioculatum* Breb. (Fig. 1)**

Suxena & Venkateswarlu, 1968, p. 33, figs. 25a,b.

Cells: 14-18 μm broad, 12-15 μm long, isthmus 5-6 μm broad.

Habitat: Planktonic in Kali river, September, 1999 (DK-83).

***C. botrytis* Menegh. var. *gemmiferum* (Breb.) Nordst. (Fig. 9)**

Croasdale, 1965, p. 316, pl. 6, fig.1.

Cells: 57-60 μm broad, 70-75 μm long, isthmus 15-18 μm broad.

Habitat: Planktonic in a small pond, Govt. P.G. College, Ranikhet, July, 1998 (DK-95). The present alga is narrower than the type species.

C. ***conspersum*** **Ralfs var.** ***scottii*** **Croasdale et Scott**

Croasdale & Scott, 1976, p. 531, pl. 9, fig. 11.

Cells: 50-56 µm broad, 60-65 µm long, isthmus 20-23 µm broad.

Habitat: Planktonic in a small pond near Choubattiya Garden, June, 1999 (DK-87).

C. ***difficle*** **Lutkem**

Agarkar & Agarkar, 1977, p. 29, fig. 50.

Cells: 18-20 µm broad, 32-35 µm long, isthmus 6-8 µm broad.

Habitat: Planktonic in a road side ditches near House Riding Ground, August, 1998 (DK-76).

C. ***globosum*** **var.** ***wollei*** **West et West**

Scott & Prescott, 1961, p. 59, pl. 26, figs. 10-12.

Cells: 25-27 µm broad, 30-33 µm long.

Habitat: Planktonic in Kali river, September, 1999 (DK-66). The present desmid resembles the type but for the slightly large diameter of the cells.

C. ***granatum*** **Breb. var.** ***ocellatum*** **West et West. (Fig. 8)**

Prasad & Mehrotra, 1977b, p. 57, pl. 1, fig. 24

Cells: 23-28 µm broad, 40-44 µm long, isthmus 8-10 µm broad.

Habitat: Planktonic in a puddle near Forest Department, May, 1998 (DK-67). The present taxon agrees well with the type in all essential characters but for slightly broader cells.

C. ***impressulum*** **Elfv. (Fig. 6)**

Prasad & Mehrotra, 1977b, p. 58, pl. 2, fig. 61.

Cells: 18-20 µm broad, 20-25 µm long, isthmus 3-5 µm broad.

Habitat: Planktonic in a puddle near T. V. station, June, 1999 (DK-77)

C. ***lundellii*** **Delp. var.** ***corruptum*** **(Turn.) West et West (Fig. 3)**

Scott & Prescott, 1961, p. 61, pl. 5, fig. 9.

Cells: 40-45 µm broad, 50-55 µm long, isthmus 10-15 µm broad.

Habitat: Planktonic in a small pond near Apple Garden, Ranikhet, September, 1998 (DK-79). The present form resembles the type but for smaller size.

C. phaseolus **Breb. var.** ***elevatum*** **Nordst.**

Agarkar & Agarkar, 1973, p. 13, pl. 6, fig. 66

Cells: 20-23 μm broad, 23-25 μm long, isthmus 3-5 μm broad.

Habitat: Planktonic in a puddle, near Ramnagar Road, May, 1999 (DK-71). The present alga agrees well type in all essential characters but for slightly broader cells.

C. platydesmium **(Nordst.) Nordst. et. Schmidle**

Gronbland & Croasdale, 1971, p. 16, pl. 6, fig. 79.

Cells: 40-45 μm broad, 44-47.5 μm long, isthmus 18-23 μm broad.

Habitat: Planktonic in a small pond near Pine forest, Ranikhet, December, 1998 (DK-27).

C. portianum **var.** ***nephroideum*** **Wittrock**

Scott & Prescott, 1961, p. 66, pl. 28, fig. 10.

Cells: 20-25 μm broad, 30-35 μm long, isthmus 10-12 μm broad.

Habitat: Planktonic in a ditch near College Road, August, 1998 (DK-43). The present alga resembles the type but for smaller size.

C. punctulatum **Breb. var. seriatum Krieger (Fig. 4)**

Gronblad & Croasdale, 1971, p. 17, pl. 7, fig. 79

Cells: 18-20 μm broad, 14-17 μm long, isthmus 6-8 μm broad.

Habitat: Planktonic in Kali river, September, 1999 (DK-88).

C. pyramidatum **Breb. var.** ***transitorium*** **Heimerl.**

Bharati & Bongale, 1975, p. 137, pl. 5, fig. 39.

Cells: 38-45 μm broad, 70-90 μm long, isthmus 10-12 μm broad.

Habitat: Planktonic in a small pond near Apple Garden, July, 1999 (DK-53). The present form resembles the type but for smaller size.

C. quadrifarium **var.** ***simplex*** **Krieger**

Scott & Prescott, 1961, p. 67, pl. 30, fig. 11.

Cells: 20-23 μm broad, 22-25 μm long, isthmus 5-6.5 μm broad.

Habitat: Planktonic in a small pond near Govt. P.G. College, Ranikhet, December, 1998 (DK-13).

C. *quinarium* Lund.

Gronblad & Croasdale, 1971, p. 17, fig. 98.

Cells: 25-28 μm broad, 30-32 μm long, isthmus 10-12 μm broad.

Habitat: Planktonic in a small pond near Apple Garden, Ranikhet, August, 1999 (DK-19).

C. *retusum* (Perty) Rebenh.

West & West, 1908, p. 265, pl. 91, figs. 1 & 2.

Cells: 20-22 μm broad, 30-33 μm long, isthmus 4-6 μm wide.

Habitat: Planktonic in a small pond near Deodar forest, September, 1998 (DK-21). The present alga agrees well with the type description but for the slightly broader cells.

C. *sexonatum* Gutw. (Fig. 2)

Prasad & Misra, 1985, p. 344, pl. 1 fig. 8.

Cells: 45-48 μm broad, 50-55 μm long, isthmus 7-9 μm broad.

Habitat: Planktonic in a small pond near Horse Riding Ground, March, 1999 (DK-23).

C. *subauriculatum* var. *bogoriense* Bern.

Gronblnd & Croasdale, 1971, p. 17, pl. 5, fig. 57.

Cells: 45-52 μm broad, 48-55 μm long, isthmus 25-30 μm broad.

Habitat: Planktonic in a small pond near Pine forest, Ranikhet, August, 1998 (DK-33). The present form resembles the type but for the length i.e. cells are longer than the type.

C. *trachyplerum* Lund.

Suxena *et. al.*, 1973. p. 329, figs. 54a-c.

Cells: 23-25 μm broad, 30-34 μm long, isthmus 4-6 μm broad.

Habitat: Planktonic in a small pond near Forest Department, September 1998 (DK-35). The present desmid agrees well in all essential characters with the type but for the slightly larger dimensions.

***Arthrodesmus convergens* Ehr.**

Scott & Prescott, 1961, p. 74, pl. 34, figs. 7-10.

Cells: 20-25 μm broad, 30-32.6 μm long, isthmus 5-7 μm broad.

Habitat: Planktonic in a small pond near Bus Station, Ranikhet, March, 1999 (DK-63). The present desmid agrees well in all essential characters with the type but for the slightly larger cells.

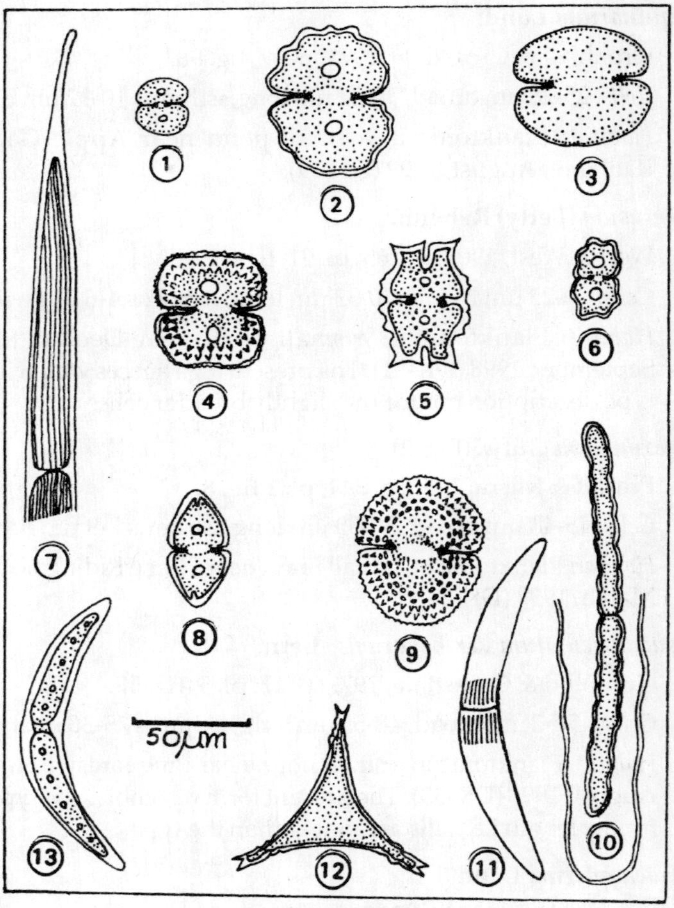

Fig. 2.1

1 : *Cosmarium bicculatum* Breb.
2 : *C. sexonatum* Gutw.
3 : *C. lundellii* Delp. var. *corruptum* (Turn.) West *et* West
4 : *C. punctulatum* Breb. var. *seriatum* Krieger
5 : *Euastrum evolutum* var. *glaziovii* (Boerg.) West *et* West
6 : *Cosmarium impressulum* Elfv.
7 : *Closterium kuetzingii* Breb. *vittatum* Nordst.
8 : *Cosmarium granatum* Breb. var. *ocellatum* West *et* West
9 : *C. botrytis* Menegh. var. *gemmiferum* (Breb.) Nordst.
10 : *Pleurotaenium ehrenbergii* (Breb.)De.Bary var. *undulatum* Scha.
11 : *Closterium porrectum* Nordst.
12 : *Staurastrum sebaldi* Reinsch.
13 : *Closterium calosporum* Wittrock.

Staurastrum sebaldi **Reinsch. (Fig. 12)**

Prasad & Mehrotra, 1977, p. 72, pl. 3, fig. 79.

Cells: 28-32 µm broad, 20-23 µm long, isthmus 3-4 µm broad.

Habitat: Planktonic in a puddle near Main Market, Ranikhet, September, 1999 (DK-47). The shape and structure of chloroplast tallies with the type but for the slightly larger dimensions.

Acknowledgements

The author is thankful to the Principal, Govt. P.G. College, Kashipur for providing laboratory facilities and to Dr. U. K. Chaturvedi, Head, Department of Botany, Bareilly College, Bareilly for his kind co-operation throughout the present investigation.

References

Agarkar, D. S. and Agarkar, M.S. 1973. Contribution to the desmids of M.P., India. *Proc. Acad. Biol.* **12**: 159.

Agarkar, M. S. and Agarkar, D. S. 1977. Desmids from Panchmari (M.P.) .*Hydrobiologia* **54** : 23-32.

Agarkar, M. S., Agarkar, D. S. and Dixit, R. 1979. Desmids from Bhandhavgarh, M. P. *Hydrobiologia* **65** : 213-223.

Bharati, S.G. and Bongale, U.D. 1975. Systematic account of freshwater algae of Raichur, Karnataka State, India. *J. Karnataka Univ.* **1**: 97-102.

Croasdale, H. 1956. Freshwater algae of Alaska 2: *Actinotaenium*, *Mixosterias* and *Cosmarium*. *Trans. Amer. Microsc. Soc.* **75**: 1-70.

Croasdale, H. 1965. Desmids of Devon is NWT Canada. *Trans. Amer. Microsc. Soc.* **80**: 301-305.

Croasdale, H. and Gronblad, R. 1964. Desmids of Lebrador. I. Desmids of the south eastern coastal area. *Trans. Amer. Microsc. Soc.* **83**: 142-212.

Croasdale, H. and Scott, A. M. 1976. New or otherwise interesting desmids from Northern Australia. *Nova Hedw.* **7**: 501-596.

Gronblad, R. and Croasdale. H. 1971. Desmids from Uganda (S.W. Africa). *Acta Bot. Fenn.* **66**: 1-40.

Habib, l. 1990a. Desmids from Bombay (Maharashtra), India. *New Bot.* **17**: 143-147.

Habib, I. 1990b. On some taxa of *Cosmarium* Corda new to Indian desmid flora. *J. Indian Bot. Soc.* **69**: 275-276.

Habib, l. 1993a. On some desmids new to Indian algal flora. *Acta Bot. Ind.* **21**: 267-269.

Habib, l. 1993b. On some desmids new to Indian algal flora. *Acta Bot. Ind.* **21**: 300-302.

Habib, l. 1993c. Contribution to the desmids of Rohilkhand division, U. P., India. *Geophytology* **23**: 177-179.

Habib, l. 1994. Desmids of Bareilly XIV (U.P.) India. *New Bot.* **20**: 43-45.

Habib, I. and Chaturvedi, U.K. 1995. On some desmids from Nepal. *J. Indian Bot. Soc.* **74**: 277-282.

Habib. I. , Pandey, U.C. and Chaturvedi, U. K. 1989. Desmids of Bareilly District-XIII, U.P., India. *Geophytology* **20**: 54-55.

Habib, l., Ansari, A. R., Mehrotra, D. K. and Chaturvedi, U. K. 1998. Some desmids from the foot hills zone of Kumaun Himalaya. *Nature Biosph.* **3** : 36-40.

Isaacs, S. W. and Hegde, G. R. 1986. Fresh water algae of Uttara Kannada District, Karnataka state. *Phykos* **25**: 102-107.

Iyengar, M.O.P. and Vimala Bai, B. 1941. Desmids from Kodakanal, South India. *J. Indian Bot. Soc.* **20**: 73-103.

Nurul Islam, A.K.M. 1970. Contribution to the knowledge of desmids of East Pakistan-I. *Nova Hedw.* **20**: 903-983.

Pandey, U.C. and Pandey, D. C. 1983. Desmids of Allahabad. *J. Indian Bot. Soc.* **62**: 166-169.

Pandey, U.C. and Habib, I. 1987. Desmids of Bareilly-I. *J. Indian Bot. Soc.* **66**: 291-296.

Pandey, U. C., Chaturvedi, U.K., Habib, I., Shukla, H. M. and Agnihotri, A. K. 1988. Some desmids new to Bareilly. *J. Indian Bot. Soc.* **67**: 71-73.

Patel, R. J. and Kumar, C.K.S. 1979. Desmids of Gujarat, India-I. Genus *Closterium* Nitzsch. *Phykos* **18** : 111-124.

Prasad, B.N. and Mehrotra , R.K . 1977a. Some desmids new to Indian flora. *J. Indian Bot. Soc.* **56** : 343-350.

Prasad, B.N. and Mehrotra, R. K. 1977b. Desmid flora of North Indian paddy fields. *New Biol.* **4**: 49-74.

Prasad, B.N. and Misra, P.K. 1985. Some taxa of *Cosmarium* Link new to Indian desmids flora. *J. Indian Bot. Soc.* **64**: 343-347.

Scott, A.M. and Prescott, G.W. 1961. Indonesian Desmids. *Hydrobiologia* **17**: 1-132.

Scott, A.M., Gronblad, R. and Croasdale, H. 1965. Desmids from Amazon Basin, Brazil. *Acta Bot. Fenn.* **69**: 1-94.

Suxena, M.R. and Venkateswarlu, V. 1968a. Desmids of Andhra Pradesh-II. From Dharamsagar lake, Warangal. *J. Indian Bot. Soc.* **47**: 23-45.

Suxena, M.R. and Venkateswarlu, V. 1968b. Desmids from Kashmir. *Phykos* **7**: 165-185.

Suxena, M.R., Venkateswarlu, V., Subbaraju, N. and Rao, V.S. 1973. The algae and the testacea of Cranganore, Kerala state, India. *J Indian Bot. Soc.* **52**: 316-341.

West, W. and West, G. S. 1904-12. A Monograph of the British Desmidiaceae. *Royal Soc. London* **1**: 1-224, **2**: 1-204, **3**: 1-272, **4**: 1-191.

Chapter 3

DIVERSITY OF ALGAL FLORA IN RELATION TO MAJOR CROPS, SOURCE OF WATER, SOIL TYPES AND FERTILIZERS IN CULTIVATED SOILS OF BIDAR AND GULBARGA DISTRICTS OF KARNATAKA, INDIA

S. B. Angadi, M.K. Santosh, V.G. Uttam and D.G. Maheshkumar

Abstract

The algal floristic diversity in relation to major crops, source of water, soil types and fertilizers has been made. It was observed that wet lands (sugarcane and paddy) had richer algal flora than the dry lands (jawar, bajra, wheat and zea mays). Lateritic and red sandy loam soils had richer algal flora than other soil types. High level of organic carbon and nitrogen appeared to be favourable for richer algal growth. Addition of manure to soils could also influence the algal flora. Members of cyanophyceae appeared to be more resistant than those of chlorophyceae and euglenophyceae, to desiccation.

Key words: Biofertilizer, cyanobacteria, diversity, floristic, soil algae

Introduction

The occurrence and succession of algae growing on the soil surface is regulated by the soil properties. Agronomically, the soil algae have been assessed from the standpoint of nitrogen content of soils. It is now generally accepted that cyanobacteria fix much more nitrogen than other heterotrophic microorganisms. Cyanobacteria being phototrophic microbes, are restricted to the photic zone and are usually located within the upper few millimeters in fine soil or few centimeters in the coarse soil (Goyal, 1997). Viable propagules of algae have, however, been reported upto a depth of one metre (Vaidya, 1964; Kamat and Patel, 1973; Friedmann and Gallan, 1974). They are carried down by agricultural activity, root growth and soil fauna (John, 1942; Lund, 1967). In India, a good deal of work has been done on the algal flora of paddy fields (Singh, 1939; Mitra, 1951; Tiwari, 1972; Sardeshpande and Goyal 1981; Kolte and Goyal, 1985; Sharma and Komi, 1992; Nayak *et. al.*, 1996; Singh *et. al.*, 1997; Rout and Dey, 1999). However, the information available on the algal flora of other crops is very meagre (Marathe, 1967; Marathe and Ananthani, 1972; Bongale and Bharati, 1980).

During the study on the algal flora of cultivated soils of Gulbarga and Bidar districts, 60 soil samples were collected and studied. Observations on the algal species in relation to major crops, source of water, soil types and fertilizers are discussed in the present paper.

Materials and methods

A total of 60 soil samples were collected from 8 different places of Gulbarga and Bidar districts of Karnataka for this study. For collecting a sample from a field, soil from 5-6 spots of a field was mixed thoroughly and about 500-600 gms of this soil was taken to the laboratory for further study. Enrichment cultures were raised from these soils using De's modified Beneck's medium (Singh, 1961). Both moist and liquid cultures were raised. Petridishes were used for the moist cultures and half litre milk bottles for the liquid cultures. The algae were collected every fortnight for about 3 months and preserved in 4% formalin for further observations.

Soils thus studied were classified into different soil types (Narayan and Shah, 1966). The information about the conditions of manuring, irrigation facilities and crop rotation etc. of all the places of field collection were noted down. Soils were analysed for their organic carbon(Piper, 1947) and nitrogen (Jackson, 1973).

Results and discussion

Correlation of major crops and source of water with algal composition

Seventy per cent of the total field soils during the present study were having one or the other irrigation facility (Table 3.1). In these soils, major crops like paddy, sugarcane, fruits and vegetables were grown. These soils harboured higher number of algal organisms. Remaining 30% of the soils were dry and received low rain. Major drought tolerant crops like jawar, bajra, maize, pulses and wheat were grown in these soils. These soils harboured a much lower number of algae.

Therefore, paddy, sugarcane and vegetable field soils had significantly greater number of total algae than the jawar, bajra, maize and wheat field soils. In all types of crop field soils, cyanobacteria were found in higher number than the chlorophyceae and euglenophyceae (Table 3.2). Similar observations have also been made by Mitra (1951),Pandey (1965), Tiwari (1972), Sardeshpande and Goyal (1981) and Angadi (1982). Cyanobacteria appear to be more resistant to desiccation than other groups of algae due to morphological and physiological adaptations (Singh, 1961). It is known that members of chlorophyceae and euglenophyceae are not resistant to desiccation (Lund, 1945). High organic carbon and nitrogen content of field soils appear to support the higher algal populations.

Correlation of soil types and algal composition

It was observed that lateritic soil and red sandy loam soil harboured more algae than the black soil (Table 3.3). However, chlorophycean and euglenophycean members represented very poorly in all the soil types. The richer algal floristic composition of lateritic and red sandy loam soils are due to their porous, friable, well drained nature and high content of organic carbon and nitrogen. The poorer algal flora of black soils is due to clayey nature and poor aeration. Similar observations have also been made by Mitra (1951), Marathe (1967) and Bongale (1981).

Correlation of fertilizers and algal composition

Soils manured with chemical fertilizers and both chemical and farm yard manure were observed to have richer algal flora while

Table 3.1: Algal species in relation to different water sources.

Source of water	*Total number of algal species*						
	No.of soil samples studied	*Cyano-phyceae*	*Chloro-phyceae*	*Eugleno-phyceae*	*Total*	*Organic carbon (%)*	*Nitrogen (%)*
Low rain water	18	21	02	02	25	1.05	0.096
Well water	23	26	02	01	29	0.82	0.071
Borewell water	12	18	02	03	23	1.18	0.087
Pond/Lake water	07	20	01	-	21	1.34	0.085

Table 3.2: Algal species in relation to major crops.

Major crops	*Total number of algal species*						
	No.of soil samples studied	*Cyano-phyceae*	*Chloro-phyceae*	*Eugleno-phyceae*	*Total*	*Organic carbon (%)*	*Nitrogen (%)*
Jawar, Wheat, Bajra, *Zea mays*	07	11	-	-	11	0.73	0.069
Grapes, Pomegranate, Citrus, Coconut	04	04	-	-	04	0.83	0.091
Green gram, Pigeon pea, Black gram, Fodder crop, Onion, Garlic, Brinjal	04	21	02	01	24	0.98	0.090
Sugarcane	20	23	02	03	28	0.91	0.071
Paddy	15	21	01	01	23	1.40	0.098

Table 3.3: Algal species in relation to soil types.

Soil types	*Total number of algal species*						
	No.of soil samples studied	*Cyano-phyceae*	*Chloro-phyceae*	*Eugleno-phyceae*	*Total*	*Organic carbon (%)*	*Nitrogen (%)*
Lateritic soils							
a) Lateritic clay loam	13	21	02	02	25	1.03	0.082
b) Lateritic sandy loam	04	12	02	03	17	1.35	0.083
Black soils							
a) Medium black sandy loam	10	11	–	–	11	0.72	0.074
b) Medium black sandy clay loam	03	09	–	–	09	0.64	0.063
c) Deep black clay soils	11	12	–	–	12	0.60	0.071
Red sandy loam	19	24	02	03	29	1.42	0.100

Table 3.4: Algal species in relation to fertilizers.

Fertilizers	*Total number of algal species*						
	No.of soil samples studied	*Cyano-phyceae*	*Chloro-phyceae*	*Eugleno-phyceae*	*Total*	*Organic carbon (%)*	*Nitrogen (%)*
Urea, D.A.P., Superphosphate	22	28	02	03	33	1.05	0.080
Field yard manure and chemicals	13	19	00	03	22	1.01	0.082
Field yard manure	15	13	02	01	16	1.02	0.092
Unmannured	10	15	02	01	18	0.995	0.081

soils manured with farm yard manure and unmanured soils had lesser algal flora (Table 3.4).

Gistl (1933) found higher number of algal species in soils with various combinations of chemical fertilizers than the unmanured soils. Similar observations have been made by Mitra (1951), Laloraya and Mitra (1973) and Angadi (1982). Higher level of organic carbon and nitrogen in fertilized fields favours high algal diversity as compared to unmanured soils.

Therefore, the soils with irrigation facilities favour the abundant occurrence of algae by providing prevailing water logged condition and increased availability of moisture for longer period. Lateritic and red sandy soils had richer algal flora as compared to other type of soils. Cyanophyceae dominated over chlorophyceae and euglenophyceae in all the soil types and major crop groups under investigation. Cyanophyceae appear to be more resistant than the chlorophyceae and euglenophyceae. Besides the physico-chemical properties of soils, supplementation of manure to soils determine the quality and quantity of algal flora.

Acknowledgement

Thanks to the Chairman, Department of Botany, Gulbarga University, Gulbarga for providing necessary facilities.

References

Angadi, S.B. 1982. *Studies on Soil Algae*. Ph.D. Thesis, Karnataka University, Dharwad.

Bongale, U.D. 1981. Abundance of algal species in relation to soil types and major crops in cultivated soils of Karnataka state (India). *Plant Soil* **60**: 293-298.

Bongale, U.D. and Bharati, S.G. 1980. On the algal flora of cultivated soils of Karnataka State, India. *Phykos* **19**: 95-109.

Friedmann, E.I. and Gallan, M. 1974. Desert algae, lichens and fungi, In: *Desert Biology II* Edited by Brown, G.W. Jr., Academic Press, New York.

Gistl, R.1933. Erdalgen and Dungung, Erdalgen and Anionlen. *Arch. Microbiol*. **4**: 318-378.

Goyal, S.K. 1997. Algae and the soil environment. *Phykos* **36**: 1-13.

Jackson, M.L. 1973. *Soil Chemical Analysis*. Prentice Hall of India, New Delhi.

Rout, J. and Dey, A. 1999. A study of algal flora from rice fields of Irongmara (Barak valley, Assam). *Phykos* **38**: 19-25.

John, R.P. 1942. An ecological and taxonomic study of the algae of British soils. 1. The distribution of surface growing algae. *Ann. Bot.* **6**: 323-349.

Kamat, N.D. and Patel, M.Z. 1973. Soil algae of rice field at different depths. *Botanique* **4**: 101-106.

Kolte, O.S. and Goyal, S.K. 1985. Distribution pattern of blue-green algae in rice fields soils of Vidharbha region of Maharashtra state. *Phykos* **24**: 156-162.

Laloraya, V.K. and Mitra, A.K. 1973. Studies on the blue-green algae of paddy fields of India, Part-I. Cultural studies, general consideration and distribution pattern of the blue-green algae in the paddy fields of India. *Nowa Hedw.* **47**: 227-262.

Lund, J.G.W. 1945. Observation of soil algae. I. The ecology, size and taxonomy of British soil diatoms. *New Phytol.* **44**: 196-219.

Lund, J.G.W. 1967. Soil algae. In: *Soil Biology*. Edited by Burges, A. and Raw, F., Academic Press, New York.

Marathe, K.U. 1967. Studies on soil algae of India. I. Soil algae from cultivated fields of Dhulia (Maharashtra state). *J. Univ. Bombay* **35**: 100-105.

Marathe, K.U. and Ananthani, Y.S. 1972. Observations on the algae of some Indian arid soils. *Botanique* **3**: 13-19.

Mitra, A.K. 1951. The algal flora of certain Indian soils. *Indian J.Agric. Sci.* **21**: 357-373.

Narayan, N. and Shah, C.C. 1966. *Physical Properties of Soils*, P.C. Manaktala and Sons Pvt. Ltd., Bombay.

Nayak, H., Sahu, J.K. and Adhikary, S.P. 1996. Blue-green algae of rice fields of Orissa state II. Growth and nitrogen fixing potential. *Phykos* **35**: 93-110.

Pandey, D.C. 1965. A study on the algae from paddy field soils of Ballia and Ghazipur districts of Uttar Pradesh, India, Part-I, Cultural and ecological considerations, *Nowa Hedw.* **10**: 299-334.

Piper, C.S. 1947. *Soil and Plant Analysis*. Univ. of Adelaide, Adelaide.

Sardeshpande, J.S. and Goyal, S.K. 1981. Distributional pattern of blue-green algae in rice fields soils of Konkan region of Maharashtra state. *Phykos* **20**: 102-106.

Sharma, B.M. and Komi, P.N. 1992. Nitrogen fixing blue-green algae from rice farms of Jammu and Kashmir. In: *Biofertilizer Technology Transfer*. Edited by Gangawane, V., pp. 187-190.

Singh, N.I., Dorycanta, H., Devi, G.A., Singh, N.S. and Singh M.S. 1997. Blue-green algae from rice field soils of Nagaland. *Phykos* **36**: 115-120.

Singh, R.N. 1939. An investigation into the algal flora of paddy fields soils of the United provinces. *Indian J. Agric.Sci.* **9**: 55-77.

Singh, R.N. 1961. *Role of Blue-Green Algae in Nitrogen Economy of Indian Agriculture*, ICAR, New Delhi. pp. 175.

Tiwari, G.L. 1972. A study of blue-green algae from paddy field soils of India. *Hydrobiologia* **39**: 335-350.

Vaidya, S.M. 1964. *Ecology of algae in Indian soils with special reference to light factor*. Ph.D. Thesis, University of London, London.

Chapter 4

CURRENT STATUS OF *AZOLLA* LAM. TAXONOMY

A. Pabby, A. S. Ahluwalia and S. Dua

Abstract

Exact taxonomic position of an experimental organisum is of utmost significance for a fruitful research investigation. Taxonomic classification of *Azolla* even upto species or ecotype level has remained more or less inconclusive. However, with the amalgamation of old (morphological) and new (biochemical and/or molecular) features /techniques, researchers have tried to solve the problem. Efforts are still required to be made to streamline its position so that improved strains/species can be worked out to exploit the agronomic potential of this versatile association. In this review, an attempt has been made to compile the available information concerning taxonomy of this water fern with the aim to further facilitate the interest of scientific community for resolving the exact taxonomic status of *Azolla-Anabaena* association.

Key words: *Azolla*, Azollaceae, classification, molecular features, morphology, taxonomy.

Introduction

Taxonomy of an organism forms its basis for other sciences. The biologists must know what organism they are working on before passing on more information about that to the scientific community. Scientists, investigating in certain disciplines of biology other than taxonomy cannot publish their observations without the knowledge of the proper taxonomic position of the experimental organism. Many of the experiments fail on repetition because the organism is not correctly identified taxonomically. Every biologist, working with one or the other species, and their findings, even at molecular level, are largely influenced by the choice of a particular species (Mayr, 1975). However, the importance of the correct identification of biological specimens under experimentation and its role in many biological researches has not been given due recognition. This is also true, particularly, in case of research programmes on the utilization and propagation of the water fern, *Azolla*.

Azolla Lam. is a eukaryotic, heterosporous, leptosporangiate aquatic fern which have symbiotic association with the prokaryotic, heterocystous, nitrogen-fixing *Anabaena azolla*e and eubiont of unknown function (Padhya, 1997; Wagner, 1997). *Azolla* is an economically important pteridophyte which has long been used the world over for its diversified applications (Lumpkin and Plucknett, 1980; Braun-Howland and Nierzwicki-Bauer, 1990; Ahluwalia *et. al.*, 2002). Presently collections of *Azolla* cultures are maintained at International Rice Research Institute (IRRI), Philippines and at the Catholic University of Louvain, Belgium, where accessions are evaluated for their productivity and nitrogen-fixing potential under various ecological conditions. The agronomic values of the *Azolla-Anabaena azolla*e symbiosis have shown to be improved by genetic manipulation of both the macrosymbiont and microsymbiont *via* sexual crossing of *Azolla* and cross-inoculation of the microsymbionts.

Azolla consists of seven species, which are largely classified based on the morphology of their reproductive organs (Lumpkin and Plucknett, 1980; Tan *et. al.*, 1986; Dunham and Fowler, 1987). Unfortunately, most of *Azolla* accessions never become fertile under maintenance conditions of germplasm collections. In the field and green houses, one frequently observes mixed populations of supposed pure "stock" of *Azolla*. The fragments can easily multiply vegetatively and establish themselves in a mixed population which generally escape the attention of the scientist. Thus, the identity of many accessions became suspect after entering a germplasm collection because of cross contamination by other organisms. Therefore, it is necessary for users of *Azolla* to have knowledge of *Azolla* taxonomy.

Keeping in view the above problem, an attempt has been made to review the literature available from the studies conducted both at morphological and molecular /biochemical level to throw some light on the systematic classification of *Azolla*.

Taxonomic history

The genus *Azolla* is a conjugation of two Greek words *azo* (to dry) and *allyo* (to kill) reflecting the inability of the plants to survive dry conditions (Moore, 1969; Ashton and Walmsely, 1976; Lumpkin and Plucknett, 1980). It was established by Lamarck in 1783 with the description of *Azolla filiculoides*. *Azolla* has been variously placed in Marsiliaceae (Brown, 1810), Salviniaceae (Dumortier, 1829; Smith, 1938; Black, 1948; Bailey, 1949; Lawrence, 1951; Benson, 1957) and later on to a monotypic family Azollaceae (Wettstein, 1903, Melchior and Werdermann, 1954; Eichler, 1965; Konar and Kapoor, 1974). The current classification scheme of the genus *Azolla* is:

Divison - Pteridophytes
Class - Filicopsida
Order - Salviniales
Family - Azollaceae

Fossilized *Azolla* specimens have been dated back to around 2-65 million years ago (West, 1953). Fossil records show evidence of at least 25 extinct species (Reed, 1954; Hills and Gopal, 1976) although, Fowler (1975) indicated that the number is as high as 48, with seven extant species namely *Azolla filiculoides* Lamarck, *A. caroliniana* Willdenow, *A. rubra* R. Brown, *A. microphylla* Kaulfuss, *A. mexicana* Presl, *A. pinnata* R. Brown and *A. nilotica* De Caisne. In addition, *A. japonica* Franch is considered to be a variety of *A. filiculoides* and *A. africana* Desv. as a synonym of *A. nilotica* (Fowler and Stennett-Wilson, 1978).

General morphology and life cycle

Fully grown *Azolla* macrophyte called 'frond' ranges from 1cm for *A. pinnata, A. mexicana, A. microphylla*, and *A. caroliniana*, to 5-7cm for *A. filiculoides* and 40cm or more for A. nilotica. The sporophyte of *Azolla* consists of multibranched, prostrate, floating stem (rhizome) that bear deeply bilobed leaves and determinant, adventitious roots. The aerial dorsal lobe is cholrophyllous, and contains a leaf cavity which harbours the symbiotic *Anabaena azollae* (Peters and Meeks,

1989; Braun-Howland and Nierzwicki-Bauer, 1990; Plazinski, 1990) alongwith small populations of bacteria *i.e. Pseudomonas* species, *Azobactor* species, *Alcalgenes faecalis* and *Caulobactor fusiformis* (Plazinski *et. al.*, 1990; Malliga and Subramanian, 1995).

Azolla exhibits both sexual (involving a complex cycle) and asexual or vegetative reproduction. During sexual reproduction, instead of forming a leaf lobe at the first ventral lobe initial of lateral branch, the fern produces two types of sporocarps, the microsporocarps and the megasporocarps. The sporocarps at the ventral side may either be of the same type or mixed. The microsporocarps produce 8-130 microsporangia each containing 32-64 spores aggregated into 3-10 massulae (Moore, 1969; Shi and Hall, 1988), whereas megasporocarp produces a single megasporangium containing a single megaspore. Every megaspore contains a small colony of *Anabaena azolla*e filaments bearing akinetes.

After a period of dormancy, liberated microspores and megaspores germinate and grow into respective prothalli that produce male gametes (antherozoids) and female gametes (oosphere). Embryo develops after fertilization of oosphere with antherozoides, germinates and pushes itself out of the sporocarp to form a plantlet.

Taxonomic characters

Various morphological characters; vegetative and reproductive, were used from time to time by scientists for classification of *Azolla* upto section and species level. They are:

Vegetative characters

(i) Sporophyte morphology

The vegetative sporophyte consists of main stem with several lateral branches that in turn produce branchlets. Each frond acquires a particular shape (deltoid or triangular, feather-like, circular or fan shaped) due to the apical dormancy of the main stem. The sporophyte of *A. pinnata* exhibits triangular shape whereas *A. nilotica* assumes a feather-like appearance at maturity (Mettenius, 1867; Strasburger, 1873; Baker, 1886, 1887; Sadebeck, 1902; Christensen, 1906; Tan *et. al.*, 1986; Saunders and Fowler, 1992), while all other species – *A.*

filiculoides, A. caroliniana, A. rubra, A. microphylla and *A. mexicana* have similar morphological outlook. They appear circular to fan shaped (Svenson, 1944; Dunham and Fowler, 1987).

(ii) Frond characters

The leaf characters such as shape and dimensions of dorsal lobe of leaf; variation in shape of hyaline papillae; their arrangement on the frond and number of cells forming the papillae are known to possess taxonomic value. The length of upper leaf lobe in *A. filiculoides* varies from 1.5-2mm (Pieterse *et. al.*, 1977), in *A. caroliniana* it is 0.5mm (Svenson, 1944) and in *A. pinnata*, 0.8-1.9mm x 0.5-1.1mm (Sweet and Hills, 1971). Lawalree (1946) and Van Ooststroom (1948) described the shape of dorsal lobe in *A. filiculoides* as ovate or broadly ovàte and in *A. caroliniana* as ovate narrowing towards the apex, obtuse or subacute asymmetrical.

The dorsal lobe of leaves of *Azolla* bear hyaline papillae (also called trichomes) that are made up of different number of cells. Interestingly, Section-Rhizosperma develops papillae on the stem and branches, a feature not observed among all the species of Section-Azolla which have papillae present only on the leaves (Cardenas *et. al.*, 1986). These leaf trichomes are unicellular in *A. filiculoides* but other taxa of Section-Azolla possess trichomes of atleast bicellular nature (Strasburger, 1873; Bernard, 1904; Pieterse *et. al.*, 1977; Lumpkin and Plucknett, 1982; Zimmerman *et. al.*, 1989b). Scanning electron microscopy showed that species of Section-Azolla have round nipples on the surface of the dorsal lobe and those in Section-Rhizosperma have prolate ones (Shi *et. al.*, 1984). Stomata have two guard cells in *A. filiculoides* whereas *A. pinnata* has only one (Seto and Nasu, 1975).

(iii) Root characters

The arrangement of root hair on the root, presence of root cap or sheath, number of root epidermal cells and outer cortical cells have some taxonomic potential. Cardenas *et. al.* (1986) observed regular rows of clusters of 5-6 root hair arranged planarly in *A. pinnata*. All these arise from the same level of nodal position of the root axis. Other taxa such as *A. mexicana, A. caroliniana* and *A. filiculoides* produce root hair in clusters of 2, 3, and 4 respectively, while irregular arrangement of root hair along the root axis has been observed in *A.*

nilotica. A coiled root apex without a root cap or root sheath developed at maturity in *A. filiculoides* (Cardenas *et. al.*, 1986).

(iv) Coloration

The formation of reddish pigments in the leaves and stems of few species of *Azolla* has been given taxonomic meaning by some workers. Fujiwara and co-workers (1947) distinguished red-duckweed (*A. pinnata*) from blue-duckweed (*A. filiculoides*). Becking (1976) also reported that populations of *A. filiculoides* and *A. caroliniana* develop a brick red colour. Similarly colour of *A. caroliniana* is dark green and it is more glaucouse in *A. filiculoides* (Pieterse *et. al.*, 1977).

As root characters can be seen only in fresh material, and development of coloration can be demonstrated as an ecological response on exposure to strong light (stress conditions) and growing at high altitudes and / or in certain mineral deficient culture media, therefore, these vegetative characters cannot be considered alone for distinguishing various species / ecotypes of *Azolla* due to their phenotypic plasticity.

Reproductive characters

By and large, the basic classification of *Azolla* since the time of Baker (1886) and Svenson (1944) lies in the sporocarp characters such as number of sporocarps, number of massulae / microsporangia, structure and position of glochidia, number of floats / megasporocarp, float texture, surface ornamentation of megasporoderm, breadth of collar and density of megasporoderm filosum. Sporocarp occur in all species of *Azolla* in pairs (Moore, 1969) except *A. nilotica* in which they are present in tetrads (Demalsey, 1953; Lumpkin, 1981). Van Ooststroom (1948) reported 6 massulae / microsporangium in *A. caroliniana* and 4-8 in *A. filiculoides*. Their number was 4 in *A. mexicana* (Svenson, 1944) and 3-8 in *A. pinnata* (Sweet and Hills, 1971).

The presence or location of glochidia on massulae can also be used as distinguishing character (Konar and Kapoor, 1974). The species belonging to Section Azolla have glochidia positioned on the total surface (Svenson, 1944) while those of Section Rhizosperma are located on the inner surface (Konar and Kapoor, 1974) and are absent in *A. nilotica* (Demalsey, 1958). These glochidia are septate in

A. microphylla and *A. mexicana*, and unseptate in *A. caroliniana* and *A. filiculoides* (Svenson, 1944; Pieterse *et. al.*, 1977).

The New World species (of Section Azolla) exhibit three floats on the megaspore apparatus, whereas Old World species (of Section Rhizosperma) has nine such structures (floats), arranged in two groups, with three above and six below (Hills and Gopal, 1967; Moore, 1969; Sweet and Hills, 1971; Lumpkin and Plucknett, 1980, Tan *et. al.*, 1986; Peters and Meeks, 1989; Saunders and Fowler, 1992). SEM observations of *Azolla* megaspores showed two types of float structure – the smooth type observed in *A. pinnata*, *A. mexicana* and *A. nilotica* and punctured type has been seen in *A. microphylla* and *A. filiculoides* (Tan *et. al.*, 1986).

Perkins *et. al.* (1985) reported that application of SEM revealed distinct perine organization which was constant within taxa and consistently different between taxa. Martin (1976) suggested that other features such as nature of collar, column and float retention mechanism might also be used as the important taxonomic characteristics. Many accessions do not develop sporocarps in culture collections, therefore, without fruiting specimens, species identification can only be tentative.

Cytological studies

The role of cytological studies has been highlighted by a number of workers indicating that chromosome number observed during meiosis can be a useful criterion for distinguishing *Azolla* species (Loyal *et. al.*, 1982; Singh *et. al.*, 1984; Thanh, 1984; Lin and Sleep, 1988; Nayak and Singh, 1989; Stergianou and Fowler, 1989,1990; Saunders and Fowler, 1993).

In *A. pinnata*, the chromosome number was reported to be n = 22 (Loyal *et. al.*, 1982). Thanh (1984) indicated 2n = 66, while Singh *et. al.* (1984) and Nayak and Singh (1989) reported 2n = 44 and 66 respectively. Stergianou and Fowler (1989,1990) confirmed 2n = 44 for *A. pinnata*. A mitotic count of 2n = 44 for *A. filiculoides* was reported by Thanh (1984) and Stergianou and Fowler (1990). 2n = 40 for this species of *Azolla* has also been observed by many workers (Singh *et. al.*, 1984; Nayak and Singh, 1989).

A. caroliniana was found to have 2n = 44 by Stergianou and Fowler (1990) but Litardiere (1921) reported 2n = 48. A count of 2n =

42 has also been recorded by Thanh and Hang (1988). Likewise *A. mexicana* have 2n = 44 (Thanh and Hang, 1988; Stergianou and Fowler, 1990) though earlier workers recorded 2n = 48 (Singh *et. al.*, 1984; Nayak and Singh, 1989) and 40 (Thanh, 1984) for this species.

The diploid plants of *A. microphylla* were found to have 2n = 44 (Stergianou and Fowler, 1990) while 2n = 48 was also recorded for this species (Thanh and Hang, 1988). A triploid of Mexican origin with 2n = 66 was also identified. The differences in chromosome number recorded by various workers are because of much smaller size of the chromosomes and their higher number.

Classification

Saunders and Fowler (1993) proposed a phylogenetically revised classification for *Azolla* on the basis of published morphological data and studies involving cytological and cladistic analysis.

Family: Azollaceae

Genus: *Azolla*

Subgenera: Azolla Section: Azolla

Species: *A. filiculoides* Lam., *A. rubra* R. Br., *A. mexicana* Presl., *A. caroliniana* auct. non

Willd., and *A. microphylla* auct. non Kaulf.

Subgenera: Azolla Section: Rhizosperma

Species: *A. pinnata* R. Br.

Subgenera: Teterasporocarpia

Species: *A. nilotica* Decen. ex Mett.

Description

1. Subgenera – Azolla

Stem branching pattern dorsoventrally flattened, protostelic stem vascular system, ventral leaf lobe unistratose and translucent (except at base), lack stomata and trichomes, plants become red under adverse environmental conditions due to anthocyanin pigment, roots solitary and growing from stem branching point, sporocarps arranged in pairs, basic chromosome number n =22.

1a. Section - Azolla

Type species – *A. filiculoides* Lam.

Dichotomous branching pattern of stem is giving rise to round plants , occasionally elongate in mature specimens of some species, stem glaborous, megaspore apparatus with three floats and a large collar, megasporoderm with infrafilosum, massulae with trichomes having barbed apices.

1b. Section – Rhizosperma

Type species – *A. pinnata* R. Br.

Main stem with alternatively arranged lateral branches, branching pattern dorsoventrally flattened, stem with bicellular trichomes at maturity (occasionally absent in juvenile plants, megaspore apparatus with nine floats and a small collar, megasporoderm without infrafilosum, massulae with simple spiniform trichomes.

2. Subgenera – Tetrasporocarpia

Type species – *A. nilotica* Decne. ex Mett.

Main stem with alternate lateral branches, plants are not dorsoventrally flattened, with assurgent leaf arrangement, stem with multicellular trichomes (occasionally unicellular or absent in juvenile plants), amphiphloic siphonostelic vascular system, ventral leaf lobe more than one cell thick, chlorophyllous, with stomata and trichomes, plants never red, roots in large fasicles from stem branching points, sporocarps arranged in fours, megaspore collar small, megasporoderm without infrafilosum, massulae with simple filiform trichomes or occasionally absent, basic chromosome number n = 26.

Biochemical/molecular studies

The traditional classification system is based primarily on the morphology of very limited number of reproductive and vegetative structures. Reproductive structures are rarely found and vegetative organs show a high phenotypic difference depending upon the various environmental conditions. Cytological studies in *Azolla* have also contributed to our understanding of the taxonomy of the genus, but there is lack of sufficient variation in the chromosome size and number to allow for high resolution discrimination of the species

(Stergianou and Fowler,1990). This problem is particularly evident in Section – Azolla, which contains five of the seven extant species, that are morphological similar.

Recent applications of biochemical and molecular technologies such as flavonoid analysis, isozyme characterization, restriction fragment legnth polymorphism (RFLP) analysis and DNA amplification, have suggested some rearrangements in taxonomy of *Azolla* particularly in Section – Azolla (Zimmerman *et. al.*, 1989a ,b; 1991a,b; Coppenolle *et. al.*, 1993; Eskew *et. al.*, 1993; Kim *et. al.*,1997).

Van Hove *et. al.* (1983) conducted flavonoid analysis with one collection of *Azolla*, but the use of products of secondary chemistry for systematics offer limited information concerning the genes producing them. Preliminary examination *via* enzyme electrophoresis of the selected isolates indicated that allozyme diversity is evident in Section – Azolla, where species level classification by morphological means is difficult.

Isozyme analysis of nine enzymes showing 19 polymorphic loci indicated that *A. filiculoides* and *A. rubra* are biochemically unique and recognizable as species but principle component analysis did not reflect any distinctiveness in *A. microphylla, A. mexicana* and *A. caroliniana* to be considered as distinct species (Zimmerman *et. al.*, 1989b). These three neotropical species of Section – Azolla were found to be very similar in their DNA fingerprinting from hybridization to heterologous and random homologous gene probes (Zimmerman *et. al.*, 1991b). Results obtained through isozyme and restriction fragment length polymorphism analysis are more consistent and similar, thus further strengthen the similarity of these three species (Zimmerman *et. al.*, 1991a,b).

Coppenolle and co workers (1993) used polymerase chain reaction (PCR) to amplify random sequences of DNA from 25 accessions of *Azolla* to evaluate the usefulness of this technique for identification and phylogenetic analysis of the aquatic fern. They found high frequency of polymorphism among accessions. They used information to construct a dendrogram and principle component analysis indicated that genus *Azolla* has three distinct groups: Group 1 contains five species of Section Azolla ; Group 2 contains cluster of *A. pinnata* species with in Rhizosperma Section ; and Group 3 contains the cluster of *A. nilotica* species within Rhizosperma Section.

Eskew and co workers (1993) developed DNA amplification fingerprinting (DAF) to solve the problem of identifying *Azolla* accessions that offers several advantages over isozyme and RFLP analyses.

Concluding remarks

Reviewing of taxonomic position of *Azolla* indicates that proper usage of this association, especially in genetic manipulation, is limited because of poor identification of accessions. Therefore, further collaborative efforts in research are required in this direction to make the best out of this important natural resource for the ecological sustainability and stability of this planet.

References

Ahluwalia, A.S., Pabby, A. and Dua, S. 2002. *Azolla*: A green gold mine with diversified applications. *Indian Fern J.* (in press).

Ashton, P.J. and Walmsley, R.D. 1976. The aquatic fern *Azolla* and its *Anabaena* symbiont. *Endeavour* **19**: 39–43.

Bailey, I.H. 1949. *Manual of cultivated plants*, 2nd ed. Macmillan Congress. New York.

Baker, J.G. 1886. A synopsis of Rhizocarpeae. *J. Bot.* **25**: 97–101.

Baker, J.G. 1887. *Azolla* Lam. In: *Handbook of fern allies*, West, Newton, London. pp. 137–138.

Becking, J.H. 1976. Contributions of plant-algal associations. In: *Proc. 1st Inter. Sym. Nitrogen Fixation*. Edited by Newton, W. E. and Nyman, C. J. Vol. 2, Washington State University Press, Pullman. pp. 556-580.

Benson, L. 1957. *Plant classification*. Health and Co., Boston.

Bernard, C. 1904. A propos d'*Azolla*. *Rev. Tran. Bot. Neerl.* **1**: 1–13.

Black, J.M. 1948. *Flora of South Australia*. (Part 1) South Australian Govt. Printer, Adelaide.

Braun–Howland, E.B. and. Nierzwicki–Bauer, S.A. 1990. *Azolla–Anabaena* symbiosis: Biochemistry, physiology and molecular biology. In: *Handbook of symbiotic cyanobacteria*. Edited by Rai, A.N. CRC Press, Florida. Pp. 65–118 .

Brown, R. 1810. *Prodromus Florae Novae Mollandiae.* Insulae Van–Dieman. London. Johnson.

Cardenas, L.B., Tan, B.C. and Torreta, N.K. 1986. Some overlooked taxonomic characters of the root in *Azolla*. *Phil. Agr.* **69**: 43–51.

Christensen, C. 1906. *Index Filocum*. Hafnicre.

Coppenolle, B. Van, Watanabe, I., Van Hove, C. and. McCouch, S.R 1993. Genetic diversity and phylogeny analysis of *Azolla* based on DNA amplification by arbitrary primers. *Genome*. **36**: 686–693.

Demalsey, P. 1953. Le Sporophyte d' *Azolla nilotica*. *Cellule* **56**: 5–60.

Demalsey, P. 1958. Nouvelles researcher sur le sporophyte d' *Azolla*. *Cellule* **59**: 235–268.

Dumortier, B.C. J. 1829. *Analyse des Familles-des plantes*. Tournary.

Dunham, D.G. and Fowler, K. 1987. Taxonomy and species recognition in *Azolla* Lam. In: *Azolla utilization*. Fuzhou, Fujian, China. pp. 7–16.

Eichler, H. 1965. *Supplement to J. M. Black's flora of South Australia*. South Australian Govt. Printer, Adelaide.

Eskew, D.J., Caetano, A., Bassam, B. J. and Gresshoff, P. M. 1993. DNA amplification fingerprinting of the *Azolla–Anabaena* symbiosis. *Plant Mol. Biol.* **21**: 363–373.

Fowler, K. 1975. An escape mechanism for spermatozoids in *Azolla* massulae. *Amer. Fern. J.* **65**: 7–10.

Fowler, K. and Stennett–Wilson, J. 1978. Sporoderm architecture in modern *Azolla .Fern Gaz.* **11**: 405–412.

Fujiwara, A., Tsuboe, I. and Yoshida, F. 1947. On free living nitrogen fixation of Akaukikusa *Noyaku (Agr. Sci.)* **1**: 361–363.

Hills, D.L.V. and Gopal, B. 1967. *Azolla primaeva* and its phylogenetic significance. *Can. J. Bot.* **45**: 1179–1191.

Kim, J. Ji–Hee, Krawczyk, K., Lorentz, W.P. and Zimmerman, W.J. 1997. Fingerprinting cyanobionts and hosts of *Azolla* symbiosis by DNA amplification. *World J. Micro. Biotech.* **13**: 97–101.

Konar, R.N. and Kapoor, R.J. 1974. Embryology of *Azolla pinnata*. *Phytomorphology* **24**: 224–261.

Lamarck, J.B.A.P.M. DE. 1783. *Encyclopedia methodique botanique*. **1**: 343.

Lawalree, A. 1946. Azollaceae. In: *Flora*. Edited by Tutin, T.G., Heywood, V.H. , Burger, N.A., Valentine, D.H. , Walters, S.M. and Webb, D.A. Cambridge University Press. pp.25.

Lawrence, G. H. 1951. *Taxonomy of Vascular Plants*. Macmillan Congress, New York.

Lin, Y. X. and Sleep, A. 1988. A chromosome count for *Azolla filiculoides* (Azollaceae: Pteridophyta). *Fern Gaz.* **13**: 193–198.

Litardiere, R. de. 1921. Recherchis sur element chromosomique dans la caryocinese somatique des Filicineees. *Cellule* **31**: 225–274.

Loyal, D.S., Gollen, A.K. and Ratra, R. 1982. Morphological and cytotaxonomic observations on *Azolla pinnata*. *Fern Gaz.* **12**: 230–232.

Lumpkin, T.A. 1981. An introduction to *Azolla nitotica*. *Acta Bot. Sin.* **23**: 492–495.

Lumpkin, T.A. and Plucknett, D.L. 1982. *Azolla as a green manure: Use and management in crop production*. Westview Press, Boulder, Colorado.

Lumpkin, T.A. and Plucknett, D.L. 1980. *Azolla*: Botany, physiology and use as a green manure. *Econ. Bot*. **34**: 111–153.

Malliga, P. and Subramanian, G. 1995. Bacteria associated with leaf cavities of *Azolla pinnata* R. Br. *Indian J. Microbiol*. **35**: 21–24.

Martin, A. R.H. 1976. Some structures in *Azolla* megaspore and an anomalous form. *Rev. Paleobot. Palynol*. **12**: 141–169.

Mayr, E. 1975. Difficulties and importance of the biological species concept. In: *The Species Problem*. Amer. Asoc. Adv. Sci. Publ. No. 50. pp. 371–388.

Melchior, H. and Werdermann, E. 1954. Engler Sallabus der Pflanzenfamilien. 12th edition Vol. 1 Gebruder Borntraeger, Berlin-Nikolassee.

Mettenius, G.H. 1867. Filicinae. In: *Plantae*. Edited by Kotschy, T. and Peyritsch, J. Tinneanal., Wien. pp. 51–54.

Moore, A.W. 1969. *Azolla:* Biology and agronomic significance. *Bot. Rev*. **37**: 17–34.

Nayak, S.K. and Singh, P.K. 1989. Cytological studies in genus *Azolla*. *Cytologia*. **54:** 275–286.

Padhya, M.A. 1997. Efficacy of *Azolla pinnata* R. Br. as a biofertilizer. *Adv. Plant Sci. Res.* **5**: 154–158.

Perkins, S.A., Peters, G.A., Lumpkin, T.A. and Calvert, H.E. 1985. Scanning electron microscopy of perine architecture as a taxonomic tool in the genus *Azolla* Lamarck. *Scan. Elect. Micros.* **4**: 1719–1734.

Peters, G.A. and Meeks, J.C. 1989. The *Azolla –Anabaena* symbiosis: Basic biology. *Ann. Rev. Plant Physiol. Plant Mol. Biol.* **40**: 193–210.

Pieterse, A.H., Delange, L. and Vanvliet, J.P. 1977. A comparative study of *Azolla* in the Netherlands. *Acta. Bot. Neerl.* **26**: 433–449.

Plazinski, J. 1990. The *Azolla–Anabaena* symbiosis In: *Molecular biology of symbiotic nitrogen fixation.* Edited by Gresshoff, P.M. CRC Press, New York. pp. 52–71.

Plazinski, J., Taylor, R., Shaw, W., Croft, L., Rolfe, B.G. and Gunning, B.E.S. 1990. Isolation of *Agrobacterium* sp., strain from the *Azolla* leaf cavity. *FEMS Microbiol. Lett.* **70**: 55–60.

Reed, C.F. 1954. Index marsileata *et. al.* Salviniata. *Bol. Soc. Brot. Ser. 2a.* **28**: 5–61.

Sadebeck, R. 1902. Salviniaceae . In: *Naturl. Pflanzenfam.* Edited by Engler and Prantl. **1**: 383–402.

Saunders, R.M. K. and Fowler, K. 1992. A morphological taxonomic revision of *Azolla* Lam. Section Rhizosperma (Mey) Mett. (Azollaceae). *Bot. J. Linn. Soc.* **109**: 329–357.

Saunders, R.M.K. and Fowler, K. 1993. The superspecific taxonomy and evolution of the fern genus *Azolla* (Azollaceae). *P1. Syst. Evol.* **184**: 175–193.

Seto, K. and Nasu, T. 1975. Discovery of fossil *Azolla* massulae from Japan and some notes on recent Japanese species. *Bull. Osaka Mus. Nat. Hist.* **29**: 51–60.

Shi, D.J., and Hall, D.O. 1988. The *Azolla–Anabaena* association: Historical perspective, symbiosis and energy metabolism. *Bot. Rev.* **54**: 353–386.

Shi, D.J., Li, S.Q. and Chang, Y.Z. 1984. Studies on the microstructure and ultra–structure of photosynthetic apparatuses in *Azolla. Acta Phytotaxon. Sin.* **2**: 32–37.

Singh, P.K., Patra, R.N. and Nayak, S.K. 1984. Sporocarp germination, cytology and mineral nutrition of *Azolla* species. In: *Practical applications of Azolla for rice production*. Edited by Silver, W.S. and Schroder, E.C. International Workshop, Mayaguez, Puerto Rico. The Hague. pp. 55–65.

Smith, G.M. 1938. Salviniaceae. In: *Cryptogamic Botany Vol. 11. Bryophytes and Pteridophytes*. McGraw Hill Inc., New York. pp. 353–362.

Stergianou, K.K. and Fowler, K. 1989. Preliminary report of chromosome counts in genus *Azolla* (Pteriodphyta). *Fern Gaz.* **13**: 317–319.

Stergianou, K.K. and Fowler, K. 1990. Chromosome number and taxonomic implications in the fern genus *Azolla* (Azollaceae). *P1ant Syst. Evol.* **173**: 223–239.

Strasburger, E. 1873. *Ueber Azolla*. Verlag von Ambr. Abel. Jena, Leizig.

Svenson, H.K. 1944. The New World species of *Azolla*. *Amer. Fern J.* **34:** 69–84.

Sweet, A. and Hills, L.V. 1971. A study of *Azolla pinnata* R. Brown. *Amer. Fern J.* **71**: 1–13.

Tan, B.C., Payawal, P., Watanabe, I., Lacdan, N. and Ramirez, C. 1986. Modern taxonomy of *Azolla*: A review. *Phil. Agr.* **69**: 491–512.

Thanh, L.D. 1984. Cytology and morphogenesis of sexual organs of *Azolla*. In: *Advances in nitrogen fixation Research*. Edited by Weeger, C. and Newton, W. E.. The Hague. pp. 509.

Thanh, L. D. and Hang, D. T. 1988. Chromosome number in *Azolla*. In: *Proceedings of 14th International Congress Genetics*, Toronto, Canada.

Van Hove, C., Diarra, H.F. and Godard, P. 1983. *Azolla* In: *West Africa Azolla project*, West Africa Rice Development Association.

Van Ooststroom, S.J. 1948. Azollaceae. *Flora Neerl.* **I**: 79–80.

Wagnar, G.M. 1997. *Azolla*: A review on its biology and utilization. *Bot. Rev.* **63**: 1–26.

West, R.G. 1953. The occurrence of *Azolla* in British interglacial deposits. *New Phytol.* **52**: 267–271.

Wettstein, R. von. 1903. *Handbuch der systematischen Botanik*. 2. Leipzig, Wien Deuticke.

Zimmerman, W.J., Lumpkin, T.A. and Watanabe, I. 1989a. Isozyme differentiation of *Azolla* Lam. *Euphytica*. **42**: 163–170.

Zimmerman, W.J., Lumpkin, T.A. and Watanabe, I. 1989b. Classification of *Azolla* spp., section Azolla. *Euphytica* **43**: 223–237.

Zimmerman, W.J., Watanabe, I. and Plucknett T.A. 1991a. The *Azolla* – *Anabaena* symbiosis: Diversity and relatedness of neotropical host taxa. *Plant Soil*. **137**: 167–170.

Zimmerman, W.J., Watanabe, I., Ventura, T., Payawal, P. and Lumpkin, T.A. 1991b. Aspects of the genetic and botanical status of neotropical *Azolla* species. *New Phytol*. **119**: 561–566.

Section II
PHYSIOLOGY, BIOCHEMISTRY AND ECOLOGY

Chapter 5

NITROGEN METABOLISM IN CYANOBACTERIA

Surendra Singh, Pramod K. Pandey, Vinay S. Chauhan, Bhanumati Singh, Rishi K. Saxena and Prakash S. Bisen

Abstract

Nitrogen constitutes 10 – 11% of the dry-weight of cyanobacterial cell which mainly use inorganic from of nitrogen such as nitrate, nitrite, dinitrogen and ammonium to fulfil their nitrogen demand, but urea and other organic sources of nitrogen can also be assimilated by some cyanobacteria. Most of the nitrogen in biosphere is present in the atmosphere in the form of dinitrogen (N_2) and the majority of organisms cannot use this most abundant form of nitrogen (N_2). Only nitrogen fixing organisms or diazotrophs are capable of using N_2. All these organisms possess nitrogenase enzyme. When combined nitrogen source such as NO_3^-, NH_4^+ and urea is absent, some members of cyanobacterial group differentiate specialized cells termed heterocysts, for aerobic nitrogen fixation. However, NH_4^+ is preferred inorganic nitrogen source which gets incorporated into glutamine and glutmate *via* glutamine synthetase/ glutamine oxoglutarate aminotransferase; (GS/GOGAT) pathway, the primary NH_4^+ assimilating pathway in cyanobacteria. NH_4^+ can be taken up from the external medium by the cyanobacterial cells, but it can also be derived from other nutrients, essentially N_2, NO_3^- and urea. NO_3^-, NO_2^- and NH_4^+ transport is energy-dependent process which occurs through a membrane bound permease system. Photophosphorylation supports the energy requirement for the transport of above nutrients. Photosynthetically reduced ferredoxin acts as an electron donor for NO_3^- reduction. NO_2^- efflux system in cyanobacteria is NO_3^- assimilation and energy – dependent which is regulated at NO_3^- uptake and nitrate reductase

(NR) level. Further, NO_2^- efflux is NH_4^+- repressible, however, the product of NH_4^+ assimilation via GS is required for NH_4^+ repression of NO_2^- efflux. The NR and NiR (nitrite reductase) are substrate – inducible and NH_4^+, arginine and proline-repressible. Molybdenum (Mo) is required as a part of low molecular weight cofactor for the NR activity and the enzyme is rich with sulfhydryl (-SH) group. Upon cyanophage infection, the Mo – cofactor synthesis is elevated and the resulting elevated level of Mo-cofactor finally reconstitutes the large cellular pool of apoprotein into active enzyme. Some amino acids like glutamine, arginine and proline are taken up and concentrated within the cell by means of energy dependent transport system. In cyanobacteria, the assimilation of urea and amino acids have been poorly investigated. However, some observations revealed that the transport system of glutamine, arginine and proline is biphasic, consisting of a rapid phase followed by a slower second phase. It is further evidenced that intracellular glutamine pool regulates the glutamine transport in cyanobacteria. ^{14}C arginine transport and arginase systems are also NH_4^+- repressible and NH_4^+ is the initial repressor. Proline oxidase is a membrane-bound enzyme as maximum activity is found in EDTA – lysozyme extract. Glutamate acts as an inhibitor of heterocyst differentiation and nitrogenase activity than as a N-source and the availability of NO_3^- and NH_4^+ alleviates the glutamate toxicity by inhibiting its uptake. This glutamate toxicity may be the reason for the localization of glutamate – forming enzyme (GOGAT) within vegetative cells and not within heterocysts. The future direction on cyanobacterial research is focussed on the sensing of the N-status of the cell and their transduction to the protein(s) effecting regulation of gene expression.

Key words: ^{14}C – Arginine, arginase, ^{14}C – glutamine, glutamine synthetase, glutamate synthase, ^{14}C -proline uptake, proline oxidase, NO_3^-uptake, nitrate reductase, nitrite reductase, NH_4^+ uptake.

Introduction

Cyanobacteria are a diverse and widely distributed group of prokaryotes which are booned to carry out oxygenic photosynthesis, making a large contribution to global nitrogen fixation by unicellular, filamentous nonheterocystous and filamentous heterocystous forms (Stanier and Cohen-Bazire, 1977; Haselkorn, 1978; Bothe *et. al.*, 1980; Gallon, 1980, 1992; Fay, 1992).

Nitrogen constitutes as much as about 10 – 11% of the total dry weight of an exponentially growing cyanobacterial cell (Wolk, 1991;

Flores and Herrero, 1994; Stal, 2000). Availability of nitrogen has been identified in many instances as the factor limiting productivity in natural habitats. Cyanobacteria mainly use the inogranic sources of N such as dinotrogen, nitrate and ammonium, in decreasing order of availability. In general, they can use nitrate or ammonium as the sole N source for growth and many have additional ability to fix N_2. However, urea and other organic sources of N such as some amino acids, can also be assimilated by some cyanobacteria (Tandeau de Marsac and Houmard, 1993; Singh and Bisen, 1993, 1994 b,d; Flores and Herrero, 1994; Reyes *et. al.*, 1997; Stal, 2000).

Under autotrophic conditions, the utilization of any form of nitrogen by cyanobacteria is strictly dependent on light and CO_2. The light dependence corresponds to the requirement for reductant and ATP in the assimilatory process involved. The CO_2 requirement results from a set of interactions between carbon and nitrogen metabolism, some of which are regulatory in nature and determine that the assimilation of C and N are balanced.

The fact that inorganic nitrogen sources other than NH_4^+ is first metabolised to NH_4^+ prior to assimilation of their nitrogen atoms into nucleic acids, cell wall, porphyrins, polyamines and proteins. Some amino acids, like arginine, are assimilated by the production of NH_4^+ among their catabolic products. Glutamine synthetase/ glutamate synthase (GS/GOGAT) plays a central role in cyanobacterial nitrogen metabolism particularly in NH_4^+ assimilation. Glutamate produced in GS/GOGAT pathway is not only the major nitrogen donor for the biosynthesis of other nitrogen containing metabolites but is itself a precursor for some amino acids and 5-aminolevulinate, the immediate precursor for porphyrin, phycobilin and chlorophyll biosynthesis. Glutamine also donates nitrogen to some metabolites. A general scheme representing major routes in nitrogen-assimilatory metabolism is shown in Fig. 5.1. It is evident that amino acids arginine and aspartate together make up cyanophycin, a unique storage reservoir of nitrogen (and carbon) in many cyanobacteria.

Utilization of different N sources requires their entry through the cell wall with the aid of specialized transport systems followed by their hydrolysis-reduction. Nitrogen-fixing cyanobacteria possess a central position in the nutrient cycling largely due to their inherent capacity to fix atmospheric nitrogen directly into ammonium, making

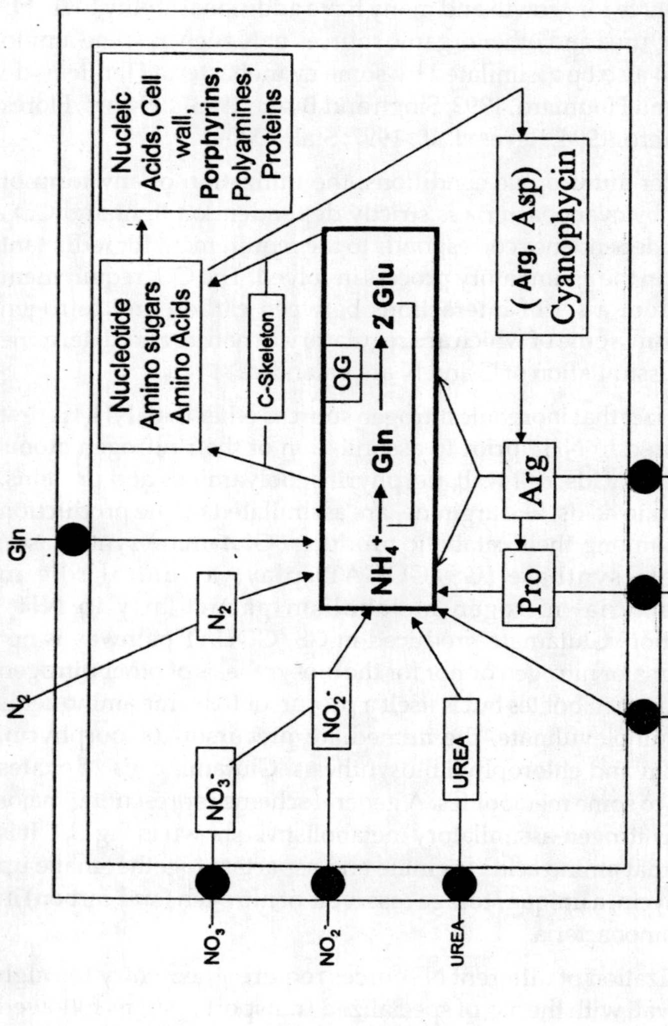

Fig. 5.1: Major routes in cynobacterial N-metabolism. The participation of permeases in the transport of some N-source is shown with red closed circle, OG= 2-Oxoglutarate (After Flores and Herrero, 1994)

it available to higher plants (Kumar and Kumar, 1988; Kumar *et. al.*, 1996). According to one assumption, cyanobacteria fix over 35 million tonnes of nitrogen annually (Hader *et. al.*, 1987; Kumar *et. al.*, 1996).

Several reviews on this topic are available (Stewart, 1980; Wolk, 1980, 1982; Bothe, 1982; Houchins, 1984) however, a better understanding of cyanobacterial nitrogen metabolism is prerequisite for enhancing their potential use as biofertilizer (Hader *et. al.*, 1987; Kumar and Kaushik, 1992; Sinha and Kumar, 1992; Kumar *et. al.*, 1996). In this article we attempt to summarize and discuss some aspects of the biochemistry and physiology of nitrogen assimilation in cyanobacteria.

Nitrogen fixation

Many cyanobacteria are able to grow at the expense of atmospheric nitrogen under aerobic conditions and many more are able to perform N_2 fixation when anaerobic conditions are provided experimentally. Nitrogenase is the conserved enzyme complex that catalyses the ATP – dependent transfer of six electrons from ferredoxin or flavodoxin to N_2 (Howard, 1994) . Hence the requirement for both reductant (as reduced ferredoxin) and energy (ATP) links N_2 fixation to photosynthesis. Nitrogenase is irreversibly inactivated by O_2, therefore all N_2-fixing organisms have evolved methods to protect the enzyme from the damaging O_2 rich atmosphere. This problem is exacerbated in cyanobacteria since they generate O_2 intracellularly during photosynthesis. Cyanobacteria have evolved a number of strategies to overcome the problem of O_2 toxicity to the nitrogenase (Fay, 1992). Filamentous cyanobacteria such as *Nostoc – Anabaena* group form specialized cells termed as heterocysts (hetero = others) that contains N_2 fixing machinery. A unique cell wall that serves a barrier to atmospheric O_2 and elimination of O_2 evolving PS II activity prevent the inactivation of nitrogenase in the heterocyst (Wolk *et. al.*, 1994, 1996). Further details of aerobic and microaerobic fixation as well as other facts of nitrogen fixation like its genetic regulation mechanism involved in heterocyst differentiation etc. have been admirably reviewed (Haselkorn and Builkema, 1992; Bothe, 1982; Wolk *et. al.*, 1994). We discuss here the reaction of cyanobacterial nitrogen fixation in relation to oxygen protection of nitrogenase.

In heterocystous cyanobacteria, the nitrogen fixation is restricted in heterocyst only. An experimental procedure has been established for screening, nonheterocystous cyanobacteria for their ability to perform N_2 fixation in the complete absence of O_2 (Rippka and Waterbury, 1977). The procedure involves subjecting the cells to a period of N starvation under Ar-CO_2 in the light (which could allow the cells to accumulate glycogen); the cells are subsequently treated with 3-(3,4-dicholorophenyl) 1, 1-dimethylurea (DCMU; to avoid photosynthetic O_2 evolution) under an atmosphere of Ar-CO_2. Under these conditions, nitrogenase activity supported by reducing power derived from the metabolism of the previously accumulated glycogen is observable in cyanobacteria. The diffusion of O_2 and the penetration of light to cells near the center of a *Trichodesmium* colony may be extremely limited, which may preserve nitrogenase activity (Carpenter and Price, 1976). However , not all non-heterocystous can fix N_2 in aerobic or microaerobic conditions, for instance, cyanobacteria *Plectonema boryanum* fixes N_2 only under conditions of anaerobiosis (Stewart and Lex, 1970).

Unicellular cyanobacteria from various environmental niche have been shown to fix N_2 (Wyatt and Silvey, 1969; Carpenter and Price, 1976; Reddy *et. al.*, 1993). These organisms are ubiquitous, and unique in their ability to fix N_2 within a cell that maintains a functional O_2 evolving photosynthetic apparatus. In certain strains of *Gloeothece* and *Synechococcus* grown under a light/dark cycle, there is a diurnal pattern of N_2-fixation that peaks in the dark (Huang and Chow, 1988; Rai *et. al.*, 1992). The temporal separation of photosynthesis and N_2-fixation ensures that photosynthetic O_2 evolution is not concurrent with N_2-fixation (Mitsui *et. al.*, 1986). It has been shown that in continuous cultures that were well-mixed and well-aerated, thus preventing O_2 super saturation in the light, N_2-fixation occurs predominantly in the light (Ortega-Calvo and Stal, 1991). This means that more efficient provision of ATP and/or reductant can be supplied to nitrogenase in the light than in dark.

Ferredoxins are low molecule iron proteins, functioning as electron carriers with a low (-400 mV) reduction/oxidation potential. In N_2-fixing cyanobacteria two functionally distinct types of ferredoxins Pet F and FdxH were identified (Böhme and Haselkorn, 1988). Recently, the role of amino acids on oxygen sensitivity to ferredoxin FdxH in *Anabaena variabilis* was discovered (Singh *et. al.*,

1999, 2001). It was observed by protein modeling that exchange of valine to leucine at position 77 of heterocystous ferredoxin $FdxH_2$ was critical in oxygen sensitivity of ferredoxin. Further, the exchange of leucine to valine in FdxH1 makes it O_2 unstable. Thus, overall leucine in ferredoxin at position 77 is critical for its O_2 sensitivity and that protein modeling can be used to identify the role of specific amino acids in O_2 sensitive proteins.

NO_3^- and NO_2^- assimilation

NO_3^- is probably the most abundant source of combined nitrogen for cyanobacterial nutrition (Flores and Herrero, 1994). The assimilation of NO_3^- by cyanobacteria involves NO_3^- uptake and reduction of intracellular NO_3^- (*via* NO_2^-) to NH_4^+, which is incorporated into organic compound. NO_2^- which can also fulfill the nitrogen requirement of cyanobacteria, is taken up into the cell and then reduced to NH_4^+ (Rodrigues *et. al.*, 1992; Flores and Herrero, 1994; Singh *et. al.*, 1999).

NO_3^-/NO_2^- uptake and NO_2^- efflux

In cyanobacteria, the uptake of NO_3^-/NO_2^- is an ATP dependent process which takes place through a membrane bound permease system (Flores *et. al.*, 1983; Singh, 1992b; Luque *et. al.*, 1994; Singh *et. al.*, 1996a).

Nitrate transport in cyanobacteria reflected high affinity for nitrate (Flores *et. al.*, 1983) with Km values of 1-4 µm having been determined in *Synechococcus* sp and 26-30 µm in *Anabaena* sp. (Meeks *et. al.*, 1983; Rodriguez *et. al.*, 1992). The NO_3^- uptake system is mainly dependent on light as the dark incubated cyanobacterial cells showed drastic decline in its NO_3^- accumulating ability as compared to its light grown counterparts (Flores *et. al.*, 1983; Singh, 1992 b; Singh *et. al.*, 1996 b). The competition for uptake between NO_3^- and NO_2^- suggests that the same system operates the active transport of both the substrates (Madueno *et. al.*, 1987; Singh, 1992 b). Furthermore, NO_2^--efflux in dark incubated cells was negligible presumably due to less NO_3^- uptake by such cells and addition of DCMU, an inhibitor of photosystem II (PS II), resulted a six fold decline in NO_3^- uptake activity, suggesting the vital role (s) of PS II mediated energy generation to support the active transport of nutrients across the cyanobacterial membrane. The inhibition of

NO_3^-, NO_2^- uptake and NO_2^- efflux by dicyclohexyl carbodiimide (DCCD), an inhibitor of bacterial ATPase, also suggests the involvement of ATP as a link between energy metabolism and the nutrients uptake/efflux processes (Singh *et. al.*, 1996b). Carbonyl cyanide–m-cholorphenyl hydrazone (CCCP), an uncoupler of photophosphorylation, significantly inhibited NO_3^- uptake, thus in turn inhibiting the NO_2^- efflux activity and suggesting the vital role of cyclic electron flow in the regulation of NO_3^- uptake and NO_2^- efflux processes in cyanobacteria (Table 5.1). PCMB (paracholoromercury benzoate), which is known to interact with sulfhydryl group (-SH) also abolished the NO_3^- uptake/NO_2^- efflux systems in *Nostoc* MAC (Table 5.2). Glutamine, arginine and urea inhibited the NO_2^- efflux system significantly in *Nostoc* MAC. However, the cells lacking functional urease (acetohydroxamic acid treated cells) could not show the urea inhibition of NO_2^- efflux, suggesting the involvement of NH_4^+, resulting from hydrolysis of urea, in the regulation of NO_2^- efflux in *Nostoc* MAC (Singh *et. al.*, 1997). Glucose, on the other hand stimulated the NO_2^- efflux activity, however, this stimulation was inhibited by the simultaneous presence of DCMU. This further confirms the role of PS II generated energy in the stimulation of NO_2^- efflux system in *Nostoc* MAC (Singh *et. al.*, 1997).

The detailed research work on NO_2^- efflux in cyanobacteria was carried out in our laboratory and for the first time it was suggested that: (i) NO_2^- efflux system in *Nosoc* MAC is NO_3^- assimilation and energy dependent process which can be regulated at the level of NO_3^- uptake and nitrate reductase (NR), (ii) similar to NO_3^- uptake,NO_2^- efflux system is NH_4^+ repressible, however, the product of NH_4^+ assimilation *via* GS is required for repression to occur, (iii) *de novo* protein synthesis is required for derepression of the NO_2^- efflux system, (iv) the catalytic activity of NR also seems to play an important role in the regulation of NO_2^- efflux system, (v) the catalytic function of GS does not appear to be involved in NO_3^- assimilation dependent NO_2^- efflux and (vi) the optimum pH, temperature and illumination for maximum NO_2^- efflux were 8.0, 30°C and 50 μmol m^{-2} s^{-1} respectively (Pandey *et. al.*, 1995; Singh *et. al.*, 1996a). Further , it was observed that Na^+ influences the level of accumulation of NO_3^- in *Nostoc* MAC (Singh *et. al.*, 1996b) which is due to the effect of

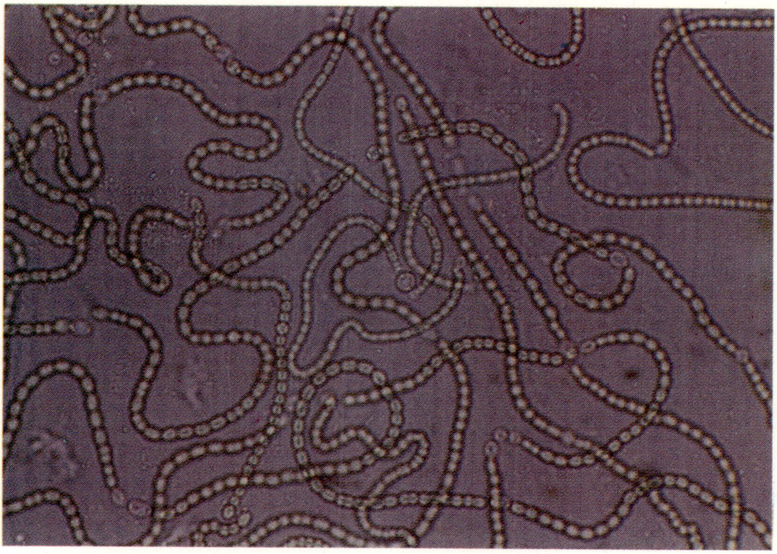

Fig. 2 (a) Wild-type strain of *Anabaena variabilis* showing heterocyst and vegetative cells.

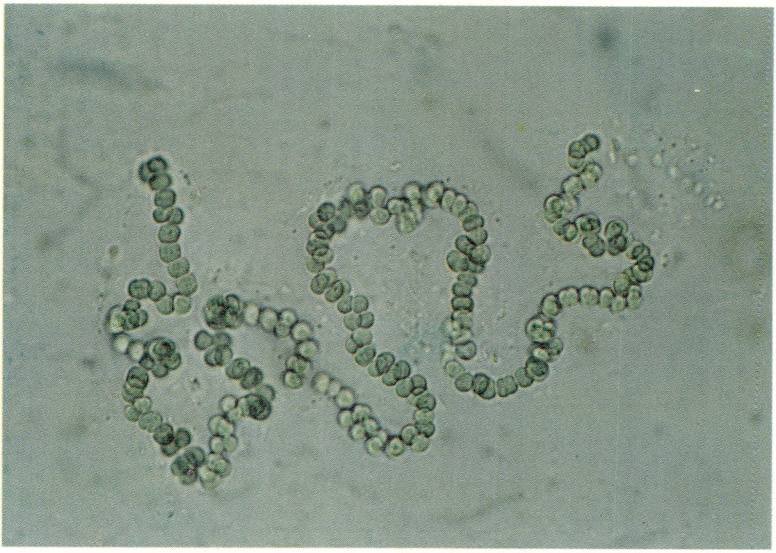

(b) Het–Fix-mutant stratin showing coiled structure

Na^+ on membrane potential affecting a process distinct from NO_3^- inward transport for instance NO_2^- efflux (Singh *et. al.*, 1996a).

Table 5.1: Effect of illumination, darkness, ATP, DCMU and various metabolic inhibitors on nitrate uptake and nitrite efflux activities in *Nostoc* MAC (after Singh *et. al.*, 1996 b), ND – Not detected

Treatments	*Nitrate uptake (nmol. NO_3^- mg^{-1} protein)*	*Nitrite efflux (nmol.NO_2^- mg^{-1} protein)*
Control	133.0	90.2
Dark	13.0	ND
Dark+ ATP	96.2	70.1
Light + DCMU	20.25	3.75
DCCD	6.36	ND
CCCP	10.21	1.02
PCMB	ND	ND
Tungstate	118.0	3.39

Table 5.2: Effect of various inhibitors on percentage inhibition of NR enzyme of *Anabaena doliolum* (after Bisen and Shanthy, 1991)

Inhibitors	*Per cent Inhibition*
Control	0
Erythromycin	96
Chloramphenicol	95
Potassium chlorate	100
Dark	100
Sodium azide	8
PCMB	92
DCMU	89

Recently, an UV-B heat-shock tolerant mutant (UV-HS^t) of the unicellular cyanobacterium *Anacystis nidulans* was isolated (Gour *et. al.*, 1997) and experiments were conducted on the nitrogen metabolism in the mutant and its wild-type counterpart. The UV-HS^t multant strain had lower level of NO_3^- uptake and NO_2^- efflux activities as compared to its wild-type counterpart. The addition of MSX (L-methionine – DL- sulfoximine), a glutamate analogue and an irreversible inhibitor of GS, permitted the development of NO_3^- uptake systems in both wild type and its UV- HS[t] mutant. However, it could not do so in case of the NO_2^- efflux system when NH_4^+ incubated cells were transferred to NO_3^- medium, suggesting that NH_4^+ metabolism through GS is required for regulation of NO_2^- efflux system. It is, however, not required for the regulation of NO_3^- uptake system in *A. nidulans* (Gour *et. al.*, 1997). Similar results on the repression/derepression of NO_2^- uptake has also been reported in *Nostoc* ANTH (Singh, 1993d). A NaCI-tolerant mutant strain of diazotrophic cyanobacterium *Anabaena variabilis* was isolated by nitrosoguanidine (NTG) mutagenesis which has impaired GS activity and liberates NH_4^+ in external medium (Chauhan, 1999 a, b). The mutant strain also takes NO_3^- and NO_2^- more rapidly as compared to its wild-type counterpart in presence and absence of NaCI thus suggesting that under NaCI stress, both the strains take up more NO_3^- and NO_2^- to counteract the deleterious effect of NaCI on N_2 fixing machinery. However, due to NaCI induced inhibition of nitrogen metabolizing enzyme (NR), wild-type strain could not assimilate NO_3^- and hence could not grow and survive under such conditions. Since, NaCI-tolerant mutant having impaired GS showed higher uptake of NO_3^-, NO_2^- as compared to its wild-type counterpart, a role of reduced Na^+ influx and increased Na+ efflux in regulation of NO_3^- and NO_2^- uptake and NaCI tolerance was suggested in mutant strain (Chauhan *et. al.*, 2001). Het^- Fix^- mutant strain of *Anabaena variabilis* showed low rate of nitrate and nitrite uptake as compared to its wild-type counterpart (Singh *et. al.*, 2001). Mutant strain is incapable of heterocyst formation and nitrogenase synthesis, therefore it grows on medium having combined nitrogen. It shows high level of chromatic adaptation with coiled morphology (Fig. 5.2).

NO_3^- reduction

Intracellular NO_3^- is reduced to NO_2^- in a two electron reaction catalysed by nitrate reductase (NR). The resulting NO_2^- is then reduced to NH_4^+ in a six electron reaction catalyzed by nitrite reductase (NiR). Both NR and NiR use reduced ferredoxin as the physiological electron donor (Manzano *et. al.*, 1976; Mentez *et. al.*, 1981; Arizmendi and Serra, 1990). In the reaction catalyzed by NR, two moles of ferredoxin are oxidized per mole of NO_3^- reduced to NO_2^-, whereas in the reaction catalyzed by NiR, six moles of ferredoxin are oxidized per mole of NO_2^- reduced to NH_4^+ . Flavodoxin can serve as an electron donor alternative to ferredoxin during growth under iron limited conditions.

$$NO_3^- + 2Fd\,(red) + 2H^+ \text{---------------} NO_2^- + 2Fd\,(ox) + H_2O$$

$$NO_2^- + 6\,Fd\,(red) + 8H^+ \text{---------------} NH_4^+ + 6Fd(ox) + 2H_2O$$

NR and NiR

NR and NiR catayse the reduction of NO_3^- to NO_2^- and NO_2^- to NH_4^+, respectively. NR from heterocystous cyanobacteria exhibits a biphasic kinetics behaviour with NO_3^- as the variable substrate that is generally not observed with NR from non-heterocystous strains (Martin-Nieto *et. al.*, 1992). Ferredoxin-NR and NiR from several cyanobacteria appear to be tightly attached to thylakoid membranes, and prolonged sonication or treatments with detergents are being required to release the enzyme in soluble form (Hattori and Myers, 1967; Ortega *et. al.*, 1976; Mikami and Ida, 1984). The NR in *Anabaena doliolum* and NiR in *A. cycadae* were reported to be substrate inducible and NH_4^+repressible (Bisen and Shanthy, 1991; Singh, 1992c). Chlorate, which is found to be a competitive inhibitor of NO_3^- assimilation has been used to isolate different class of chlorate resistant mutants to understand the mechanism of regulation of NR in *A. doliolum* (Bisen and Shanthy, 1992a). Molybdenum (Mo) is required as a part of low molecular weight cofactor for the NR activity and the enzyme is rich with sulfhydryl (-SH) groups. DCMU, an inhibitor of PS II, inhibited the NR activity further suggesting the involvement of PS II generated energy in the regulation of NR activity in *A. doliolum* (Table 5.2). An increased NR activity in *Phormidium uncinatum* and *Nostoc muscorum* after cyanophage LPP-1 and N-1 infection, respectively was reported for the first time from our laboratory (Bisen *et. al.*, 1986; Bagchi *et. al.*, 1987). It was observed that NR activity after cyanophage N-1 infection in *N. muscorum* is controlled by two independent factors, which function jointly. It is suggested that upon infection, the Mo cofactor synthesis is elevated and the resulting elevated level of Mo cofactor finally reconstitutes the large cellular pool of apoprotein into active enzyme (Bagchi *el al.*, 1987). While working with a glutamine auxotrophic mutant of *Anabaena cycadeae*, we found that similar to *A. doliolum*, the NR activity is also NO_3^- inducible and arginine repressible (Singh and Bisen, 1994c) and GS activity is not required for arginine inhibition of NR

Table 5.3: ^{14}C labelled arginine uptake and arginase activity of the wild-type and mutant strain of the cyanobacterium *Anabaena cycadeae* (after Singh and Bisen, 1994b)

Nitrogen source	*^{14}C arginine uptake* [a]		*Arginase activity* [b]	
	Wild type	*Mutant*	*Wild type*	*Mutant*
N_2	225.0	232.0	218.6	230.2
NO_3^-	227.6	239.2	219.1	235.7
NH_4^+	33.5	35.3	24.7	27.6
Glutamine	217.7	226.9	215.5	228.9

[a] n mol. mg $^{-1}$ chla^{-1} min^{-1} [b] n mol. urea produced mg^{-1} chla^{-1} min^{-1}

activity. The NR activity in UVHS^t mutant strain of *A. nidulans* is relatively lower as compared to its wild type counterpart (Gour *et. al.*, 1997). UV-B treatment inhibited the NR activity by more than 50% in the wild type, however, it could cause the marginal reduction in NR activity in the UV-HS^t strain. Using glutamine auxotrophic mutant of *A. cycadeae*, it has been reported that NiR is arginine and proline repressible and the catalytic function of GS is necessarily required for the arginine and proline inhibition of NiR (Singh and Bisen, 1994d). NH_4^+ resulting from the catabolism of arginine and proline should be metabolized via GS for repression of NiR to occur. Similar results were also reported on NiR cellular levels in *Nostoc* ANTH in response to NH_4^+ (Singh, 1992d).

NH_4^+ assimilation

NH_4^+ can also be directly taken up from the external medium and cyanobacteria in general are able to grow with NH_4^+ as nitrogen source. NH_4^+ ions are transported by carrier mediated energy dependent transport system in *A. doliolum* (Bisen and Shanthy, 1992b). In general, cellular suspensions of several cyanobacteria have been shown to efficiently take up low concentrations (0.25mM) of NH_4^+ from the outer medium (Healey, 1977; Boussiba, 1989).NH_4^+ transport increased as a function of time in illuminated cells upto 600 µM and then exhibited saturation at 800 µM. However, no NH_4^+ transport was recorded under the dark conditions. The inhibition of NH_4^+ transport in dark or in the presence of DCMU, strongly suggests for a dependence of the NH_4^+ transport on the light energy generated by PS II. Secondly, the inhibition of the NH_4^+ transport by MSX, which is inhibitory to GS, suggests for a coupling of the NH_4^+ transport system with the GS/GOGAT pathway (Bisen and Shanthy, 1992). [^{14}C]- methylammonium can be used as a probe to study the NH_4^+ transport system. Available data suggest that [^{14}C]-methylammonium uptake is mediated by a permease which catalyses the active transport that is dependent on membrane potential, NaCl^r strain of *Anabaena variabilis* in its free-living state assimilated NH_4^+ at a slower rate as compared to its wild counterpart in the absence of NaCI. The addition of NaCI significantly inhibited NH_4^+ uptake and GS activities in the wild type *A. variabilis*, however, it could not do so in NaCl^r mutant strain (Chauhan *et. al.*, 1999c). NH_4^+ upake and GS activity thus showed parallel behaviour in both wild type *A. variabilis* and its NaCl^r mutant. Immobilization of cells further decreased the NH_4^+ uptake and GS activity in both the strains, indicating that the reduction in NH_4^+ uptake and GS activity may be the main cause for increased NH_4^+ production in NaCl^r mutant strain.

There are two major pathways for NH_4^+ incorporation into organic constituents of the cell; directly into glutamate by means of GDH or by a cycle of reactions catalyzed by GS and GOGAT (Fig. 5.1). GDH catalyses the reductive amination of 2-oxoglutarate to glutamate. GS catalyses the ATP dependent ligation of glutamate and NH_4^+ rendering glutamine, and GOGAT transfers the amido group of glutamine to 2-oxoglutarate rendering two glutamate molecules. The concerted action of GS and GOGAT yields one molecule of glutamate per molecule of NH_4^+. However, the GS/GOGAT pathway appears to be a major pathway for NH_4^+ assimilation not only in N_2-fixing, but also in non-N_2-fixing cyanobacteria (Bisen and Shanthy, 1992c; Singh, 1992c) as GDH activity is either very low or even undetectable in some cyanobacteria (Hoare *et. al.*, 1967; Neilson and Doudoroff, 1973). The GS inhibitor, MSX was observed to depress severely the assimilation of newly fixed NH_4^+ in N_2 fixing cultures of *A. cylindrica* (Stewart and Rowell, 1975) , and GOGAT was detected in two heterocyst forming cyanobacteria (Lea and Miflin, 1975). Characterization and regulatory mechanism of GS in *Anabaena doliolum* suggested that the enzyme was synthesized maximally under N_2-fixing conditions than in presence of NH_4^+ as nitrogen source (Meeks *et. al.*, 1977;

Vega-Palas *et. al.*, 1990; Bisen and Shanthy, 1992c). Biosynthetic and γ-glutamyl transferase activities of GS showed specificity towards divalent cations Mg^{2+} and Mn^{2+}, respectively. The enzyme also required these cations to stabilize the structure and function under extreme conditions such as high and low temperature and denaturation (Bisen and Shanthy, 1992c).

Assimilation of urea

Urea appears to be taken up intact and actively by the cells and then metabolized intracellularly (Healey, 1977). Intracellular urea is degraded in cyanobacteria by the enzyme urease releasing one molecule of CO_2 and two molecules of ammonium (Berns *et. al.*, 1966; Mackerras and Smith, 1986; Singh, 1990). Urease has been shown to require the presence of Ni^{2+} in the medium (Singh, 1990). Further studies on the regulation and genetics of urea assimilation in cyanobacteria would be worth while.

Assimilation of amino acids

Although a considerable amount of research work has been done on inorganic nitrogen metabolism, little attention has been paid to the amino acid metabolism in cyanobacteria. Only few amino acids (glutamine, arginine and ornithine) were found to serve as sole nitrogen source for some cyanobacteria (Rawson, 1985; Thiel and Laone, 1986; Herrero and Flores, 1990; Singh, 1993 d). In a detailed study , the ability of nine cyanobacterial strains (both unicellular and filamentous, as well as N_2 fixers and non N_2-fixers) to take up various amino acids was examined (Montesinos *et. al.*, 1997). All the strains had a transport system for neutral amino acids, while some also had systems specific for basic amino acids, confirming earlier work on amino acid transport (Labarre *et. al.*, 1987; Flores and Muro-Paster, 1988). We studied the glutamine, arginine and proline assimilation in various cyanobacteria (Singh and Bisen, 1993, 1994 a, b). Glutamine, arginine and proline were utilized as N-source and were taken up and concentrated within the cyanobacterial cells by means of energy-dependent transport system (Singh, 1992a). The transport system of glutamine, arginine and proline were found to be biphasic consisting of a rapid phase followed by a slower second phase (Singh and Bisen, 1993, 1994a, b). Although glutaminase activity has been detected (Haystead *et. al.*, 1973; Chen *et. al.*, 1987), GOGAT appears to be the main enzyme converting exogenously supplied glutamine to glutamate. Azaserine (an inhibitor of GOGAT) which could not inhibit *in vitro* and *in vivo* GS activity effectively inhibited the ^{14}C glutamine transport. ^{14}C glutamine transport was also not significantly affected when glutamate, methylglutamate, aspartate, arginine, lysine, hydroxylysine, ornithine and MSX were simultaneously available during transport assay, thus suggesting that glutamine uptake takes place *via* a general amino acid permease, which does not, however, transport basic and acidic amino acids (Singh and Bisen, 1994a). Furthermore, azaserine treated cells had increased and decreased levels of glutamine and glutamate respectively, suggesting that the increased intracellular glutamine level is responsible for the inhibition of ^{14}C glutamine uptake. It was the first report for the role of intracellular glutamine pool in the regulation of glutamine transport *in Anabaena* 7120 (Singh and Bisen, 1994a). The catabolism of arginine in cyanobacteria occurs by two enzymes, arginase and arginine deaminase (Hood and Carr, 1971; Gupta and Carr, 1981; Martel *et. al.*, 1993). The former enzymatic process catalyses the catabolism of arginine to ornithine and urea (Weathers *et. al.*, 1978). The observations on nitrogen regulation of arginine uptake and arginase activity in *A. cycadeae* and its gutamine auxotrophic mutant (Singh and Bisen, 1994b) suggest that the development of ^{14}C- arginine uptake and arginase systems started simultaneously and both the systems were independent. ^{14}C - arginine uptake and arginase systems were NH_4^+ repressible, and NH_4^+ was the initial repressor signal (Table 5.3). However, glutamine could not inhibit the ^{14}C arginine uptake and arginase activity, suggesting that repression of ^{14}C arginine uptake and arginase by NH_4^+ does not require the operation of GS and that NH_4^+ itself is the repressor signal (Singh, 1993b; Singh and Bisen, 1994b).

Proline, which acts as an osmoprotectant in cyanobacteria is utilized as sole source of nitrogen by *Nostoc* ANTH (Singh and Bisen, 1993). GS activity is necessarily required for the assimilation of proline as sole nitrogen source. It is, however, not required for the uptake of proline. Similar results were also observed in *A. cycadeae* (Singh, 1993b). ^{14}C- proline uptake is constitutive as the uptake was similar in N_2, NH_4^+ and glutamine grown cells and could not be induced by incubating the cells with proline. ^{14}C- proline uptake was also an energy dependent process as the metabolic inhibitors and energy uncouplers, DCMU, HOQNO (2-heptyl-4-

hydroxyquinoline), KCN (potassium cyanide), FCCP (p-trifluoromethoxy phenyl hydrazone), CCCP, DCCD inhibited the ^{14}C- proline uptake completely. Intracellular glutamine pool does not regulate the proline uptake (Singh and Bisen, 1993). The proline oxidase activity was rather similar in N_2, NH_4^+ and glutamine grown cells, however, NO_3^- grown cells had reduced level of proline oxidase activity. Proline oxidase is a membrane bound enzyme as maximum enzyme activity was observed in EDTA-lysozyme extract of *Nostoc* ANTH (Singh and Bisen,1993). Glutamate acts an inhibitor of heterocyst differentiation and nitrogenase activity rather than as a nitrogen source in *Nostoc* ANTH and that availability of NO_3^- and NH_4^+ alleviates the glutamate toxicity by inhibiting its uptake (Singh, 1993c). This glutamate toxicity may be the reason for the localization of glutamate forming enzyme, GOGAT within vegetative cells and not within heterocysts.

References

Arizmendi, J.M. and Serra,. J.L. 1990. Purification of some properties of the nitrite reductase from cyanobacterium *Phormidium liminosum*. *Biochim Biophys. Acta.* **1040**: 237-244.

Bagchi, S.N., Kaloya, P. and Bisen, P.S. 1987. Effect of cyanophage N-1 infection on the synthesis and stability of *Nostoc muscorum* nitrate reductase. *Curr. Microbiol.* **15**: 61-66.

Berns, D.S., Holohan, P. and Scott., E. 1996. Urease activity in blue-green algae. *Science* **152**: 1077-1078.

Bisen, P.S., Audholia, S. and Bagchi, S.N. 1986. Nitrate reductase activity of a cyanobacterium *Phormidium uncinatum* after cyanophage LPP-1 infection. *FEMS Microbiol. Lett.* **33**: 69-72.

Bisen, P.S. and Shanthy, S. 1991. Regulation of assimilatory nitrate reductase in the cyanobacterium *Anabaena doliolum*. *Curr. Microbiol.* **23**: 239-244.

Bisen, P.S. and Shanthy, S. 1992 a. Physiological and biochemical characterization of chlorate resistant mutants of *Anabaena doliolum*. *Curr. Microbiol.* **25**: 353-358.

Bisen, P.S. and Shanthy, S. 1992b. Modifiers of ammonium transport in the nitrogen-fixing cyanobacterium *Anabaena doliolum*. In: *National Symposium on Cyanobacterial Nitrogen fixation.* Edited by Kaushik, B.D. Associated Publ., New Delhi, pp. 41-47.

Bisen, P.S. and Shanthy, S. 1992c. Biochemical characterization of glutamine synthetase from the diazotrophic cyanobacterium *Anabaena doliolum*. *Curr. Microbiol.* **25**: 69-75.

Böhme, H. and Haselkorn, R. 1988. Molecular cloning and nucleotide sequence analysis of the gene coding for heterocyst ferredoxin from the cyanobacterium *Anabaena* sp strain PCC 7120. *Mol. Gen. Genet*. **214**: 278-285.

Bothe, H. 1982. Nitrogen fixation. In: *The Biology of Cyanobacteria*. Edited by Carr, N.G. and Whitton, B.A. Blackwell Sci. Publ., Oxford. pp. 87-104.

Bothe, H., Neuer, G., Kalbe, I. and Eisbrenner, G. 1980. Electron donors and hydrogenase in nitrogen-fixing micro-organisms. In: *Nitrogen-Fixation*. Edited by Stewart, W.D.P. and Gallon, J.R., Academic Press., London. 83-112.

Boussiba, S. 1989. Ammonia uptake in the alkalophilic cyanobacterium *Spirulina platensis*. *Plant Cell Physiol.* **30**: 303-308.

Carpenter, E. and Price, C.C. 1976. Marine *Oscillatoria (Trichodesmium)* explanation for aerobic nitrogen fixation without heterocyst: *Science* **191**: 1278-1280.

Chauhan, V.S., Singh, B., Singh, S. and Bisen, P.S. 2001. Regulation of Sodium influx in the NaCI-resistant (NaCI^r) mutant strain of the cyanboacterium *Anabaena variabilis*. *Curr. Microbiol.* **42**: 100-105.

Chauhan, V.S., Singh, S. and Bisen, P.S. 1999a. Development of NaCI resistant mutant strain of the diazotrophic cyanobacterium *Anabaena variabilis* having derepressed nitrogenase and impaired glutamine synthetase activities. *Physiol.Mol. Biol. Plants* **5**: 33-36.

Chauhan, V.S., Singh, S., Pandey, P.K. and Bisen, P.S. 1999b. Isolation and partial characterization of NaCI tolerant mutant strain of *Anabaena variabilis* with impaired glutamine synthetase ctivity. *J. Basic Microbiol.* **39**: 219-226.

Chauhan, V.S., Singh, S., Gour, R.K. and Bisen, P.S. 1999c. Immobilization caused physiological and biochemical changes in NaCI resistant (NaCI^r) mutant strain of diazotrophic cyanobacterium *Anabaena variabilis*. *Indian J. Expl. Biol.* **32**: 696-700.

Chen, C., Van Baalen, C. and Tabita, F.R. 1987. DL-7- azqtryptophan and citrulline metabolism in the cyanobacterium *Anabaena* sp.strain IF. *J. Bacteriol.* **169**: 1114-1119.

Fay, P. 1992. Oxygen relation to nitrogen fixation in cyanobacteria. *Microbiol. Rev.* **560**: 340-373.

Flores, E., Guerrero, M.G. and Losada, M. 1983. Photosynthetic nature of nitrate uptake and reduction in cyanobacteria *Anacystis nidulans*. *Biochim. Biophys. Acta* **722**: 408-418.

Flores, E. and Herrero, A. 1994. Assimilatory nitrogen metabolism and its regulation. In: *The Molecular Biology of Cyanobacteria.* Edited by Bryant, D.A. Kluwer Academic Publ., The Netherlands. pp. 487-517.

Flores, E. and Muro-Pastor, M.I. 1991. Mutational and kinetic amino acid analysis of basic amino acid transport in the cyanobacterium *Synechocystis* sp. PCC 6803. *Arch. Microbiol.* **154**: 521-527.

Gallon, J. R. 1992. Reconciling the incompatible: N_2 fixation and CO_2. *New Phytol.* **122**: 571-609.

Gallon, J.R. 1980. Nitrogen fixation by photoautotrophs. In: *International Symposium on Nitrogen Fixation*. Edited by Stewart, W.D.P. and Gallon, J.R., Academic Press , N.Y. **451**: 197-238.

Gour, R.K., Singh, S., Pandey, P.K. and Bisen, P.S. 1997.UV-B and heat shock induced changes in the wild type and UV-B heat shock tolerant (UV-HS^t) strain of the unicellular cyanobacterium *Anacystis nidulans*. *J. Basic Microbiol.* **37**: 259-267.

Gupta, M. and Carr, N.G. 1981. Enzymology of arginine metabolism in heterocyst forming cyanobacteria. *FEMS. Microbiol. Lett.* **12**: 179-181.

Häder, D.P., Worrest, R.C. and Kumar, H.D. 1987. Aquatic ecosystems, UNEP Environmental Effects Panel Report , pp. 39-48.

Haselkorn, R. 1978. Heterocysts. *Ann. Rev. Plant Physiol.* **29**: 319-344.

Hattori, A. and Myers, J. 1967. Reduction of nitrate and nitrite by subcellular preparation of *Anabaena cylindrica*. II Reduction of nitrate to nitrite. *Plant Cell Physiol.* **8**: 327-337.

Haystead, A., Dharmawardene, M.W.N. and Stewart, W.D.P. 1973. Ammonia assimilation in a nitrogen-fixing blue green algae. *Plant Sci.Lett.* **1**: 439-445.

Healey, F.P. 1977. Ammonium and urea uptake by some fresh water algae. *Can. J. Bot.* **55**: 61-65.

Herrero, A and Flores, E. 1990. Transport of basic amino acids by the nitrogen fixing cyanobacterium *Anabaena* PCC 7120. *J. Biol. Chem.* **265**: 3931-3935.

Hoare, D.S., Hoare, S.L. and Moore, R.B. 1967. The photoassimilation of organic compounds by autotrophic blue green algae. *J. Gen. Microbiol.***73**: 317-326.

Hood, W. and Carr, N.G. 1971. Apparent lack of control by repression of arginine metabolism in blue green algae. *J. Bacteriol.* **107**: 365-367.

Houchins, J.P. 1984. The physiology and biochemistry of hydrogen metabolism in cyanobacteria. *Biochim. Biophys. Acta* **768**: 227-265.

Huang, T. C.V. and Chow, T.J. 1988. Comparative studies of some nitrogen fixing unicellular cyanobacteria isolated from rice field. *J. Gen. Microbiol.* **134:** 3089-3097.

Kumar, A. and Kumar, H.D. 1988. Nitrogen-fixation by blue-green algae, In: *Plant Physiology Research. Society for Plant Physiology and Biochemistry*. Edited by Sen, S.P. Ist International Congress of Plant Physiology, New Delhi, India. pp. 85-103.

Kumar, A., Sinha, R.P. and Hader, D.P. 1996. Effect of UV-B on enzyme of nitrogen metabolism in the cyanobacterium *Nostoc calcicola*. *J. Plant Physiol.* **148**: 86-91.

Kumar, H. and Kaushik, B.D. 1992. Standardization of algalization technology for amelioration of salt affected soils. In: *National Symposium on Cyanobacterial Nitrogen-fixation.* Edited by Kaushik, B.D. Associate Publ.Co., New Delhi, India. pp. 343-352.

Labarre, J., Thuriaux, P. and Chauvat, F. 1987. Genetic analysis of amino acid transport in the facultatively heterotrophic cyanobacterium *Synechocystis* sp. strain 6803.*J.Bacteriol.* **169**: 4668-4673.

Lea, P.J. and Miflin, B.J. 1975. Glutamate synthase in blue-green algae. *Biochem. Soc. Trans.* **3**: 381-384.

Luque, I., Flores, E. and Herrero, A. 1994. Nitrate and nitrite transport in the *cyanobacterium Synechococcus* sp.strain PCC 7942 are mediated by same permease. *Biochim. Biophys. Acta* **1184**: 296-298.

Mackerras, A.H. and Smith, G.D. 1986. Urease activity of the cyanobacterium *Anabaena cylindrica. J. Gen. Microbiol.* **132**: 2749-2752.

Madueno, F., Flores, E. and Guerrero, M.G. 1987. Competition between nitrate and nitrite uptake in the cynobacterium *Anacystis nidulans. Biochim. Biophys. Acta* **896**: 109-112.

Manzano, C., Candau, P., Gomez Moreno, C., Relimpio, A.M. and Losada, M. 1976. Ferredoxin-dependent photosynthetic reduction of nitrate and nitrite by particles of *Anacystis nidulans. Mol. Cell Biochem.* **10**: 161-169.

Martein-Nieto, J., Flores, E. and Herrero, A. 1992. Biphasic kinetic behaviour of nitrate reductase from heterocystous, nitrogen-fixing cyanobacterium. *Plant Physiol.* **100**: 157-163.

Martel. A., Jansson, E. Gracia-Reina, G. and Lindblad, P. 1993. Ornithine cycle in *Nostoc* PCC 73102. Arginase OCT and arginine deaminase and the effect of the addition of external arginine, ornithine or citrulline. *Arch. Microbiol.* **151**: 475-478.

Meeks, J.C., Wolk, C.P., Thomas, J., Lockau,W. and Shaffer, P.W. 1977. The pathways of assimilation of $^{13}NH_4^+$ by the cyanobacterium *Anabaena cylindrica. J. Biol. Chem.* **252**: 7894-7900.

Meeks, J.C., Wykoff, K.C., Chapman, J.S. and Enderlin, C.S.C. 1983. Regulation of expression of nitrate and dinitrogen assimilation by *Anabaena* species. *Appl. Environ. Microbiol.* **45**: 1351-1359.

Mendez, J., Herrero, A. and Vega, J.M. 1981. Characterization of catalytic properties of nitrite reductase from *Anabaena* sp. 7119. *Z. Pfanzenphysiol.* **103**: 305-315.

Mikami, B. and Ida, S. 1984. Purification and properties of ferredoxin-nitrate reductase from the cyanobacterium *Plectonema boryanum. Biochim. Biophys. Acta* **791**: 294-304.

Mitsui, A., Kumazawa, S., Takahasi, A., Ikemoto, H.S. and Arai, T. 1986. Strategy by which nitrogen fixing unicellular cyanobacteria grow photoautotrophically. *Nature* **323**: 720-722.

Montesinos, M.L., Herrero, A. and Flores., E. 1997. Amino acid transport systems in taxonomically diverse cyanobacteria and identification of two genes encoding elements of a neutral amino acid permease putatively involved in recapture of leaked hydrophobic amino acids. *J. Bacteriol.* **179**: 853-862.

Neilson, A.H. and Doudoroff, M. 1973. Ammonia assimilation in blue-green algae. *Arch Microbiol.* **89**: 15-22.

Ortega-Calvo, J.J. and Stal, L.J. 1991. Diazotrophic growth of the unicellular cyanobacterium *Gloeothece* sp PCC 690 in continuous culture. *J. Gen. Microbiol.* **137**: 1789-1797.

Ortega, T., Castillo, F. and Cardenas, J. 1976. Photolysis of water coupled to nitrate reduction by *Nostoc muscorum* subcellular particles. Biochem. Biophys. Res. Commun. **71**: 885-891.

Pandey, P.K., Singh, B.B., Singh, S. and Bisen, P.S. 1995. NO_2^- efflux and its regulation in cyanobacterium *Nostoc* MAC. *Curr. Microbiol.* **31**: 119-123.

Rai, A.N.,Borthakur, M. and Bergman, B. 1992. Nitrogenase deprivation, its regulation and metabolic changes associated with diazotrophy in the non-heterocystous cyanobacterium *Plectonema boryanum* PCC 73110. *J. Gen Microbiol.* **138**: 481-491.

Reddy, K.J., Haskell, J.B., Sherman, D.M. and Sherman, L.M. 1993. Unicellular aerobic nitrogen fixing cyanobacteria of the genus *Cyanothece. J. Bacteriol.* **175**: 1284-1292.

Reyes, J.C., Muro-Pastor, M.I. and Florencio, F.J. 1997. Transcription of glutamine synthetase gene (*gln A* and *gln N*) from the cyanobacterium *Synechocystis* sp. strain PCC 6803 is differently regulated in response to nitrogen availability. *J. Bacteriol.* **179**: 2678-2689.

Rippka, R. and Waterbury, J.B. 1979. The synthesis of nitrogenase by non-heterocystous cyanobacteria. *FEMS. Microbiol. Lett.* **2**: 83-86.

Rodriguez, R., Lara, C. and Guerrero, G.C. 1992. Nitrate transport in the cyanobacterium *Anacystis nidulans* R2: kinetic and energetic aspects. *Biochem. J.* **282**: 639-643.

Rowson, D.M. 1985. The effect of exogenous amino acids on growth and nitrogenase activity in the cyanobacterium *Anabaena cylindrica* PCC 7122. *J. Gen. Microbiol.* **131**: 2549-2554.

Singh, B., Chauhan, V.S., Gour, R.K., Singh, S. and Bisen, P.S. 1999. Heterocysts: Current status and future prospects. *J. Indian Bot. Soc.* **78**: 25-32.

Singh, B., Chauhan, V.S., Singh, S. and Bisen,P.S. 2001. Isolation and partial characterization of Het Fix mutant strain of the diazotrophic cyanobacterium *Anabaena variabilis* showing chromatic adaptation. *Curr. Microbiol.* **42**: 321-327.

Singh, B.B., Curdt, I., Jakobs, C., Shomberg, D., Bisen, P.S. and Bohme, H. 1999. Identification of amino acids responsible for the oxygen sensitivity of ferredoxin from *Anabaena variabilis* using site-directed mutagenesis. *Biochim. Biophy. Acta* **1412**: 288-294.

Singh, B.B., Curdt, I., Shomberg, D., Bisen, P.S. and Böhme, H. 2001. Valine 77 of heterocystous ferredoxin FdxH2 in *Anabaena variabilis* strain ATCC 29413 is critical for its oxygen sensitivity. *Mol. Cell. Biochem.* **217**: 137-142.

Singh, B.B. , Pandey, P.K., Singh, S. and Bisen, P.S. 1996a. Evidence for the nitrate assimilation-dependent nitrate excretion in cyanobacterium *Nostoc* MAC. *World J. Microbiol. Biotechnol.* **12**: 285-287.

Singh, B.B., Pandey, P.K., Singh, S. and Bisen, P.S. 1996b. Regulation of nitrate uptake and nitrite efflux in the cyanobacterium *Nostoc* MAC. *J. Basic Microbiol.* **36**: 433-438.

Singh, S. 1992a. Regulation of glutamine uptake in the cyanobiont *Nostoc* ANTH.*FEMS. Microbiol. Lett.* **98**: 245-248.

Singh, S. 1992b. Regulation of nitrite uptake in the cyanobiont *Nostoc* ANTH. *Indian J. Exp. Biol.* **30**: 288-291.

Singh, S. 1992c Nitrite metabolism in the cyanobacterium *Anabaena cycadeae*: Regulation of nitrite uptake and nitrite reductase by ammonia. *FEMS. Microbiol. Lett.* **98**: 249-254.

Singh, S. 1992d.Regulation of nitrite reductase cellular levels in the cyanobiont *Nostoc* ANTH. *Biochem. Physiol. Pflanzen.* **188**: 241-246.

Singh, S. 1993 a. Role of glutamine synthetase, glutamine and NH_4^+ in the regulation of glutamine uptake in the cyanobacterium *Anabaena cycadeae. J. Gen. Appl. Microbiol.* **39**: 57-64.

Singh, S. 1993 b. Role of glutamine synthetase activity in the uptake and metabolism of arginine and proline in the cyanobacterium *Anabaena cycadeae. FEMS. Microbiol. Lett.* **106**: 335-340.

Singh, S. 1993 c. Regulation of glutamate metabolism in the cyanobiont *Nostoc* ANTH by nitrogen sources. *J. Basic Microbiol.* **33**: 41-45.

Singh, S. 1993d. Role of glutamine synthetase, glutamine and NH_4^+ in the regulation of NO_2^- uptake in the cyanobiont *Nostoc* ANTH. *J. Plant Physiol.* **142**: 403-406.

Singh, S. and Bisen, P.S. 1993. Proline metabolism and its regulation in the cyanobiont *Nostoc* ANTH. *J. Plant Sci. Res.* **9**: 1-6.

Singh, S. and Bisen, P.S. 1994a. Evidence for the role of intracellular glutamine level in the regulation of glutamine uptake in the *cyanobacteirum Anabaena* 7120. *Curr. Microbiol.* **29**: 319-322.

Singh, S. and Bisen, P.S. 1994b. Arginine metabolism in cyanobacterium *Anabaena cycadeae*: regulation of arginine uptake and arginase by ammonia. *Curr. Microbiol.* **29**: 49-52.

Singh, S. and Bisen, P.S. 1994 c. Glutamine synthetase and arginine inhibition of nitrate reductase activity in *Anabaena cycadeae. World. J. Microbiol. Biotechnol.* **10**: 191-193.

Singh, S. and Bisen, P.S. 1994d. Inhibition of nitrite reductase and urease by arginine and proline in the cyanobacterium *Anabaena cycadeae. J. Basic Microbiol.* **34:** 401-404.

Singh, S.,Chauhan, V.S. and Bisen, P.S. 1997. Nitrogen and carbon regulation of NO_2^- efflux in the cyanobacterium *Nostoc* MAC. *Physiol. Mol. Biol. Plants.* **3**: 171-172.

Sinha, R.P. and Kumar, A. 1992. Screening of blue-green algae for biofertilizer. In: *National Seminar on Organic Farming*. Edited by Patil, P.S., Pune, India. pp. 95-97.

Stal, L.J. 2000. Cyanobacterial mats and stromalites. In: *The Ecology of Cyanobacteria*. Edited by Whitton, B.A. and Potts, M. Kluwer Academic Publ,, Dordrecht, London. pp.61-120 .

Stanier, R.Y. and Cohen-Bazire, G. 1977. Phototrophic prokaryotes: The Cyanobacteria. *Ann. Rev. Microbiol.* **31**: 225-274.

Stewart, W.D.P. 1980. Some aspects of structure and function in N_2 fixing cyanobacteria. *Ann. Rev. Microbiol.* **34**: 497-536.

Stewart, W.D.P. and Lex, M. 1970. Nitrogenase activity in the blue green algae *Plectonema boryanum* strain 594. *Arch.Microbiol.* **73**: 250-256.

Stewart, W.D. P. and Rowell, P. 1975. Effect of L-methionine-DL-sulfoximine on the assimilation of newly fixed NH_3, acetylene reduction and heterocyst production in *Anabaena cylindrica*. *Biochem. Biophys. Res. Commun.* **65**: 846-856.

Tandeau de Marsac, N. and Houmard, J. 1993. Adaptation of cyanobacteria to environmental stimuli: new steps towards molecular mechanisms. *FEMS Microbiol. Rev.* **104**: 119-190.

Theil, T. and Leone, M. 1986. Effect of glutamine on growth and heterocysts differentiation in the cyanobacterium *Anabaena variabilis*. *J. Bacteriol.* **168**: 769-774.

Vega-Palas, M.A., Madueno, F., Herrero, A. and Flores, E. 1990. Identification and cloning of a regulatory gene for nitrogen assimilation in the cyanobacterium *Synechococcus sp.* strain PCC 7942. *J. Bacteriol.* **172**: 643-647.

Weathers, P.J., Chee, H.L. and Allen, M.M. 1978. Arginine catabolism in *Aphanocapsa* 6308. *Arch. Microbiol.* **118**: 1-6.

Wolk, C.P. 1973. Physiology and cytological chemistry of blue-green algae. *Bact. Rev.* **37**: 32-101.

Wolk, C.P. 1980. Heterocysts ^{13}N and N_2 fixing plants. In: *Nitrogen Fixation*. Edited by Newton, W.E. and Orme-Johnson. University Park Press, Baltimore, Vol. II, pp. 279-292.

Wolk, C.P. 1982. Heterocysts. In: *The Biology of Cyanobacteria*. Edited by Carr, N.G. and Whitton, B.A. Blackwell Science Publ., Oxford, pp. 359-386.

Wolk, C.P. 1991. Genetic analysis of cyanobacterial development. *Curr. Opin. Gene. Develop.* **1**: 336-341.

Wyatt, J.T. and Silvey, J.K.G. 1969. Nitrogen fixation by *Gloeocapsa*. *Science*. **165**: 908-909.

Chapter 6

IMPACT OF Cu, Zn AND Cd ON CERTAIN PHYSIOLOGICAL AND BIOCHEMICAL CHARACTERISTICS OF *MICROCYSTIS* SP.

Subhashree Pradhan and L.C. Rai

Abstract

This paper provides information on the growth, nutrient (NO^-_3, NH^+_4, PO^{3-}_4 and urea) uptake, activities of urease NR, ATPase, Apase, O_2 evolution, ATP content and electron transport chain of the less studied nuisance cyanobacterium *Microcystis*. In contrast to NO_3^-, PO^{3-}_4 and urea, NH^+_4 uptake showed the highest sensitivity to all the metals. However, NO^-_3 uptake was maximally inhibited by Cu. The inhibition of nutrient was generally a function of metal concentration. In the case of enzyme activities, Cd showed maximum toxicity followed by Cu and Zn except for NR which showed greater inhibition to Cu than other metals tested. Like those of nutrients uptake, the enzyme activities also depicted a concentration dependent inhibition.

Though all the reactions of photosynthetic electron transport showed significant inhibition by Cu, Zn and Cd, the reaction II ($H_2O \rightarrow PBQ$) showed the highest inhibition. PSII also registered greater inhibition than PSI and whole chain electron transport. This study suggests a generalized toxicity hierarchy of Cd> Cu> Zn on different variables of the test cyanobacterium and the toxicity was generally concentration dependent.

Key words: Algal growth, Cyanobacteria, heavy metal, *Microcystis*, nutrient, toxicity.

Introduction

Increased anthropogenic loading on earth coupled with rapid industrialization, intensive agriculture, fast depletion of non-renewable resources and high demand for various synthetic chemicals has accelerated the release of toxic chemicals into the environment thus polluting our aquatic and terrestrial ecosystems. Unlike other chemical contaminants heavy metals are non-biodegradable, hence they persist in the sediment. Following their gradual release into water bodies, metals tend to penetrate the biological membrane that separates organisms from their surrounding milieu, and affect organism in various ways. A number of metals in traces are essential for various metabolic activities of algae, cyanobacteria and higher plants (Rai *et. al.*, 1981; De Filippis and Pallaghy, 1994). However, these metals become toxic when present at high concentration. Heavy metals known to be involved in the stabilization of a host of metabolic processes from cell walls to protein conformation. They are often highly effective catalysts of a range of diverse biochemical processes and in some cases are able to trigger or inhibit reactions.

Cu is an essential redox-active micronutrient that functions as an enzyme catalyst and forms an important part of prosthetic groups of many enzymes. It acts as an essential component of plastocyanin, PS I and enzyme amine oxidase. It stimulates growth of *Chlorococcus, Scenedesmus, Ankistrodesmus* (Devi Prasad and Devi Prasad, 1982). Likewise Zn is another essential trace metal for algae. Zn acts as a prosthetic group of enzymes like alkaline phosphatase, dipeptidase, aldolase, alcohol dehydrogenase, carbonic anhydrase and lactic dehydrogenase (Coleman, 1984; Harmens *et. al.*, 1993; Pinton *et. al.*, 1994). It also plays an important role in the maintenance of integrity of ribosomes (Rai *et. al.*, 1981) and transport and assimilation of nutrients. Cd is unique among metals and is always found in association with Zn. In sharp contrast to Zn, essentiality of Cd has not been fully demonstrated for biological species. However, some processes where Cd is known to be involved include the stimulation of protease, RNAase, acid and alkaline phosphatase in various plants (Lee *et. al.*, 1976; Jana and Choudhari, 1982).

Most studies of heavy metal toxicity have merely determined the concentration of a metal that envokes inhibitory and lethal responses in the test organisms under controlled conditions. A number of reviews on metal toxicity to algae have been published during the last decade (Rai *et. al.*, 1981; De Filippis and Pallaghy, 1994). Algae being primary producers of aquatic ecosystem are the most suitable materials for toxicity bioassay because they possess high accumulation potential due to their vast surface exchange area exposed to the environment (Davies, 1978). A vast literature has accumulated on inhibition of algal growth by heavy metals like Cu (Rachlin *et. al.*, 1982a; Rai *et. al.*, 1994). Excess Cu may replace other metals in metalloproteins or may interact directly with -SH groups of proteins (Uribe and Stark, 1982). Excess Cu inhibits a large number of enzymes and several aspects of plant processes, including photosynthesis (Vavilin *et. al.*, 1995; Yruela *et. al.*, 1996), pigment synthesis (Chettri *et al.*, 1998), permeability of plasma membrane causing loss of potassium from the cells (De Filippis, 1979), reduce pigment concentration (Rijstenbil *et. al.*, 1994), inhibit NO^-_3 uptake and nitrate reductase (Florence and Stauber, 1986).

Cadmium has been reported to inhibit growth of a variety of algae (Devi Prasad and Devi Prasad, 1982; Husaini and Rai, 1991; Payne and Price, 1999). De Filippis *et. al.* (1981) while studying the effects of Cd on *Euglena* found that this metal inhibited NADP oxidoreductase, there by significantly lowering the intracellular supply of NADPH. This result is an outcome of the inhibition of the enzyme which requires sulphydryl groups for its activity and Cd is a strong sulphydryl antagonist (Vallee and Ulmer, 1972). Exposure to Cd did result in a significant increase in the surface area of the thylakoids and coalescence of cellular lipid inclusions of treated *Plectonema boryanum* cells when compared to the control cells (Rachlin *et. al.*, 1982c). Rachlin *et. al.* (1982a) also reported that Cd exposure produced a significant enlargement in the polyphosphate bodies of *Plectonema boryanum*. Cd is known to inhibit the nutrient uptake (Rai *et. al.*, 1994) and enzyme activities like alkaline phosphatase in *Nostoc linckia* (Husaini and Rai, 1991) nitrate reductase in *Scenedesmus obliquus* (Hofslagare, 1985).

Cyanobacteria have simple prokaryotic cell organization. They carry oxygenic photosynthesis, a process presumably the *most* sensitive indicator of heavy metal toxicity (De Filippis and Pallaghy, 1976). Impact of heavy metals on natural phytoplankton (Davies, 1983), algae (Sorentino, 1979; Rai *et. al.*, 1981), and cyanobacteria (Stratton and Corke, 1979; Les and Walker, 1984) has been the subject of numerous reports. Cyanobacterium *Microcystis* sp. is often dominant in the plankton of many ponds, lakes, and reservoirs of tropical as well as temperate regions. It is known, to proliferate in nutrient rich water bodies and produce toxins in temperate lakes. Because of prevalence in natural water bodies, there is every possibility for this organism to be exposed to metal insults. Surprisingly nothing is known about the toxicity of the test metals on this organism.

Attempts have therefore been made to investigate the effects of different concentrations of Cu, Zn and Cd on O_2 evolution, ATP content, nutrient uptake, enzyme activity of bloom forming cyanobacterium *Microcystis* sp.

Materials and methods

The test organism *Microcystis* sp. collected from Laxmi kund in Varanasi, was isolated in pure form and grown in Parker's medium (1982) pH 9.2 at 29±2°C under continuous illumination of 72 μ mol photon $m^{-2}s^{-1}$ light intensity. Stock solutions of $CuCl_2 \cdot 2H_2O$, $ZnCl_2$ and $CdCl_2$ were prepared in Milli-Q water and sterilized by passing through millipore membrane filter (0.22 μm) before supplementing to the culture medium.

Per cent survival and growth measurement

The per cent survival of the test organism against test metals was scored by the plate colony count method. Colonies so developed were counted after 15 days with the help of colony counter, and per cent survival was subsequently scored. Since the test organism forms a homogenous suspension in liquid growth medium, a direct optical density reading gives a reliable measure of growth yield. On every 2nd day, the culture suspension was aseptically withdrawn and the optical density of the suspension was estimated at 663 nm in a Bausch and Lomb Spectronic-20 colorimeter. Observations were recorded up to 15th day using reference blank of basal culture medium.

Protein was estimated following the method of Lowry *et. al.* (1951) as modified by Herbert *et. al.* (1971).

Nutrient uptake

The uptake of NO_3^- was estimated colorimetrically by brucine-sulphuric acid method (Nicholas and Nason, 1957) by measuring the depletion of NO_3^- from the external medium. The ammonium left in the external medium after algal consumption was estimated using Nessler's reagent as described by Herbert *et. al.* (1971). The uptake of PO_4^{3-} and urea following 24h of treatment was estimated colorimetrically by stannous chloride method (APHA, 1985), and by diacetyl method using thiosemicarbazide (Wootton, 1974).

Assay of enzyme activity

Whole-cell urease activity was assayed following the method of Mackerras and Smith (1986). The estimation of *in vivo* nitrate reductase activity was done following the method of Camm and Stein (1974). The activity is based on the total nitrite formed. Alkaline phosphatase activity was assayed by the method of Ihlenfeldt and Gibson (1975). The activity is based on total p-nitrophenol formed in a known volume of culture suspension. ATPase activity of the test alga was assayed according to the method of Lockau and Pfeffer (1982) with minor modification.

Oxygen evolution, ATP content and electron transport chain

Photosynthetic O_2-evolution of exponentially. growing *Micracystis* sp. was measured in terms of µmole O_2-evolved mg^{-1} protein h^{-1} by a calibrated Clark type polarographic oxygen electrode connected to an oxygen analyser (Digital oxygen system, model 10, Rank Brothers, UK). A 10 ml algal suspension was taken in a reaction vessel. The temperature of the vessel was maintained by a thermocirculator. ATP was measured by luciferin-luciferase assay using LKB-1250 Luminometer (Larsson and Olsson,1979). The electron transport activity was determined by three basic assays (Lien 1978) and expressed in terms of either evolution or consumption of O_2. Activity of PSI was determined in presence of DCMU by measuring O_2 consumption after addition of ascorbate (5 µM) and methyl viologen (MV,1.4mM). Likewise the PSII was measured in terms of O_2 evolution in presence of p-benzoquinone (PBQ, 2mM). The whole chain electron transport was measured in the presence of potassium ferricyanide as electron accepter and water as electron donor. The overall reaction can be expressed as:

Reaction I Ascorbate $\rightarrow$ DCPIP$\rightarrow$ MV (PS I)

Reaction II $H_20 \rightarrow$ PBQ (PS II)

Reaction III $H_20 \rightarrow$ Ferricyanide (Redox coupling between PSI and PS II)

Statistical analysis

The results were analysed using analysis of variance and student's t-test. The data were subjected to analysis of variance (multifactorial) to determine the differences in inhibition of various physiological and biochemical characteristics due to metals and its different concentrations.

Results

Per cent survival and growth behaviour

Effect of different concentrations of Cd, Cu and Zn on per cent survival of *Microcystis* sp. is summarized in Figure 6.1. About 50 per cent survival of *Microcystis* was found at 0.02, 0.2 and 0.2 µg ml^{-1} concentration of Cd, Cu and Zn respectively. Approximately 25 per cent and 75 per cent survival of the test cyanobacterium was noticed at 0.01 and 0.03 µg ml^{-1} of Cd; 0.1 and 0.4 µg ml^{-1} of Cu, and 0.1 and 0.4 µg ml^{-1} of Zn respectively. However, the lethal doses for Cd, Cu and Zn were 0.06, 1.5 and 2.0 µg ml^{-1} respectively.

Figure 6.2 summarizes the effect of LC_{25}, LC_{50} and LC_{75} concentrations of test metals on growth behaviour of *Microcystis*. Alga treated with different dosage of metals depicted an increase in the lag phase as well as fast retardation of growth. After a lag phase of four days the alga continued to grow exponentially for 12 days. A gradual reduction in growth in terms of specific growth rate and final yield was observed at increasing concentration of the test metals. Approximately 65, 60 and 48 per cent reduction in final yield was obtained at sublethal concentration of Cd, Cu and Zn respectively.

Nutrient uptake

Table 6.1 presents data on the effects of Cd, Cu and Zn on nutrient uptake (NO_3^-, NH_4^+, PO_4^{3-} and urea) by the test cyanobacterium after 24h of treatment. Reduction in the NO_3^- uptake was more pronounced at increasing concentrations of metals. NO_3^- uptake was maximally inhibited by Cu followed by Cd and Zn. Exposure of the cyanobacterium to LC_{25} dose of metals (Cd, Cu and Zn) produced respectively 10, 23 and 5 per cent inhibition of NO_3^- uptake. This inhibition was more pronounced at higher concentration. At LC_{50} concentration of the test metals, a decrease in NO_3^- uptake from 1.40 to 1.13, 0.80 and 1.13 µg NO_3^- μg^{-1} protein respectively for Cd, Cu and Zn was observed. Approximafely 19 and 33; 43 and 62 and 19 and 24 per cent inhibition of NO_3^- uptake

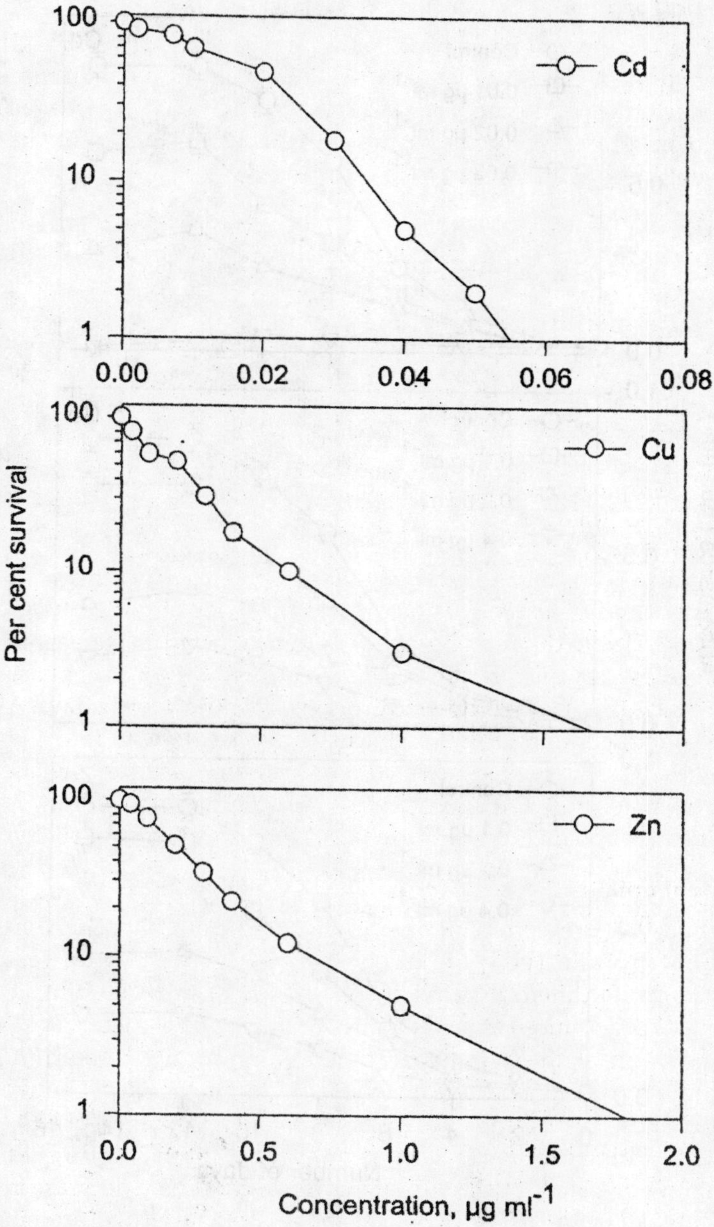

Fig. 6.1: Per cent survival of *Microcystis* sp. at different concentration of Cd, Cu and Zn

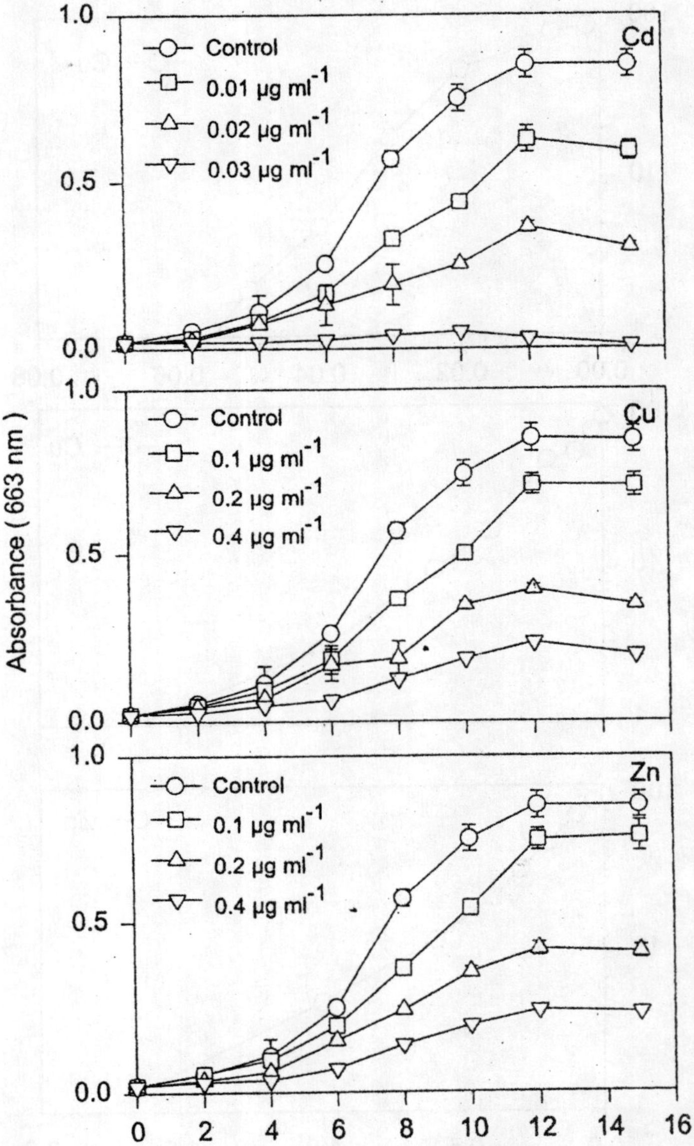

Fig. 6.2: Growth behaviour of *Microcystis* sp. at different concentration of Cd, Cu and Zn. Vertical bars show S.D.

was noticed at LC_{50} and LC_{75} doses of Cd, Cu and Zn respectively. Data pertaining to the effect of all the three metals on NH_4^+ uptake by *Microcystis* has been compiled in Table 6.1. A concentration dependent reduction of NH_4^+ uptake was clearly noticed in this case. There was approximately 65, 68 and 44 per cent inhibition of NH_4^+ uptake at LC_{50} concentration of Cd, Cu and Zn respectively. NH_4^+ uptake in control cells was 0.191 µg NH_4^+ μg^{-1} protein. With supplementation of LC_{50} doses of Cd, Cu and Zn, uptake dropped down respectively to 0.067, 0.060 and 0.105 µg NH_4^+ µg $protein^{-1}$.

Table 6.1: Cd, Cu and Zn toxicity on nutrient (NO_3^+, NH_4^+, PO_4^{3-} and Urea) uptake after 24 h of treatment

Metal	*Concentration (µg ml^{-1})*	*NO_3^- uptake (µg NO_3^- μg^{-1} protein)*	*NH_4^+ uptake (µg NH_4^+ μg^{-1} protein)*	*PO_4^{3-} uptake µg PO_4^{3-} μg^{-1} protein)*	*Urea uptake (µg urea μg^{-1} protein)*
Control		1.40± 0.05	0.191±0.03	0.36±0.06	0.027 (0.009)
	0.01	1.26 ±0.08 (10)	0.113±0.025 (61)	0.16 ±0.05 (56)	0.015±0.008 (44)
Cd	0.03	1.133±0.1 (19)	0.067±0.005 (65)	0.114±0.06 (68)	0.012±0.007 (56)
	0.04	0.933±0.09 (33)	0.006±0.001 (97)	0.038±0.005 (89)	0.010±0.007 (63)
	0.1	1.06±0.09 (23)	0.128±0.02 (32)	0.270±0.08 (25)	0.021±0.009 (22)
Cu	0.2	0.80±0.15 (43)	0.060±0.008 (68)	0.209±0.095 (58)	0.018±0.01 (33)
	0.4	0.15±0.13 (62)	0.030±0.001 (84)	0.140±0.07 (61)	0.014±0.007 (48)
	0.1	1.33±0.085 (5)	0.135 ± 0.03 (28)	0.342±0.08 (5)	0.022±0.008 (18.5)
Zn	0.2	1.13±0.087 (19)	0.105±0.02 (44)	0.287±0.09 (21)	0.019±0.006 (30)
	0.4	1.06±0.09 (24)	0.060±0.008 (68)	0.232±0.1 (36)	0.017±0.005 (37)

All the values are mean ± SE, Data in parentheses denote per cent inhibition. All the treatments are significantly different (P<0.05) from their respective controls according to student's 't' test.

Inhibition of PO_4^{3-} and urea uptake also followed the trend of NH_4^+ uptake; the uptake of all these nutrients was decreased with increasing concentration of test metals. Supplementation of 0.03, 0.2, 0.2 μg ml^{-1} Cd, Cu and Zn produced a decline in PO_4^{3-} uptake from 0.36 μg PO_4^{3-} μg^{-1} protein to 0.114, 0.209 and 0.287 μg PO_4^{3-} μg^{-1} protein respectively and at this concentration 68, 58 and 21 per cent reduction of PO_4^{3-} uptake was obtained. Further, *Microcystis* registered a decrease in urea uptake from 0.027 μg urea μg^{-1} protein to 0.015, 0.012, 0.010, 0.021, 0.018 and 0.014; and 0.022, 0.019 and 0.017 μg urea μg^{-1} protein respectively after 24 h of treatment by LC_{25}, LC_{50}, LC_{75} doses of Cd, Cu and Zn. Approximately 44, 56 and 63 per cent; 22, 33 and 48 per cent and 18.5, 30 and 37 per cent inhibition of urea uptake was obtained respectively at above concentrations.

Of the four nutrients NH_4^+ uptake demonstrated the highest sensitivity to above metals. Except NO_3^- uptake, the inhibition of NH_4^+, PO_4^{3-} and urea uptake was highly pronounced at LC_{25}, LC_{50} and LC_{75} concentrations of Cd. NO_3^- uptake by *Microcystis* was much more inhibited when the test alga was supplemented with Cu. The result of nutrient uptake analyzed by ANOVA demonstrated a highly significant ($P<0.01$) correlation between the inhibition of nutrient uptake and different concentrations of metals.

Enzyme activities

The effect of the test metals on nitrate reductase, urease, alkaline phosphatase and ATPase activities of *Microcystis* subjected to LC_{25}, LC_{50} and LC_{75} doses of Cd, Cu and Zn is summarized in Table 6.2. The results presented in this table vividly demonstrate a concentration dependent decrease in almost all enzymatic processes of *Microcystis*. It is pertinent to pinpoint that the exposure of the cyanobacterium to LC_{25}, LC_{50} and LC_{75} doses of metals, reduced the urease activity from 5.01 μg NH_4^+ μg^{-1} protein to 3.75, 2.41 and 1.44 μg NH_4^+ μg^{-1} protein by Cd; 4.04, 3.68 and 3.10 μg NH_4^+ μg^{-1} protein by Cu and 4.30, 4.11 and 3.75 μg NH_4^+ μg^{-1} protein by Zn respectively. There was 52, 27 and 18 per cent reduction of urease activity following supplementation of LC_{50} concentration of Cd, Cu and Zn respectively.

Spiking of the cyanobacterium with LC_{25}, LC_{50}, LC_{75} doses of test metals, inhibited the NR activity by approximately 19, 42 and 62 per cent by Cd, 22, 48, 69 per cent by Cu and 14, 38 and 57 per cent by Zn. At LC_{50} concentration of Cd, Cu and Zn, reduction in NR activity from 1.58 μg NO_3^+ μg^{-1} protein in control cells to 0.92, 0.82 and 0.98 μg NO_3^+ μg^{-1} protein was clearly observed. Similar to the inhibition of nitrate uptake, the reduction in the NR activity of *Microcystis* was maximum in the presence of Cu, followed by Cd and Zn. A significant decrease ('t' significant at P<0.05) in the NR activity following metal treatment was observed.

Table 6.2: Toxicity of Cd, Cu and Zn on enzyme (Urease, NR, APase, ATPase) activities after 24 h of treatment

Metal	*Concentration ($\mu g\ ml^{-1}$)*	*Urease ($\mu g\ NH_4^+\ \mu g^{-1}$ protein)*	*NR (μg NR μg^{-1} protein)*	*Apase (μg APase μg^{-1} protein)*	*ATPase ($\mu g\ PO_4^{3-}\ \mu g^{-1}$ protein)*
Control		5.01±0.043	1.58±0.06	1.90±0.039	0.33±0.045
Cd	0.01	3.75±0.32 (25.1)	1.28±0.1 (19)	0.19±0.035 (52.1)	0.13±0.038 (61)
	0.02	2.41±0.38 (52)	0.92±0.055 (42)	0.6O±0.031 (68)	0.08±0.029 (75.6)
	0.03	1.44±0.45 (71)	0.60±0.095 (62)	0.50±0.031 (74)	0.03±0.035 (90)
Cu	0.1	4.04±0.50 (19)	1.23±0.085 (22)	1.30±0.035 (2)	0.23±0.35 (31)
	0.2	3.68±0.40 (27)	0.82±0.05 (48)	1.00±0.031 (47)	0.16±0.40 (51.3)
	0.4	3.10±0.38 (38)	0.49±0.06 (69)	0.91±0.38 (52)	0.I3±0.48 (60)
Zn	0.1	4.30±0.25 (14.1)	1.36±0.07 (14)	1.60±0.030 (16)	0.28±0.038 (30)
	0.2	4.11±0.39 (18)	0.98±0.08 (37.7)	1.30±0.025 (31)	0.23±0.0385 (36.6)
	0.4	3.75±0.38 (25)	0.67±0.01 (57.4)	1.10±0.038 (42)	0.20±0.052 (39)

All the values are mean ± SE, Data in parentheses denote per cent inhibition. All the treatments are significantly different (P<0.05) from their respective controls according to student's 't' test.

An exposure of the cyanobacterium to different concentrations of metals showed a general reduction in alkaline phosphatase activity. The supplementation of LC_{25}, LC_{50}, LC_{75} dose of Cd registered respectively 52, 68 and 74 per cent inhibition. Likewise Cu registered respectively 32, 47 and 52 per cent inhibition, and Zn registered 16, 31 and 42 per cent inhibition of alkaline phosphatase activity.

It is worthwhile to indicate that the metal induced inhibition of ATPase activity was quite similar to that of other enzymatic activities. LC_{25}, LC_{50} and LC_{75} doses of Cd, Cu and Zn inhibited the ATPase activity by 61, 75 and 90 per cent; 31, 51 and 60 per cent and 30, 37 and 39 per cent respectively. The pattern of inhibition of the enzyme activity was largely similar to nutrient uptake i.e., a decrease in the enzyme activities was noticed with increase in metal concentration. Table 6.2 further revealed that all enzymatic activities of *Microcystis* were maximally inhibited by Cd followed by Cu and Zn except NR activity which registered greater inhibition by Cu. Statistical analysis showed significant ($P<0.001$) inhibition of all the enzymatic processes following supplementation of test metals.

O_2^- evolution, ATP content and electron transport chain

The data emerging from the experiments carried out to determine the effect of LC_{25}, LC_{50} and LC_{75} concentrations of Cd, Cu and Zn on oxygen evolution and ATP content are given in Table 6.3. All the metals strongly inhibited O_2 evolution in illuminated cultures after 2 hr. At 0.2 μg ml^{-1} of Cu and Zn, the inhibition was respectively 77 and 65 per cent. The photosynthetic O_2 evolution of control cells of *Microcystis* was 30.23 μmol mg^{-1} protein h^{-1}. It declined to 4.56 and 7.01 μmol mg^{-1} protein h^{-1} at LC_{75} dose of Cu and Zn respectively. At high concentration of Cd, instead of O_2 evolution O_2 consumption was obtained.

Table 6.3 shows ATP content of *Microcystis* after 2h exposure to different metals. The control cells with 3.21 μmol ATP μg^{-1} protein h^{-1} depicted consistent decline in ATP content with increasing concentration of test metals. The test cyanobacterium following supplementation of LC_{25}, LC_{50} and LC_{75} doses of Cd produced respectively 33, 41 and 64 per cent reduction in ATP content. However, the Cu and Zn induced inhibition of ATP content was approximately 20, 34 and 55 per cent and 10, 28 and 48 per cent respectively at its LC_{25}, LC_{50} and LC_{75} doses. A significant reduction in O_2 evolution and ATP content of *Microcystis* following treatment with different metals, metal doses and interaction between metals and metal doses ($P < 0.001$, ANOVA) was observed.

Table 6.4 contains data on the effects of test metals on various components of photosynthetic electron transport of *Microcystis* sp. Though all the three reactions of the electron transport system were inhibited severely by Cu, Zn and Cd, the reaction II ($H_20 \rightarrow$ PBQ) was found to be more sensitive in *Microcystis* sp. The addition of a

Table 6.3: Cd, Cu and Zn toxicity on O_2 evolution and ATP content after 2h of treatment

Metal	*Concentration (µg ml⁻¹)*	*O_2 evolution (µ mol O_2 evolved/consumed mg⁻¹ protein h⁻¹)*	*ATP content (µ mol ATP mg⁻¹ protein h⁻¹)*
Control		30.23±0.2	3.21±0.05
Cd	0.01	14.04±0.522 (54)	2.16±0.07 (33)
	0.02	21.61±0.226 (171.49)	1.88±0.08 (41)
	0.03	14.82±0.299 (148.53)	1.17±0.05 (64)
Cu	0.1	14.10±0.255 (53.30)	2.57±0.08 (20)
	0.2	6.90±0.141 (77)	2.10±0.06 (34)
	0.4	4.56±0.106 (85)	1.44±0.03 (55)
Zn	0.1	14.10±0.289 (53)	2.89±0.06 (10)
	0.2	10.75±0.393 (65)	2.32±0.07 (28)
	0.4	7.01±0.501 (77)	1.67±.0.08 (48)

All the values are mean ± SE, Data in parentheses denote per cent inhibition. All the treatments are significantly different ($P<0.05$) from their respective controls according to student's 't' test.

low concentration of metal in the intact cells caused a partial inhibition, the latter was increased with increasing metal concentration. PS II was more inhibited (60 per cent reduction) than PS I (42 per cent reduction) and whole chain electron transport (53 per cent) at LC_{50} dose of Cd. Similar trend of inhibition of electron transport system was noticed in Cu and Zn supplemented *Microcystis* culture. At LC_{50} concentration of Cu and Zn, reaction II was inhibited by 51 and 49 per cent, whereas approximately 23 and 38 per cent inhibition of reaction I and 36 and 38 per cent for reaction III was noticed. PS I activity, assayed in term of O_2 consumption, showed less inhibition at a selected concentration of metal than PS II and redox coupling between PS I and PS II. Approximately 22-53 per cent reduction in PS I activity, 23-75 per cent in PS II and 36-69 per cent reduction in redox coupling between PS I and PS II was found when Cd concentration was raised from 0.01 to 0.04 µg ml^{-1}. Likewise 20-32 per cent, 21-48 per cent decrease in PS I; 21-63 per cent, 37-56 per cent decrease in PS II and 24-51 per cent and 24-49 per cent reduction in redox coupling between PS I and PS II was observed following an increase in Cu and Zn concentration from 0.1 to 0.4 µg ml^{-1}. The data of statistical analysis (ANOVA 2 way) revealed a significant variation in the PS I, PS II and whole chain activities following supplementation of different concentrations of Cd, Cu and Zn ($P < 0.001$).

Table 6.4: Effect of test metals on electron transport chain of *Microcystis* sp.

Metal	*Concentration (µg ml^{-1})*	*PSI activity (µmol O_2– consumed mg^{-1} protein hr^{-1})*	*PS II activity (µmole O_2– evolved mg^{-1} protein h^{-1})*	*Whole chain (µmole O_2– evolved mg^{-1} protein h^{-1})*
Control		58.72±0.014	42.62±0.09	38.66±0.005
Cd	0.01	45.51±0.100 (23)	32.64±0.90 (23)	24.67±0.022 (36)
	0.02	33.97±0.266 (42)	17.06±0.090 (60)	18.09±0.007 (53)
	0.03	27.54±0.105 (53)	10.64±0.055 (75)	12.07±0.022 (69)
Cu	0.1	47.11±0.308 (20)	33.83±0.097 (21)	29.12±0.170 (24)
	0.2	45.09±0.342 (23)	20.86±0.135 (51)	24.82±0.227 (36)
	0.4	39.88±0.220 (32)	15.63±0.078 (63)	18.92±0.116 (51)
Zn	0.1	46.37±0.083 (21)	26.80±0.100 (37)	29.44±0.100 (24)
	0.2	36.49±0.100 (38)	21.62±0.020 (49)	23.58±0.019 (38)
	0.4	30.35±0.104 (48)	18.77±0.00 (56)	19.59±0.011 (49)

All the values are mean ± SE, Data in parentheses denote per cent inhibition. All the treatments are significantly different ($P<0.05$) from their respective control according to student's 't' test.

Discussion

Heavy metal toxicity to algae is one of the most debated environmental problems. Plants, algae and other organisms sensitive to metals, display metabolic disturbances and growth inhibition at metal contents in the tissues only slightly higher than the normal levels. High levels of metals inhibit a large number of enzymes and interfere with several metabolic processes including photosynthesis, pigment synthesis and membrane integrity. One most important effect is the blocking of photosynthetic electron transport leading to the production of radicals which start peroxidative chain reaction involving membrane lipids. Further, reduction in growth rate seems to be due to inhibition of *in vivo* primary processes like O_2 evolution and ^{14}C fixation (De Filippis *et. al.*, 1981).

Reduction in the per cent survival and growth of the test alga at increasing concentration of Cu, Zn and Cd (Figs. 6.1, 6.2) confirmed the toxic potentials of these metals, and agrees well with the findings of De Fillippis and Pallaghy (1976) and Rai *et. al.* (1994). The toxicity of metals may be due either to the disruption of the permeability of the cell membrane or inhibition of photosynthetic pigments and enzyme activities (Raizada *et. al.*, 1984). Several reports describe inhibition of growth and metabolism of algae and cyanobacteria by Cd (Rachlin and Grosso, 1991), Cu (Rai *et. al.*, 1994; Vavilin *et. al.*, 1995), and Zn (Rana and Kumar, 1975). Rachlin *et. al.* (1982 a) found that with increasing Cd concentration there was reduction in the growth rate of *Chlorella saccharaphila.* This is in agreement with the findings of Fisher and Frood (1980) for *Skeletonema costatum, Nitzschia closterium, Chaetoceros compressium, Asterionella japonica* and Rachlin *et. al.* (1982b) for *Chlorella saccharophila, Navicula incerta* and *Nitzschia closterium.* These results offer support to the contention that growth reduction is a credible determinant of metal toxicity and that the degree of response of the algae is dependent on the amount of metals that traverse the membrane. Another reason for reduction of growth of cyanobacterium might be the inhibition of cell division. The reduction of cell division could be due to binding of the test metals to sulphydryl group which is responsible for regulation of the cell division in plants (Visviki and Rachlin, 1991).

Concentration dependent reduction in nutrient uptake is supported by the findings of the earlier workers for Cd with *Nostoc linckia* (Husaini and Rai, 1991) and Cu and Ni with *Chlorella vulgaris*

(Rai *et. al.*, 1996). These workers also suggested that exposure to heavy metals such as Pb, Cu, Zn and Cd might stimulate uptake of phosphorus which is deposited in polyphosphate bodies of different species. The inhibition of nutrient uptake may be due to competition of Cu, Zn and Cd with nutrients for binding onto the cyanobacterial cell surface (Vallee and Ulmer, 1972), inactivation of the enzyme complex by binding with their sulfhydryl groups (Porter and Sheridan, 1981) or exhaustion of energy yielding substrates (Norris and Kelly, 1977) required for the active uptake of different nutrients.

The uptake of nutrients, critical to growth and reproduction of different plants including algae, was severely affected by heavy metals used. Nutrient uptake is an energy dependent process (Ulrich *et. al.*, 1987; Oelmuller *et. al.*, 1988). Inadequate energy supply and adsorption of metals onto the cell surface are two important factors for inhibition of nutrient uptake. For this reason metal and nutrient uptake may compete for energy. Thus the restricted availability of energy may result in the inhibition of nutrient uptake as observed in the present study. Bordons and Jafre (1987) reported that heavy metal accumulation onto the cell through adsorption to negatively charged surface, serves as selective barrier for other nutrients.

The interaction of metals with PO_4^{3-} binding sites (Husaini and Rai, 1991) and an indirect inhibition of uptake of substrates (NO_3^-, NH_4^+, PO_4^{-3}) of different enzymes may be the other important reasons for inhibition of enzyme activity. The results are quite in agreement with the earlier findings (Hofslagare *et. al.*, 1985, Price and Morel, 1990, Husaini and Rai, 1991). The results presented in Table 6.2 vividly demonstrate a concentration dependent decrease in almost all enzymatic processes of *Microcystis* sp. Several workers (Price and Morel, 1990; Rai *et. al.*, 1996) demonstrated that inhibition of these enzymes by heavy metals is either because of reduction of ATP pool or binding of the metals to enzyme protein thereby altering its structure. It is also known that nutrient uptake is primarily dependent on the activity of membrane bound ATPase. Since ATPase activity is dependent on ATP pool, it is likely that the availability of cellular ATP may slow down the ATPase activity in *in vivo* condition, which in turn reduces the uptake of nutrients leading ultimately to the inhibition of the cell growth. The inhibition of PO_4^{3-} uptake and breakdown of p-nitrophenyl phosphate (p-NPP) following supplementation of heavy metals may cause PO_4^{3-} deficiency, which might result in the inhibition of ATP synthesis there by affecting all the ATP requiring enzymatic processes i.e. NR, GS, alkaline phosphatase, active nutrient uptake process and finally growth. Though urease activity is independent of cellular ATP (Mackerras and Smith, 1986), the inhibition of this enzyme might be due to the direct action of metals on the enzyme complex. Price and Morel (1990) studied alkaline phosphatase mediated death of cell due to Cd. They suggested that inhibition may be due to the binding of Cd to enzyme thereby not allowing the conversion of substrate into product, which ultimately becomes deficient for the cell. This inhibition of enzyme activity is closely dependent on nutrient uptake. The inhibition of nutrient uptake under Cd stress seems to be due to inhibition of glutamine synthetase and nitrate reductase which are responsible for assimilation, of NH_4^+ and NO_3^- respectively (Rao and Venkateswaran, 1986). Further, a close relationship between cyanobacterial photosynthesis and nitrate reduction (Losada and Guerrero, 1979) offers a testimony to the above contention. It can be suggested that a decrease in NR activity may lead to reduced uptake and metabolism of NO_3^- hence a decrease in the availability of carbon skeleton and energy (Losada and Guerrero, 1979). Depression of nitrate reductase activity as observed in our study, is supported by the observations on *Anabaena doliolum* treated with Cr and Zn (Dubey and Rai, 1987), *Nostoc linckia* with Cd (Husaini and Rai, 1991) and *Chlorella vulgaris* with Hg and Zn (Rai *et. al.*, 1991).

Our findings on inhibition of photosynthetic oxygen evolution by Cu, Zn and Cd (Table 6.3) agree well with those of Tripathy *et. al.* (1981), Sheoran *et. al.* (1990) and Husaini and Rai (1991). Destruction of light harvesting pigments (De Filippis, 1989), and photosynthetic membrane (Sicko-Goad, 1982) may be the main cause for inhibition of O_2 evolution in metal treated cells. Since ATP and NADH are generated "mainly from electron transport chain, any disruption of thylakoid by heavy metals might interrupt the flow of electrons thereby decreasing the energy pool of the cell. In all biological systems, ATP is the primary source of energy. Therefore, a reduction in its synthesis is likely to create energy deficiency, thus causing inhibition of all energy dependent metabolic processes of the cells. The depletion in ATP content as observed for test alga may be due to inhibition of photosynthetic electron transport chain. A significant positive correlation between the two processes *i.e.* depletion in ATP content and inhibition of photosynthetic electron transport offers confirmation to the above contention.

The inhibition of photosynthesis and associated electron transport reaction of *Microcystis* in presence of Cu, Zn and Cd as compiled in Table 6.4 resemble the findings of Tripathy *et. al.* (1981); Vavilin *et. al.* (1995) and Patsikka *et. al.* (1998). According to Shioi *et. al.* (1978) metal inhibits electron transport at two different sites. One site of action appears to be localized at the ferredoxin site and the other is the blocking between the oxidizing side of the reaction center of PS II and electron donating site of 1, 5-diphenyl carbazide. Though both PS I and PS II are inhibited by metals, PS II depicted maximum sensitivity to most of the metals tested. Our results clearly agree with the findings of Mohanty *et. al.* (1989), Vavilin *et. al.* (1995) and Yruela *et. al.* (1996). Both the donor and acceptor of PS II have been proposed to be the most sensitive sites for Cu^{2+} action (Mohanty *et. al.*, 1989; Vavilin *et. al.*, 1995; Yruela *et. al.*, 1996). Further, Vavilin *et. al.* (1995) summarized that the inhibition of PS II was either by interruption of flow of electrons of PS II at multiple site such as plastocyanin site (Kotoh and Takamiya, 1964) or inhibition of donor site of cytochrome (Singh and Singh, 1987) and the donor (Hsu and Lee, 1988) and acceptor (Shioi *et. al.*, 1978) sites of PS II.

Some important points emerging from this study are (i) a generalized toxicity hierarchy for test metals is Cd > Cu > Zn, (ii) inhibition of uptake and enzymatic activities of *Microcysitis* depends on the concentration of test metals in the culture medium, and (iii) although all the reactions of *Microcystis* were inhibited severely by the metals used, reaction II ($H_2O \rightarrow PBQ$) was found to be the most sensitive.

Acknowledgements

This study was sponsored by grants from the Department of Biotechnology, Ministry of Science & Technology, New Delhi, Pitamber Pant National Environment Fellowship Award from the Ministry of Environment and Forests, New Delhi and INDO-US project from NSF, USA to L.C. Rai.

References

American Public Health Association (APHA) 1985. *Standard Methods for the Examination Water and Waste Water.* 16th ed., APHA, Washington, D.C.

Bordons, A. and Jafre, J. 1987. Extracellular adsorption of nickel by a strain of *Pseudomonas* sp. *Enzym. Microb. Technol.* **9**: 709-713.

Camm, E.L. and Stein, J.R. 1974. Some aspects of nitrogen metabolism of *Nodularia spumigena* (cyanophyceae). *Can. J. Bot.* **52**: 719-726.

Chettri, M.K., Cook, C.M., Vardika, E., Sawidis, T. and Lanaras, T. 1998. The effect of Cu, Zn and Pb on the chlorophyll content of the lichens *Cladonia convoluta* and *Cladonia rangiformis. Environ. Exp. Bot.* **39**: 1-10.

Coleman, J. E. 1984. Carbonic anhydrase: zinc and the mechanism of catalysis. In: *Biology and Chemistry of the Carbonic Anhydrogenase.* Edited by Tashian, R. E. and Hewlett-Enmett, D. Ann. New York Acad. Sci. pp. 26-48.

Davies, A. G. 1978. Pollution studies with marine phytoplankton. II. Heavy metals. *Adv. Mar. Biol.* **15**: 381-508.

Davies, A.G. 1983. The effects of heavy metals upon natural marine phytoplankton populations. In: *Progress in Phycological Research.* Edited by Round, F .E. and Chapman, D.J. Elsevier, Amsterdam. pp. 113-145.

De Filippis, L.F. 1979. The effect of heavy metal compounds on the permeability of *Chlorella* cells. Z. *Pjlanzenphysiol.* **93**: 63-68.

De Filippis, L.F. 1989. Toxicity and resistance of algae to heavy metals. In: *Plants and Pollutants in Developed and Developing Countries.* Edited by Ozturk, M. A. Ege University Press, Izmir. pp. 19-35.

De Filippis, L. F. and Pallaghy, C. K. 1976. The effect of sublethal concentrations of mercury and zinc on *Chlorella*. I. Growth characteristics and uptake of metals. Z. *Pjlanzenphysiol.* **78**: 197-207.

De Filippis, L.F. and Pallaghy, C.K., 1994. Heavy metals: sources and biological effects. In: *Advances in Limnology.* Edited by Rai, L.C., Gaur, J.P. and Soeder, C.J. Stuttgart, Germany. pp. 31-77.

De Filippis, L.F., Harnp, P. R. and Ziegler, H. 1981. The effect of sublethal concentrations of zinc, cadmium and mercury on *Euglena* growth and pigments. Z. *Pflanzenphysiol.* **101**: 37-47.

Devi Prasad, P .V. and Devi Prasad, P. S. 1982. Effect of cadmium, lead and nickel on three fresh water green algae. *Water Air Soil Pollut.* **17**: 263-268.

Dubey, S.K. and Rai, L.C. 1987. Effect of chromiurn and tin on survival, growth, carbon fixation, heterocyst differentiation, nitrogenase, nitrate reductase and glutamine synthetase activities of *Anabaena doliolum*. *J. Plant Physiol.* **130**: 165-172.

Fisher, N.S. and Frood, D. 1980. Heavy metals and marine diatoms: Influence of dissolved organic compounds on toxicity and selection for metal tolerance among four species. *Mar. Biol.* **59**: 85-93.

Florence, T.M. and Stauber, J.L. 1986. Toxicity of copper complexes to the marine diatom *Nitzschia closterium*. *Aquat. Toxicol.* **8**: 11-26.

Harrnens, H., Gusmoa, N.G.C.P.B., Denhartog, P.R., Verkleij, J.A.C. and Emst, W.H.O. 1993. Uptake and transport of zinc in Zn-sensitive and Zn-tolerant *Silene vulgaris*. *J. Plant Physiol.* **141**: 309-315.

Herbert, D., Phipps, P. J. and Strange, R. E. 1971. Chemical analysis of microbial cells. In: *Methods in Microbiology*. Edited by Norris, J.R. and Ribbons, D.W. Academic Press, London, pp. 209-334.

Hofslagare, C., Sarnuelsson, G. and Sjobers, S. 1985. Cadmium effects on photosynthesis and nitrate assismilation in *Scenedesmus obliquus*. A potentiometric study in an open CO_2- system. *Environ. Exp. Bot.* **25**: 75-82.

Hsu, B.D. and Lee, J.Y. 1988. Toxic effects of copper on photosystem II of spinach chloroplasts. *J. Plant Physiol.* **87**: 116-119.

Husaini, Y. and Rai, L.C. 1991. Studies on nitrogen and phosphorus metabolism and the photosynthetic electron transport system of *Nostoc linckia* under cadmium stress. *J. Plant Physiol.* **138**: 429-435.

Ihlenfeldt, M.J.A. and Gibson, J. 1975. Phosphate utilization and alkaline phosphatase activity in *Anacystis nidulans (Synechococcus)*. *Arch. Microbiol.* **102**: 23-28.

Jana, S. and Choudhuri, M.A. 1982. Senescence in submerged aquatic angiosperms: Effects of heavy metals. *New Phytol.* **90**: 477-484.

Katoh, S. and Takamliya, A. 1964. Nature of copper protein binding in spinach plastocyanin. *J. Biochem.* **55**: 378-387.

Larsson, C. M. and Olsson, T. 1979. Firefly assay of adenine nucleotides from algae: Comparison of extraction methods. *Plant Cell Physiol.* **20**: 145-155.

Lee, K.C., Cunningham, B.A., Paulsen, G.M., Liang, G.H. and Moore, R.B. 1976. Effects of Cd on respiration rate and activities of several enzymes in soybean seedlings. *Physiol. Plant.* **36**: 4-6.

Les, A. and Walker, R. W. 1984. Toxicity and binding of Cu, Zn and Cd by blue green alga, *Chroococcus paris. Water Air Soil Pollut.* **23**: 129-139.

Lien, S. 1978. Hill reaction and phosphorylation with chloroplast preparation from *Chlamydomonas reinhardtii.* In: *Hand Book of Phycological Methods: Physiological and Biochemical Methods.* Edited by Hellebust, J.A. and Craigie, J.S., Cambridge University Press. pp. 305-315.

Lockau, W. and Pfeffer, S. 1982. ATP-dependent calcium transport in membrane vesicles of the cyanobacterium *Anabaena variabilis. Biochim Biophys. Acta* **733**: 124-132.

Losada, M. and Guerrero, M.G. 1979. The photosynthetic reduction of nitrate and its regulation. In: *Photosynthesis in relation to Model System.* Edited by Barber, J. Elsevier, North Holland Biomedical Press, Amsterdam, pp. 365-408.

Lowry, O.H., Rosebrough, N.J., Farr, A.L. and Randall, R.J. 1951. Protein measurement with the Fo1in phenol reagent. *J Biol. Chem.* **193**: 265-275.

Mackerras, A. H. and Smith, G. D. 1986. Urease activity of the cyanobacterium *Anabaena cylindrica. J. Gen. Microbiol.* **132**: 2749-2752.

Mohanty, N., Vass, L. and Demeter, S. 1989. Copper toxicity affects photosystem II electron transport at the secondary quinone acceptor, Qa: *Plant Physiol.* **90**: 175-179.

Nicholas, D.J. and Nason, A. 1957. Determination of nitrate and nitrite. *Meth. EnzymoL.* **111**: 320-343.

Norris, P.R. and Kelly, D.P. 1977. Accumulation of cadmium and cobalt by *Saccharomyces cerevisiae. J. Gen. Microbioi.* **99**: 317-324.

Oelmuller, R., Schuster, C. and Mohr, H. 1988. Physiological characterization of a plastidic signal required for nitrate induced appearance of nitrate and nitrite reductases. *Planta* **174**: 75-83.

Parker, D.L. 1982. Improved procedures for the cloning and purification of *Microcystis* cultures (Cyanophyta). *J. Phycol.* **18**: 471-477.

Patsikka, E., Aro, E.M. and Tyystjärvi, E. 1998 Increase in the quantum yield of photoinhibition contributes to copper toxicity *in vivo. Plant Physiol.* **117**: 619- 627.

Payne, C.D and Price, N.M. 1999. Effects of cadmium toxicity on growth and elemental composition of marine phytoplankton. *J. Phycol.* **35**: 293-302.

Pinton, R., Cammak, I. and Marschner, H. 1994. Zinc deficiency enhanced NAD (P) H- dependent superoxide radical production in plasmamembrane vesicles isolated from roots of bean plants. *J Expt. Bot.* **45**: 45-50.

Porter, J.R. and Sheridan, R.P. 1981. Inhibition of nitrogen fixation in alfalfa by arsenate, heavy metals, fluoride and simulated acid rain. *Plant Physiol.* **68**: 143-148.

Price, N.M. and Morel, F.M.M. 1990. Cadmium-substitution for zinc in a marine diatom. *Nature* **344**: 658-660.

Rachlin, J.W. and Grosso, A. 1991. The effects of pH on the growth of *Chlorella vulgaris* and its interaction with cadmium toxicity. *Arch. Environ. Contam. Toxicol.* **20**: 505-508.

Rachlin, J.W., Jensen, T.E. and Warkentine, B. 1982a. The growth response of the green alga *(Chlorella saccharophila)* to selected concentrations of the heavy metals Cd, Cu, Pb, and Zn. In: *Trace Substances in Environmental health, XVI.* Edited by Hemphill, D.D., University of Missouri, Columbia. pp. 145-154.

Rachlin, J.W., Warkentine, B. and Jensen, T.E. 1982b. The growth responses of *Chlorella saccharophila, Navicula incerta* and *Nitzschia closterium* to selected concentrations of cadmium. *Bull. Torrey Bot. Club* **109**: 129-135.

Rachlin, J.W., Jensen, T.E., Baxter, M. and Jani, V. 1982c. Utilization of morphometric analysis in evaluating response of *Plectonema boryanum* (Cyanophyceae) on exposure to eight heavy metals. *Arch. Environ. Contam. Toxicol.* **11**: 323-333.

Rai, L.C., Gaur, J.P. and Kumar, H.D. 1981. Phycology and heavy metal pollution. *Biol. Rev.* **56**: 98-151.

Rai, P .K., Mallick, N. and Rai, L.C. 1994. Effect of Cu and Ni on growth, mineral uptake, photosynthesis and enzyme activities of *Chlorella vulgaris* at different pH - values. *Biomed. Environ. Sci.* **7**: 56-67.

Rai, L.C., Rai, P.K. and Mallick, N. 1996. Regulation of heavy metal toxicity in acid tolerant *Chlorella:* Physiological and biochemical approach. *Environ. Expt. Bot.* **36**: 99-109.

Raizada, M., Rai, L.C. and Dubey, S.K. 1984. Effect of hexavalent chromium on a nitrogen fixing cyanobacterium, *Nostoc muscorum. Adv. Biol. Res.* **1&2**: 46-54.

Rana, B.C. and Kumar, H.D. 1975. Studies on chemical and biological treatment of zinc smelter-effluent and its evaluation through the growth of test algae. *Nova Hedw.* **26**: 465-471.

Rao, S. and Venkateswaran, G. 1986. Glutamine metabolism in *Neurospora crassa* under condition of copper toxicity. *J. Expt. Bot.* **31**: 947-955.

Rijstenbil, J.W., Derksen, J.W.M., Gerringa, L.J.A., Poortvliet, T.C.W., Sandee, A., , Vandenberg, M., Vandrie, J. and Wijnholds, J.A. 1994. Oxidative stress induced by copper: Defense and damage in the marine planktonic diatom *Ditylum brightwellii* grown in continuous cultures with high and low zinc levels. *Mar. Biol.* **119**: 583-590.

Sheoran, I.S., Singal. H. R. and Singh, R. 1990. Effect of Cd and Ni on photosynthesis and enzyme of photosynthetic carbon cycle in pigeon pea. *Photosyn. Res.* **23**: 345-352.

Shioi, Y., Tamai, H. and Sasa, T. 1978. Inhibition of photo system II in the green alga *Ankistrodesmus* by copper. *Physiol. Plant.* **44**: 434-438.

Sicko-Goad, L. 1982. A morphometric analysis of algal response to low dose, short term heavy metal exposure. *Protoplasma* **110**: 75-86.

Singh, D.P. and Singh, S.P. 1987. Action of heavy metals on hill activity and O_2 evolution in *Anacystis nidulans. Plant Physiol.* **83**: 12-14.

Sorentino, C. 1979. The effects of heavy metals on phytoplankton: A review. *Phykos* **18**: 149-161.

Stratton, G.W. and Corke, C.T. 1979. The effect of cadmium ion on growth, photosynthesis and nitrogenase activity of *Anabaena inaequalis*. *Chemosphere* **5**: 277-282.

Tripathy, B.C., Bhatia, B. and Mohanty, P. 1981. Inactivation of chloroplast photosynthetic electron transport activity by Ni^{2+}. *Biochim. Biophys. Acta* **638**: 217-224.

Ullrich, W.R., Aparicio, P.J., Syrett, P.J. and Castillo, F. 1987. *Inorganic Nitrogen Metabolism*. Springer-Verlag, Berlin.

Uribe, E.G. and Stark, B. 1982. Inhibition of photosynthetic energy conversion by cupric ion. Evidence for C^{2+} coupling factor interac.tion. *Plant Physiol.* **69**: 1040-1045.

Vallee, B.L. and Ulmer, D.D. 1972. Biochemical effects of mercury, cadmium and lead. *Ann. Rev. Biochem.* **41**: 91-129.

Vavilin, D. V., Polynov, V. A., Matorin, D. N. and Venediktov, P.S. 1995. Sublethal concentrations of copper stimulate photosystem II. Photoinhibition in *Chlorella pyrenoidosa*. *J. Plant Physiol.* **146**: 609-614.

Visviki, I. and Rachlin, I.W. 1991. The toxic action and interactions of copper and cadmium to the marine alga *Dunaliella minuta*, in both acute and chronic exposure. *Arch. Environ. Contam. Toxicol.* **20**: 271-275.

Wootton, I.D.P. 1974. *Microanalysis in Medical Biochemistry*. J. and A. Churchil Livingstone Edinburgh and London Ltd., pp. 79-99.

Yruela, I., Gatzen, G., Picorel, R. and Holzwart, A.R. 1996. Cu (II) – inhibitory effect on photosystem II from higher plants: A picosecond time-resolved fluorescence study. *Biochemistry* **35**: 9469-9474

Chapter 7

UV ABSORBING PIGMENTS IN EPILITHIC CYANOBACTERIA OCCURRING ON THE TEMPLES AND MONUMENTS

Amarpalli Roy and S.P.Adhikary

Abstract

The blackish-brown crusts/ mats occurring on the rock surface of several temples exposed to intense solar radiation showed prominent absorption in the UV region of the spectrum. Absorption maxima at 258-260, 312, 317 or 325 nm in the crusts/ mats was due to presence of mycosporine amino acid (MAAs) like compounds. An absorption peak at 384 nm, obtained in certain epilithic crusts was due to yellowish brown scytonemin pigment. Methanolic extract of seventeen different epilithic cyanobacterial species of which 14 were the dominant or associated components in the crusts/ tufts from temples revealed that 12 species contained one or more MAAs absorbing in the UV-B and UV-C region. Nine species belonging to *Tolypothrix* and *Calothrix*, all of which possessed distinct sheath around their trichome, absorbed strongly in the UV-A region due to extracellular sunscreen pigment scytonemin. Dark grown cultures lost their capacity to absorb in the UV-B region due to MAAs which could be reinduced in their cells only after exposure to light for prolonged periods. Absorption in the UV-A region of the spectrum due to scytonemin was retained even when the cyanobacteria were cultivated in darkness for a longer period suggesting that MAAs but not scytonemin is the major sun-screen pigment in terrestrial epilithic cyanobacteria

Key words: Cyanobacteria, *Calothrix*, epilithic, scytonemin, *Tolypothrix*, UV absorbing pigments.

Introduction

Cyanobacteria are often the primary colonizers on bare rock surfaces and soils (Whitton, 1987; Hoffmann, 1989). Some of the species possess the ability to occur in a wide variety of extremely dry habitats like arid desert rocks (Friedmann and Galun, 1974; Anagnostidis *et. al.*, 1983; Büdel and Wessels, 1991) , surface of buildings coated with cement, lime or structures covered with asbestos cement sheets in the tropics (Chadha and Pandey, 1982; Tripathi *et. al.*, 1991) and exposed rock surface of temples and monuments forming characteristic blackish brown crusts/mats (Tripahty *et. al.*,1997). Under such conditions they must therefore possess efficient mechanisms to prevent or counteract the harmful effects of strong solar radiation including those caused by the particularly deleterious UV-region of the spectrum. These mechanisms include actions of photoprotective carotenoids (Buckley and Houghton, 1976), dark DNA repair and photorepair (Levine and Thiel, 1987, Pattnaik *et. al.*, 1993) and the presence of detoxifying enzymes (Mittler and Tel-Or, 1991). Certain cyanobacterial species also prevent UV-photo damage by negative photo movements (Richardson and Castenholz, 1987; Häder, 1988). Recently Garcia-Pichel and Castenholz (1991) reported scytonemin-mediated preventive strategy to short wavelength radiation in many species that survive long periods of exposure under senescent or physiologically inactive or quiescent conditions. Scytonemin is a coloured lipid soluble pigment, occurring prominently in the sheaths of outermost parts or top layers of cyanobacterial mats, crusts or colonies, with broad absorption in the UV and violet wavelengths with an *in vivo* maximum *at ca.* 370 nm (Proteau *et. al.*, 1993). A second group of colourless, water soluble, UV absorbing, mycosporine amino acid like compounds (MAAs) with narrow absorption bands and maxima between 310 and 334 nm has also been found in 13 strains of cyanobacteria (Garcia-Pichel and Castenholz, 1993; Karsten and Garcia-Pichel, 1996). It has been suggested that significant protection from UV photo damage could be gained from possession of either MAAs or scytonemin but especially from simultaneous screening by both types of compounds (Garcia-Pichel and Castenholz, 1991; Garcia-Pichel *et. al.*, 1992, 1993). Several cyanobacterial forms have been found to inhabit the rock surface of temples and monuments forming characteristic blackish brown crusts/ tufts on their walls. Presence of UV absorbing pigments in the cyanobacterial forms surviving on the rock surface of the temples was studied. Observations were made using both field samples and culture specimens.

Materials and methods

The cyanobacterial crusts/ mats from the rock surfaces of Sun temple, Konark; Jagannath temple, Puri; Udayagiri caves, Lingaraj temple, Mukteswar temple, Brahmeswar temple, Kedar-Gouri temple, Parasurameswar temple and Rameswar temple, Bhubaneswar were used in the experiments. The blackish-brown crusts/ mats were collected from the rock surface of the temples and caves during summer months (March-May, 1994 and 1995) using sterile needles and stored in screw capped specimen bottles. In the laboratory, a pinch of each of the samples was added to 100 ml Erlenmeyer flasks containing 25 ml of sterile BG 11 medium with or without combined nitrogen (Rippka *et. al.*, 1979) and to agar plates prepared with the same mineral medium with 1% (w/v) agar agar. These cultures were incubated in a growth chamber at 25 ± 1°C under continuous light from fluorescent tubes at an light intensity of 12.5 W/m^2. The cyanobacterial forms appeared were identified following standard monographs (Tripathy *et. al.*,1997).

For studying the presence of UV absorbing pigments, the different cyanobacterial crusts/ tufts from different temples and unialgal culture of 16 species of cyanobacteria belonging to *Myxosarcina, Phormidium, Lyngbya, Plectonema, Nostoc, Calothrix, Tolypothrix* and *Fischerella* isolated from these crusts/ mats were used. All these organisms were grown in 100 ml Erlenmeyer flasks containing 25 ml of BG 11 medium (nitrogen free medium for heterocystous forms) (Rippka *et. al.*, 1979).at 25±1°C under 12.5 W/ m^2 light intensity from fluorescent tubes. Known quantity of crusts/ mats containing cyanobacterial cells at exponential growth phase were taken for extraction of UV absorbing pigments. These were extracted with 20% (v/v) methanol at 45°C, 2 h for mycosporine amino acid (MAAs) like compounds and for 3 h with 90% (v/v) methanol at 25°C for scytonemin pigment. The contents were centrifuged at 4,000 rpm for 30 min and the supernatant was used for measurement of absorption spectra in a Hitachi U 2000 UV-visible spectrophotometer in the wavelength range of 250-750 nm.

Results and discussion

UV absorbing pigments in the crusts/ tufts occurring on the temples and the corresponding cyanobacterial species

The coloured crusts and mats from eight different sites of archaeological importance comprising temples and caves, e.g. Sun temple, Konark; Kedar-Gouri temple, Udayagiri cave, Parasurameswar temple, Mukteswar temple, Brahmeswar temple, Lingaraj temple and Rameswar temple, Bhubaneswar were taken to find out the presence of MAA like substances. Absorption spectra of these extracts, depicted in Fig. 7.1, showed that all these crusts/ mats absorbed in the wavelength range 258 to 260 nm except that of the crust from Lingaraj temple where the absorption shifted to wavelength ranges of 271 and 282 nm. In addition, the crusts of Kedar-Gouri temple, Udayagiri caves and of Mukteswar temple absorbed prominently between 312 to 314 nm. Additional absorption peak at 317 nm was noted in the mats from Parasurameswar temple and Brahmeswar temple and at 325 nm in the crust of Sun temple, Konark. In certain monuments, e.g. Kedar-Gouri temple, Udayagiri cave and Rameswar temple, two different mats were encountered at each site containing different species of *Tolypothrix* as their major component. This was confirmed from the absorption spectra of their methanolic extracts and also from the morphological features of the organisms inhabiting there-in. The *Tolypothrix* species which occurred as tuft on Rameswar temple were similar to that of the organism occurring in the Udayagiri cave and Lingaraj temple of Bhubaneswar, hence, the spectra of their extracts was not depicted in Fig. 7.1. After extracting the mycosporine amino acid like substances (MAAs) from the crusts/ mats, the pellets were extracted with 90% (v/v) methanol and the absorption was measured in the wavelength range of 250 - 750 nm. Absorption peaks at 665 nm, 472 - 474 nm and at 384 nm of the extracts of the crusts/ mats from all the eight study sites were detected; the absorption at these wavelengths was due to chlorophyll-*a*, carotenoids and scytonemin pigments respectively (Table 7.1). In addition, absorption peak was also observed at 330 - 333 nm in the extracts of the crusts/ mats from Parasurameswar, Brahmeswar, Lingaraj and one of the mat from Rameswar temple. This shows that absorption in the UV-B and UV-C region between 258-325 nm could be extracted with 20% (v/v) methanol. Absorption at and above 330 nm was retained in pellets

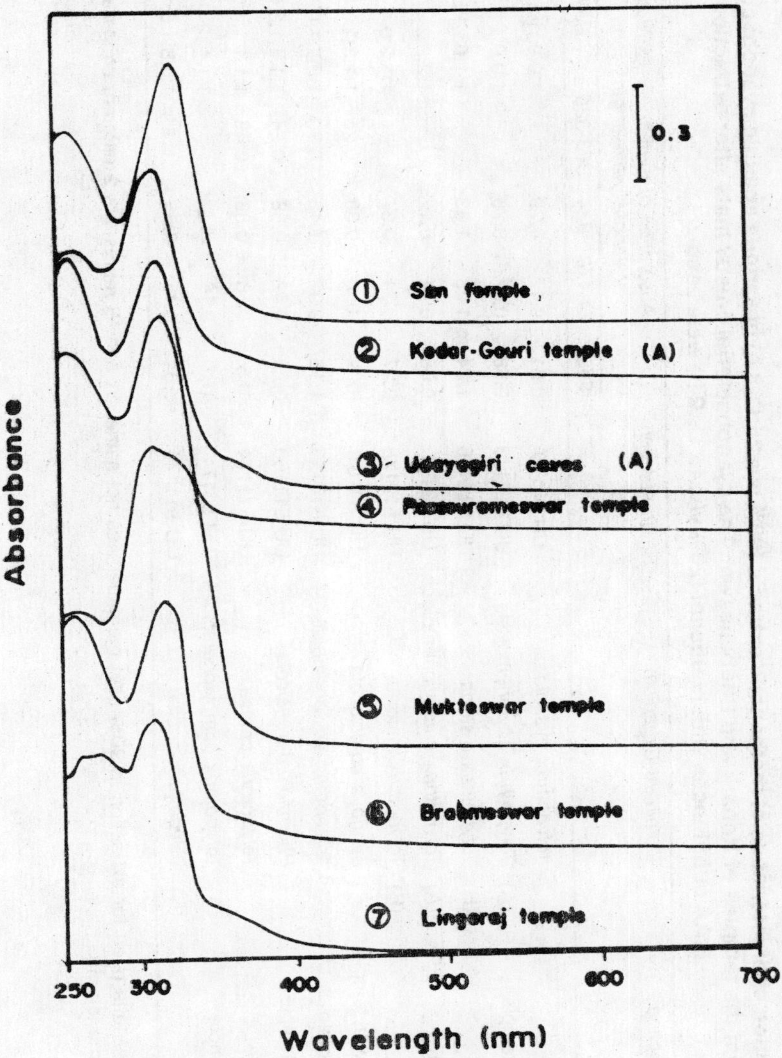

Fig. 7.1. Absorption spectra of methanolic extracts (20%, v/v, 45°C, 2h) for mycosporine amino acid – like substances (MAAs) of cyanobacterial crusts/ tufts from the exposed rock surface of seven different temples and caves.

Table 7.1: Absorption characteristics of methanolic extracts (90% v/v, 25°C, 3 h) of various epilithic cyanobacterial crusts/ tufts from the temples of Konark and Bhubaneswar. The cyanobacterial crusts/ mats after extraction of MAA substances with methanol (20% v/v, 45°C, 2 h) were used

Sl.No.	*Place of Collection*		*Dominant Organism*	*Strain number*	*Absorption of methanolic extracts at various wavelengths (nm)*				
					665	*474-470*	*384*	*333-330*	*Ratio*
1.	Sun temple		*Tolypothrix byssoidea*	UU 53170	0.51	0.47	0.79		1:0.9 :1.5
2.	Kedar-Gouri temple	(A)	*Tolypothrix scytonemoides*	UU 512173	0.58	0.57	0.75		1:0.9:1.3
		(B)	*Tolypothrix distorta*	UU 512176	1.24	1.28	1.83		1:1.03:1.5
3.	Khandagiri cave	(A)*	*Tolypothrix campylonemoides*	UU 53171	0.06	0.14	0.12		1:2.2:2
		(B)*	*Tolypothrix distorta*	UU 512176	0.27	0.34	0.37		1:1.2:0.4
4.	Parasurameswar temple		*Lyngbya corticiocola*	UU 53161	0.3	0.4	0.38	0.25	1:1.3:1.3:0.8
5.	Mukteswar temple		*Tolypothrix scytonemoides*	UU 512173	1.16	0.92	1.18	0.79	1:0.7:1.02:0.6
6.	Brahmeswar temple		*Phormidium truncicola*	UU 511160	0.29	0.42	0.38	0.24	1:1.4:1.3:0.8
7.	Lingaraj temple		*Tolypothrix crassa*	UU 511174	0.5	0.56	0.68	0.46	1:1.1:1.4:0.9
8.	Rameswar temple	(A)*	*Tolypothrix campylonemoides*	UU 53171	0.1	0.16	0.2		1:1.4 :2
		(B)*	*Tolypothrix crassa*	UU 511174	0.34	0.45	0.55	0.4	1:0.8:1.6:1.1

Quantity of the crusts/ tufts used for extraction in 90% (v/v) methanol was not same. *A and *B represents 2 different crusts/ mats collected from the same temple.

and could be extracted in the 90% (v/v) methanol. The absorption between 330 and 400 nm was due to the pigment absorbing at the UV-A region of the spectrum.

The crusts/ mats from all the study sites were also extracted with pure UV free acetone at 25°C, 3 h. Absorption spectra of the acetone extracts showed absorption peaks in the wavelength range of 660-663 nm due to chlorophyll-*a* (Table 7.2) and at 472-474 nm due to carotenoid pigments (data not shown). In addition the crusts/ mats from Sun temple, Konark; Lingaraj temple, Bhubaneswar, and one of the mats collected from Kedar-Gouri temple, Rameswar temple and Udayagiri cave, Bhubaneswar absorbed at 380-384 nm due to scytonemin. All these crusts/ mats also showed the absorption at various other wavelengths between 303-315 nm and in the wavelength range of 302-307, 312-314, 325, 338, 345-350nm. Though acetone extracts of most of these crusts/ mats showed common absorption peaks at 312-314 nm, also displayed few other additional peaks in the UV-A and UV-B region of the spectrum.

In order to find out the retention of absorption of mycosporine amino acid like substances and/ or scytonemin pigment in various epilithic cyanobacteria isolated from the crusts/ mats from the temples and grown in cultures under fluorescent light, few experiments were carried out. Seventeen different epilithic cyanobacterial species of which 14 were the components of the crusts/ mats from the temples and caves of Puri, Konark, and Bhubaneswar of Orissa state, and three others e.g. *Myxosarcina spectabilis* UU 514154, *Tolypothrix rechingeri* UU 513172 and *Fischerella muscicola* UU 58159 isolated from the crusts of Virupaksha temple, Pattadakal; Ladkhan temple, Aihole, Karnatak and Vatsala Durga temple, Bhaktapur, Nepal, respectively were used in the experiments. These organisms were routinely cultivated in BG 11 medium with or without nitrogen under fluorescent light at an intensity of 12.5 W/m^2 at 25 ± 1°C for more than an year. For experiments, the cells in their exponential growth phase were taken and extracted with 20% (v/v) methanol (45°C, 2 h) and 90% (v/v) methanol (25°C, 3 h). Absorption of the different methanolic extracts was measured in the UV and visible range of the spectrum (250-750 nm). Of the 17 epilithic cyanobacterial species used in the study, 6 species of *Tolypothrix*, 3 of *Calothrix* and 1 each of *Myxosarcina*, *Nostoc* and *Fischerella* showed absorption maxima at specific wavelengths

Table 7.2: Absorption characteristics of acetone extracts (100%, 25°C, 3 h) of various epilithic cyanobacterial crusts/ mats from the temples of Konark and Bhubaneswar

Sl.No.	Place of Collection		Dominant	Strain number	Absorption of methanolic extracts at various wavelengths (nm)							
					663-660	384-380	350-345	338	325	314-312	307-303	Ratio
1.	Sun temple		*Tolypothrix byssoidea*	UU 53170	1.4	3.65		2.86	1.58		0.8	1:2.6 :2 :1.1 :0.5
2.	Kedar-Gouri temple	(A)	*Tolypothrix scytonemoides*	UU 512173	1.58	5.53				0.7		1:3.5 :0.4
		(B)	*Tolypothrix distorta*	UU 512176	1.24	1.83				1.44		1:1.47:1.16
3.	Khandagiri cave	(A)*	*Tolypothrix campylonemoides*	UU 53171	0.28					0.3		1:1.1
		(B)*	*Tolypothrix distorta*	UU 512176	0.93	3.05				0.9		1:3.3 :0.9
4.	Parsurameswar temple		*Lyngbya corticiocola*	UU 53161	1.59		1.9			2.2		1:1.2 :1.4
5.	Mukteswar temple		*Tolypothrix scytonemoides*	UU 512173	1.26			1.13			0.92	1:0.9 :0.5
6.	Brahmeswar temple		*Phormidium truncicola*	UU 511160	1.6					2.4		1:1.5
7.	Lingaraj temple		*Tolypothrix crassa*	UU 511174	1.43	5.14	4.86				1.73	1:3.6 :3.4 :1.2
8.	Rameswar temple	(A)*	*Tolypothrix campylonemoides*	UU 53171	1.08					2.38		1:2.2
		(B)*	*Tolypothrix crassa*	UU 511174	1.98	7.32	6.73				2.37	1:3.7 :3.4 :1.2

Quantity of the crusts/ tufts used for extraction in acetone (100%) was not same. *A and *B represents 2 different crusts/ mats collected from the same temple.

in the UV-region of the spectrum (Table 7.3). In addition to their absorption at 665-666, 472-478 and 413-440 nm due to chlorophyll-*a* and carotenoid pigments, some of these organisms absorbed at 384 nm or up to 393 nm in the UV-A region due to the yellowish sheath pigment scytonemin (Garcia-Pichel and Castenholz, 1991) and at various wavelengths between 369 and 258 nm in the UV-B and UV-C region of the spectrum . The absorption peaks of methanolic extracts between 258 and 325 nm was due to mycosporine amino acid like substances (MAAs). This identification was made on the basis of absorption spectra as followed earlier (Garcia-Pichel and Castenholz, 1993; Böhm *et. al.*, 1995; Karsten and Garcia-Pichel, 1996). However, there existed a wide variation among these species with regard to their absorption maxima for MAAs and scytonemin. All those species of *Calothrix* and *Tolypothrix* which showed absorption at 384 nm possess well defined yellowish brown sheath layers around their trichome suggesting that scytonemin is located in the extracelluar sheath (Garcia-Pichel and Castenholz, 1991; Proteau *et. al.*, 1993). Further, all these *Tolypothrix* species have been found to be the dominant component of the crusts/ mats collected from temple walls of different regions of India during mid days of hot summer months exposed to bright sunlight of the tropics (Tripathy *et. al.*, 1997) suggesting the capability of these organisms to colonize the bare rock surfaces of temples even under extreme heat and dryness.

UV absorbing pigments of *Tolypothrix byssoidea* UU 53170 and *Tolypothrix scytonemoides* UU 512173 during their growth under various light/ dark regime

Since both the species of *Tolypothrix*, e.g. *Tolypothrix byssoidea* UU 53170 and *Tolypothrix scytonemoides* UU 512173 which were the prominent components of the crusts/ mats of Sun temple, Konark and Kedar-Gouri temple, Bhubaneswar respectively in the exposed rocks as well as inside the temples at darkened places, pigment composition of these species during dark incubation was studied by analysing the absorption spectrum of the methanolic extracts (90%, v/v) of their cells/ trichome. It was observed that soon after dark incubation, *Tolypothrix byssoidea* UU 53170 lost its absorption at 327 nm in the UV region due to MAAs, whereas the broad absorption at 384 nm due to scytonemin became prominent. Under low light (7.5 W/m^2), absorption due to chlorophyll-*a* pigment and MAAs was more prominent and quantitatively increased over its

Table 7.3: Absorption characteristics of methanolic extracts (20%, v/v, 45°C, 2 h and 90%, v/v, 25°C, 3 h) of various epilithic cyanobacteria isolated from the temple walls and grown in BG 11 medium with or without combined nitrogen at 25±1°C under 12.5 W/m² light intensity in pure culture. Organisms during their exponential growth phase were used in the experiments

Organism	Isolate number	Absorption of methanolic extracts at various wavelengths (nm)		
		20% methanol, v/v, 45°C, 2 h	90% methanol, v/v, 25°C, 3 h	
			UV region	visible region
Myxosarcina spectabilis	UU 514154	260,305	260,307	440,473,665
Nostoc ellipsosporum	UU 52184		265,288,334,373	434,478,666
Calothrix scytonemicola	UU 512166	258,325	340,384,393	475,665
Calothrix marchica	UU 512167	260,323	265,334	436,475,665
Calothrix ghosei	UU 518169	258,325	336	435,474,665
Tolypothrix byssoidea	UU 53170	258,325	258,327,384	434,473,665
Tolypothrix campylonemoides	UU 53171	260,313	269,313,335	433,472,665
Tolypothrix rechingeri	UU 512172	258,325	265,313,335	435,472,665
Tolypothrix scytonemoides	UU 512173	258,309	260,295,384	414,473,665
Tolypothrix crassa	UU 511174	260,305,313	258,293,331,351,369	415,473,665
Tolypothrix distorta	UU 512176	260,313	260,308,384,393	413,473,665
Fischerella muscicola	UU 58159	260,325	265,334	436,475,665

cells incubated at 12.5 W/w^2 light intensity. Absorption due to chlorophyll-*a*, scytonemin and carotenoids was prominent even when the organism was cultured up to 30 days in the inorganic medium (Table 7.4). To the contrary, in *Tolypothrix scytonemoides* UU 512173, absorption at 665 nm was decreased in the cells when the organism was cultivated for more than 15 days at 7.5 W/m^2 and kept in the dark than that of the cells grown at 12.5 W/m^2 light intensity. However, absorption at 385-389 nm due to scytonemin was more prominent within few days of cultivation of the organism in darkness. The absorption due to scytonemin in the dark grown cells was higher than those grown under light intensity of 12.5 W/m^2 and 7.5 W/m^2; absorption of chlorophyll-*a* as well as scytonemin in this organism was comparatively lower in the later light regime (Table 7.5). In order to find out whether the absorption due to MAAs can be re-induced in the dark grown cells of both the *Tolypothrix* species, the 15 days dark grown organisms were exposed to 12.5 W/m^2 light intensity for different time periods (6 h, 12 h, 24 h and 48 h) and then the absorption spectrum was measured. Within 48 h of exposure to fluorescent light after the dark growth, the absorption of MAAs in the UV range of the spectrum was not re-induced in *Tolypothrix byssoidea* UU 53170, however, the peak at 384 nm due to scytonemin became broader (Fig. 7.2). In *Tolypothrix scytonemoides* UU 512173, upon exposure to light after dark growth, any change in the spectrum of its methanolic extract was not detected (Fig. 7.3). Further experiments were carried out by incubating the cells of both the organisms in the dark for six consecutive months in the inorganic medium (without any organic substrate) and then exposing to fluorescent light or sunlight and studying the reappearance of UV absorbing pigments, if any. After exposure of the dark incubated cells of *T. byssoidea* UU 53170 to fluorescent light of 12.5 W/m^2 or to direct sunlight for 30 days (for 6 hours daily in case of exposure to sunlight), the MAAs absorbing in the UV region was regained (Fig. 7.4); chlorophyll-*a* content of the organism under exposure to direct sunlight though was reduced, absorption due to carotenoid pigments, scytonemin and MAAs was more prominent. The cells of these organisms were not disintegrated even after 6 months of exposure to darkness and synthesized yellowish brown sheath pigments after exposure to light (Figs. 7.4 and 7.5).

The present data on epilithic cyanobacteria from temple walls showed that mycosporine aminoacid like substances (MAAs) were

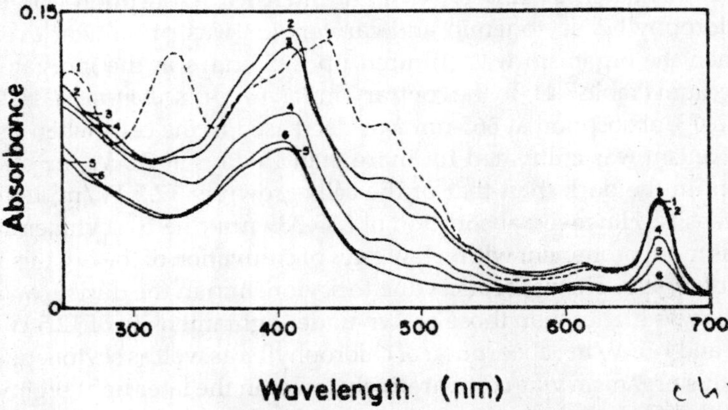

Fig. 7.2. Absorption spectra of methanolic extracts (90%, v/v, 25°C, 3 h) of *Tolypothrix byssoidea* UU 53170 exposed to fluorescence light for different time periods after growth in darkness for 15 days: 1. control (organism not pre exposed to darkness; grown at 12.5 W/m^2 light intensity); 2. pre incubated in darkness up to 15 days; 3. exposure to 12.5 W/m^2 light intensity up to 6h after dark growth; 4. exposure to 12.5 W/m^2 light intensity up to 12h after dark growth; 5. exposure to 12.5 W/m^2 light intensity up to 24h after dark growth; 6. exposure to 12.5 W/m^2 light intensity up to 48h after dark growth.

synthesized/ induced in the cells of *T. byssoidea,* which is the major component in the crusts of many temples (Tripathy *et. al.,* 1997) in response to light. Cells incubated in the dark for 5 days lost their capability to absorb in the UV-B region corresponding to the absorption due to the MAAs but of scytonemin absorbing at 384 nm was decreased only on prolonged dark incubation. Upon exposure to light after dark growth period, absorption due to MAAs was regained in the cells. Absorption spectra of the dark grown cells of the other experimental organism, *T. scytonemoides* UU 512173 isolated from the tufts of Kedar-Gouri temple which did not show prominent absorption in the UV-B range of the spectrum, was not changed much in comparison to their light grown cultures; the absorption due to scytonemin in the UV-A region was retained in the dark incubated cultures. Since the formation of MAA like substances but not of scytonemin in the cells of the *Tolypothrix* species is regulated by light regime, MAAs are possibly the major UV absorbing/sunscreen pigments in these subaerial cyanobacteria.

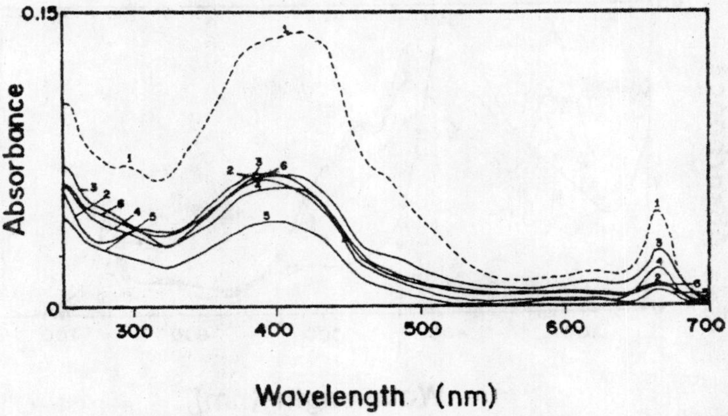

Fig. 7.3. Absorption spectra of methanolic extracts (90%, v/v, 25°C, 3 h) of *Tolypothrix scytonemoides* UU 512173 exposed to fluorescence light for different time periods after growth in darkness for 15 days: 1. control (organism not pre exposed to darkness; grown at 12.5 W/m² light intensity); 2. pre incubated in darkness up to 15 days; 3. exposure to 12.5 W/m² light intensity up to 6h after dark growth; 4. exposure to 12.5 W/m² light intensity up to 12h after dark growth; 5. exposure to 12.5 W/m² light intensity up to 24h after dark growth; 6. exposure to 12.5 W/m² light intensity up to 48h after dark growth.

Table 7.4: Absorption characteristics of 90% (v/v) methanolic extract of *Tolypothrix byssoidea* UU 53170 grown under different light intensities and in continuous dark for different time periods

Days of incubation	*Growth condition*	*Wavelength (nm)* 665	475	384	327	*Ratio* 665:475:384:327
5	12.5 W/m² light	0.025	0.021	0.05	0.067	1:0.84:2:2.68
	Diffuse light	0.045	0.025	0.08	0.102	1:0.55:1.78:2.27
	Dark	0.04	0.03	0.107		1:0.75:2.63
15	12.5 W/m² light	0.025	0.021	0.05	0.067	1:0.84:2 :2.68
	Diffuse light	0.058	0.057	0.078	0.108	1:0.98: 1.34:1.86
	Dark	0.052	0.065	0.12		1:1.25:2.31
30	12.5 W/m² light	0.025	0.021	0.05	0.067	1:0.84:2.68
	Diffuse light	0.076	0.07	0.088	0.134	1:0.92:1.16:1.76
	Dark	0.051	0.058	0.133		1:1.13:2.61

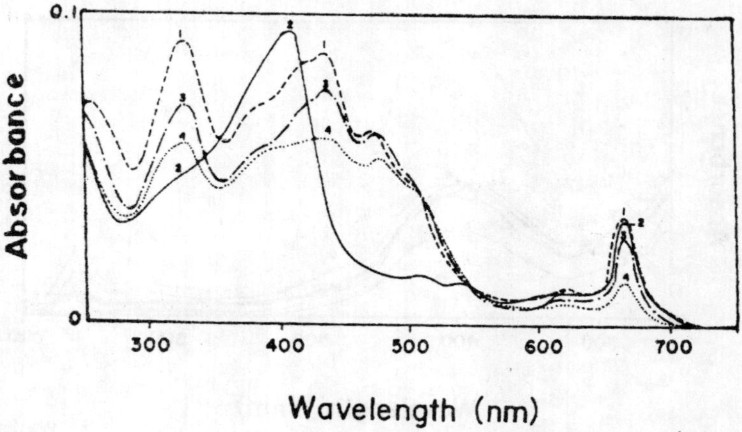

Fig. 7.4. Absorption spectra of methanolic extracts (90%, v/v, 25°C, 3 h) of *Tolypothrix byssoidea* UU 53170 grown in continuous dark up to 6 months and then exposed to fluorescent light of 12.5 W/m² and to sun light for a month: 1. control (organism grown at 12.5 W/m² light intensity); 2. growth in continuous dark in the mineral medium up to six months; 3. exposure to fluorescent light of 12.5 W/m² light intensity for a month after 6 months of dark incubation; 4. exposure to sun light (6 hours daily) for a month after 6 months of dark incubation.

Table 7.5: Absorption characteristics of 90% (v/v) methanolic extract of *Tolypothrix scytonemoides* UU 512173 grown under different light intensities and in continuous dark for different time periods

Days of incubation	*Growth condition*	*Wavelength (nm)*			*Ratio*
		665	*475*	*384*	*665:475:384*
5	12.5 W/m² light	0.007	0.01	0.025	1:1.43:3.57
	Diffuse light	0.018	0.017	0.055	1:0.94:3.06
	Dark	0.03	0.032	0.08	1:1.07:2.67
15	12.5 W/m² light	0.018	0.03	0.064	1:1.67:3.56
	Diffuse light	0.011	0.012	0.046	1:1.09:4.18
	Dark	0.012	0.017	0.068	1:1.42:5.67
30	12.5 W/m² light	0.027	0.045	0.097	1:1.67:3.59
	Diffuse light	0.02	0.022	0.07	1:1.1:3.5
	Dark	0.02	0.03	0.102	1:1.5:5.1

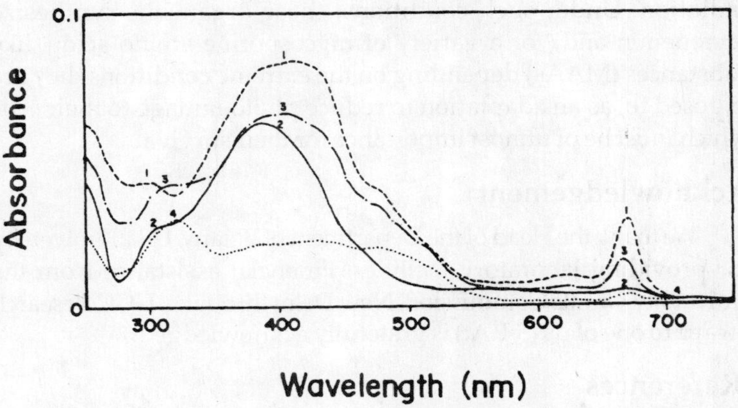

Fig. 7.5. Absorption spectra of methanolic extracts (90%, v/v, 25°C, 3 h) of *Tolypothrix scytonemoides* UU 512173 grown in continuous dark up to 6 months and then exposed to fluorescent light of 12.5 W/m² and to sun light for a month: 1. control (organism grown at 12.5 W/m² light intensity); 2. growth in continuous dark in the mineral medium up to six months; 3. exposure to fluorescent light of 12.5 W/m² light intensity for a month after 6 months of dark incubation; 4. exposure to sun light (6 hours daily) for a month after 6 months of dark incubation.

Certain cyanobacteria possess the ability to occur in a wide variety of extremely dry habitats like the exposed rock surfaces of temples and monuments of a tropical country like India. The environmental stresses of these habitats are likely to play a determinative role on the morphology and physiology of the organisms that occur in such aerophytic environments. Cyanobacterial species isolated from these habitats have been found to contain one or more UV-absorbing, mycosporine amino acid like substances and/ or extracelluar sunscreen pigment, scytonemin which has been suggested as an adaptive strategy for photo-protection against short wavelength irradiance (Garcia-Pichel and Castenholz, 1991, 1993). Certain filamentous cyanobacteria synthesized sunscreen pigments absorbing at 312 nm and at shorter wavelengths between 250 and 300 nm upon prolonged exposure to artificial solar radiation (Donkar and Hädar, 1995) and to UV (Scherer *et. al.*, 1988). The terrestrial cyanobacteria that colonize the rock surface of temples during summer months survive long periods of quiescence while simultaneously desiccated and exposed to solar radiation. Under such conditions, these organisms synthesize scytonemin and/ or a variety of mycosporine amino acid - like substances (MAAs) depending on the extreme conditions they are exposed to, as an adaptation to reduce photodamage to their cells which must be of utmost importance for their survival.

Acknowledgement

We thank the Head of the Department of Botany, Utkal University for providing laboratory facilities. Financial assistance from the University Grants Commission, New Delhi through a UGC Research award to one of us (S.P.A.) is gratefully acknowledged.

References

Anagnostidis, K., Economou-Amilli, A. and Roussomoustakaki, M. 1983. Epilithic and chasmolithic microflora (Cyanophyta, Bacillariophyta) from marbles of the Parthenon (Acropolis-Athens, Greece). *Nova Hedw.* **38**: 227-277.

Böhm, G.A., Pfleiderer, W., Böger, P. and Scherer, S. 1995. Structure of a novel oligosaccharide mycosporine amino acid ultraviolet A/B sunscreen pigment from the terrestrial cyanobacterium *Nostoc commune. J. Biol. Chem.* **270**: 8536-8539.

Buckley, C.E. and Houghton, J.A. 1976. A study of the effect of near UV-radiation on the pigmentation of the blue-green alga *Gloeocapsa alpicola. Arch. Microbiol.* **107**: 93-97.

Büdel, B. and Wessels, D.C.J. 1991. Rock inhabiting blue-green algae/ cyanobacteria from hot arid regions. *Algol. Studs.* **64**: 385-398.

Chadha, A. and Pandey, D.C. 1982. Distribution of algae on the walls of buildings. *Phykos* **21**: 122-123.

Donker,V.A. and Hader,D.P. 1995. Protective strategies of several cyanobacteria against solar radiation. *J.Plant Physiol.* **145**: 750-755.

Friedmann, E.I. and Galun, M. 1974. Desert algae, lichens and Fungi. In: *Desert Biology*. Edited by Brown, G.W. Academic Press, New York. pp. 165-212.

Garcia-Pichel, F. and Castenholz, R.W. 1991. Characterization and biological implications of scytonemin, a cyanobacterial sheath pigment. *J. Phycol.* 27: 395-409.

Garcia-Pichel, F. and Castenholz, R.W. 1993. Occurrence of UV-absorbing mycosporine-like compounds among cyanobacterial isolates and an estimate of their screening capacity. *Appl. Environ. Microbiol.* **59**: 163-169.

Garcia-Pichel, F., Sherry, N.D. and Castenholz, R.W. 1992. Evidence for a UV-sunscreen role of the extracellular pigment scytonemin in the terrestrial cyanobacterium *Chlorogloeopsis* sp. *Photochem. Photobiol.* **56**: 17-23.

Garcia-Pichel, F., Wingard, C.E. and Castenholz, R.W. 1993. Evidence regarding the UV sunscreen role of a mycosporine-like compound in the cyanobacterium *Gloeocapsa* sp. *Appl. Environ. Microbiol.* **59**: 170-176.

Häder, D.P. 1988. Ecological consequences of photomovement in microorganisms. *J. Photochem. Photobiol B. Biol.* **1**: 385-414.

Hoffmann, L. 1989. Algae of terrestrial habitats. *Bot. Rev.* **55**: 77-105.

Karsten, U. and Garcia-Pichel, F. 1996. Carotenoids and mycosporine-like amino acid compounds in members of the genus *Microcoleus* (cyanobacteria): a chemosystematic study. *Syst. Appl. Microbiol.* **19**: 285-294.

Mittler, R. and Tel-Or, E. 1991. Oxidation stress responses in the unicellular cyanobacterium *Synechococcus* PCC 7942. *Free Rad. Res. Comm.* **12-13**: 845-850.

Levine,E. and Thiel, T. 1987. UV inducible DNA repair in the cyanobacterium *Anabaena* sp. *J.Bactoriol.* **169**: 3988-3993.

Pattnaik, J., Swain, N. and Adhikary, S.P. 1993. Differential response of two species of the cyanobacterium *Anabaena* to ultraviolet (UV-C) irradiation. *J. Basic Microbiol.* **33**: 427-432.

Proteau, P.J., Gerwick, W.H., Pichel, F. and Castenholz, R. 1993. The structure of scytonemin, an ultraviolet sunscreen pigment from the sheaths of cyanobacteria. *Experientia* **49**: 825-829.

Richardson, L.L. and Castenholz, R.W. 1987. Diel vertical movements of cyanobacterium *Oscillatoria terebriformis* in a sulfide-rich hot spring microbial mat. *Appl. Environ. Microbiol.* **53**: 2142-2150.

Rippka, R., Deruellus, J., Waterbury, J.B., Herdman, M.A. and Stanier, R.Y. 1979. Generic assignments, strain histories and properties of pure cultures of cyanobacteria. *J. Gen. Microbiol.* **111**: 1-61.

Scherer, S., Chen, T.W. and Böger, P. 1988. A new UV-A/B protecting pigment in the terrestrial cyanobacterium *Nostoc commune*. *Plant. Physiol.* **88**: 1055-1057.

Tripathy, P., Roy A. and Adhikary, S.P. 1997. Survey of epilithic blue-green algae / cyanobacteria from temples of India and Nepal. *Arch. Hydrobiol. Algol. Studs.* **87**: 43-57.

Tripathi, S.N., Tiwari, B.S. and Talpasayi, E.R.S. 1991. Growth of cyanobacteria (blue-green algae) on urban buildings. *Energy Build.* **15-16**: 499-505.

Whitton, B.A. 1987. Survival and dormancy of blue-green algae. In: *Survival and Dormancy of Microorganisms*. Edited by Heins, Y. , John Wiley and Sons, New York. pp. 109-167.

Chapter 8

INFLUENCE OF PETROLEUM OILS ON ALGAE AND CYANOBACTERIA

J.P. Gaur and A.K. Singh

Abstract

Petroleum oils exert toxic effects on algae in both laboratory and field conditions. This results through inhibition of almost all physiological parameters. Differences in chemical composition are primarily responsible for variability in toxicity of oils. In general, oils rich in aromatics are more toxic than those deficient in these. Environmental factors can modify the toxicity of petroleum. The photooxidized oil is generally more toxic than the parent oil. Some oils are even stimulatory at their low concentrations. Partial degradation of aromatics and other fractions could be mediated by algae belonging to diverse groups. The toxic effects of oils on algae are highly species-specific. As a result, the composition of natural algal communities is altered by oil pollution.

Key words: Algae, cyanobacteria, oils, petroleum, pollution.

Introduction

The release of petroleum oils into aquatic systems degrades the quality of environment and threatens the inhabiting biota. Algae and cyanobacteria (blue-green algae), representing the producer component of such systems, are being increasingly exposed to oil pollutants. O'Brien and Dixon (1976) expressed concern about the lack of information pertaining to petroleum effects on algae. Vandermeulen and Ahern (1976) briefly reviewed the physiological effects of oil pollution on algae. Although Østgaard (1994) discussed the action of petroleum on marine phytoplankton and Radwan and Al-Hasan (2000) briefly reviewed the interaction of cyanobacteria with oil, no other review has recently appeared that presents a complete picture of algal interaction with petroleum. The present article is an effort in this direction.

Ecological effects of oil pollution on algae

Ecological studies on algae have been performed after accidental oil spills, simulated spills in ponds and in controlled ecosystem enclosures, and also in habitats perennially polluted with petroleum, e.g. oil refinery effluents.

Accidental oil spills

The existing reports describe qualitative changes in algal communities, whereas quantitative information is scarce. In general, the extent of damage caused to algae is related to the amount of oil spilled, the site of spill, the physical and chemical properties of oil, and the time at which the accident occurs (Maurin, 1984). The phytosociological studies after the grounding of "Torrey Canyon" revealed no long-term effects of oil spill or detergent application on algal community composition (Bellamy *et. al.*,1967). Several marine algae that had been affected by the spill started reestablishing after some time. About 10% reduction in intertidal productivity was seen, of which 80% was probably due to detergent application (Chassé *et. al.*, 1967).

Mitchell *et. al.* (1970) observed normal intertidal algal flora following "Santa Barbara" spill. However, others have demonstrated its deterimental effects on *Phyllospadix torreyi* (Foster *et. al.*, 1971), *Ulva lactuca*, *Enteromorpha intestinalis* and *Endocladia* sp. (Holmes, 1969). Several years after the spill, Nicholson and Climberg (1971) reported 78% reduction in species number. But it is unclear if the reduction in species number was due to oil or some other factor. Disturbances in algal communities after "Tampico Maru" wreck in 1957 have been described by North *et. al.* (1965). Six months after the spill, 12 intertidal and 30 sublittoral species were additionally observed. The population of the giant kelp *Macrocystis pyrifera* substantially increased due either to reduced grazing pressure or by the action of the oil itself.

A good account of the effects of "Amoco Cadiz" and "Gino" spills on macrophytic algae has been given by Maurin (1984).

Although the algae were trapped in oil, their growth and reproduction did not suffer. Peréz (1978) could not observe any significant effect of "Amoco Cadiz" spill on the population of *Laminaria digitata*. The absence of a large quantity of detritus in the habitat led him to conclude that there was no abnormal mortality of *L. digitata* in spite of high level of hydrocarbons in the milieu. He also recorded high concentrations of hydrocarbons in alginates extracted from *L. digitata* growing on polluted sites. Peréz (1978) did not observe any harmful effect on the reproduction of some seaweeds at the site of "Amoco Cadiz" accident. Peréz reported (i) dark brown spots on the fronds of *L. digitata*, (ii) fertility of stalks in *Gigartina*, *Ascophyllum*, etc., and (iii) 35% thalli of *Chondrus crispus* had cystocarp-bearing stalks. A few years after Peréz's studies, Kaas (1980,1981) observed higher growth of algae in polluted area than those in the control zone. This was probably due to eutrophication-like changes caused by organic enrichment of the environment. Algal blooms have been reported after "Tsesis" spill also (Lindén *et. al.*, 1979, Johansson *et. al.*, 1980). These workers suggested that reduced grazing pressure was the primary reason for such blooms. Boucher *et. al.* (1984) found initially high chlorophyll *a* to pheopigment ratio in a sublittoral fine sand environment polluted by "Amoco Cadiz". But the concentration of chlorophyll *a*, and the ratio of chlorophyll *a* to pheopigment remained low several years after the incident. Riaux-Gobin (1985) also made a four-year survey of microphytobenthos on a mud flat polluted with oil from "Amoco Cadiz". The algal biomass increased 7 months after the spill, declined during 1979-80, and slowly increased in 1981. Riaux-Gobin found conspicuous changes in the annual cycle of biomass and species succession; two species of *Euglena* were stimulated due either to the direct action of oil or the disappearance of grazers which were obviously oil-sensitive.

The post-spill studies in marine environment often report oil coating on algae which prevents gas exchange and exerts mechanical stress. In extreme cases of smothering, algae became excessively heavy and subjected to breakage by waves (Nelson-Smith, 1973). Many seaweeds are economically important, and oiled and tainted algae loose their commercial value. The mucilage coatings of many brown algae may act as a protective device against oils (Baker, 1970; Straughan, 1971), however, dispersants are able to penetrate these coatings. Some of the algae damaged due to oil coating may be revived by basal regrowth (Nelson-Smith, 1968). Baker *et. al.* (1980) observed disintegration of red algae upon chronic exposure to dispersants. Huge quantities of dispersants used after the grounding of "Torrey Canyon" probably accentuated the toxic effects of the spilled oil. This incident has led to a serious debate regarding the use of oil dispersants (Lindén *et. al.*, 1987). This aspect has been described in detail under a separate head. Green algae are sensitive to oils (Baker, 1971a), though they have a remarkable ability to invade the areas from where other species have become eliminated (O'Brien and Dixon, 1976).

Simulated spills

Limited efforts have been made to assess the impact of simulated spills on algae in ponds, streams and controlled ecosystem enclosures (CEEs). The latter experimental systems mimic near-natural conditions, and are being increasingly used for pollution effect studies at the ecosystem level. Lee *et. al.* (1977) studied the effect of a fuel oil on phytoplankton and other communities in a CEE. Their study revealed marked sensitivity of centric diatoms to fuel oil, whereas pennate diatoms and flagellates were relatively insensitive or even stimulated. Kuwait crude inhibited primary production of marine phytoplankton in CEE and the oil-dispersant mixture had a synergistic effect (Lacaze, 1975). Studies made by Lännergren (1978) after an oil spill in the North sea showed no harmful effects, and biomass and primary production increased at 260 ppb of hydrocarbons. Primary productivity of phytoplankton community remained decreased by 30% for 35 days in a lake exposed to 2.6 ml l^{-1} of Norman Wells crude (Hanna *et. al.*, 1975). Though *Oscillatoria* sp. increased in proportion, algal biomass remained unchanged. At 0.03 ml l^{-1} of the same oil, Snow and Scott (1975) found complete inhibition in ^{14}C uptake immediately after the spill, but recovery occurred the next day. In this study also, the proportion of *Oscillatoria* changed, but algal biomass remained unaffected. Study of Cape Simpson ponds exposed to natural oil seepage revealed increased biomass and primary productivity of Chlorophyta and Cyanophyta (Barsdate *et. al.*, 1973). Higher concentrations were nevertheless inhibitory. Miller *et. al.* (1978) studied the influence of simulated spill of Prudhoe Bay crude oil in a small tundra pond. After the spill, the primary productivity was decreased but a slight recovery occurred 6-18 days after the spill. The dominance shifted from Chrysophyta (*Chromulina*) and Cryptophyta (*Rhomomonas*) to Chlorophyta (*Chlamydomonas*), Chrysophyta (*Uroglena*) and Pyrrophyta (dinoflagellates). Throndsen (1982) observed dominance by microflagellates when Ekofisk crude (31 mg l^{-1}) and Corexit 9527 (6.4 mg l^{-1}) were added to a controlled ecosystem enclosure. Federle *et. al.* (1979) also noticed decreased algal production and shift in community structure following an experimental spill of Prudhoe Bay crude in a small tundra thaw pond near Barrow, Alaska. Though primary production recovered with time, the composition of algal community remained unchanged due to reduction in number of zooplankton grazers and the selective effect of oil. Low levels of No. 2 fuel oil also stimulated the abundance of phytoplankton due to reduced predation pressure and altered feeding behaviour (Vargo *et. al.*,1982; Dahl *et. al.*,. 1983). Hsiao *et. al.* (1978) found some oils stimulatory and the other inhibitory to phytoplankton populations. The crude oil-Corexit mixtures were more toxic than the crude oil or Corexit alone. A simulated spill of Empire mix crude oil (1.4 mg l^{-1}) inhibited algal primary productivity and respiration, but recovery occurred with the passage of time (De la Cruz, 1982). Nigerian and Arabian crude oils exerted similar effects on natural phytoplankton communities. Lindén *et. al.* (1987) reported inhibitory action of North Sea oil on periphyton, phytoplankton and bladderwrack (*Fucus vesiculosus*). They found more severe inhibition of oil-dispersant mixture than the oil alone.

Lack of information concerning the effects of oil spills on lotic algae prompted Bott and Rogenmuser (1978) and Bott *et. al.* (1978) to undertake studies on laboratory streams. Bott and Rogenmuser (1978) assessed the impact of aqueous extracts of No. 2 fuel oil, used crankcase and Nigerian crude oil on periphytic algae. Nigerian crude oil was not inhibitory to biomass accrual, whereas No. 2 fuel oil was the most toxic shifting the dominance from diatoms to blue-green algae. Extracts of used crankcase oil also depressed chlorophyll concentration, but changes in dominant algal types were less dramatic. The incorporation of ^{14}C was stimulated as well as inhibited at hydrocarbon concentrations ranging from 0.038 to 0.124 mg l^{-1}. Bott *et. al.* (1978) simulated oil spill in a running water microcosm placed in glass house conditions. They found copious increase in populations of blue-green algae (*Oscillatoria, Schizothrix, Microcoleus*) following oil exposure. *Vaucheria* was initially destroyed by the oil but regrowth occurred by the end of the experiment. Another important change was shift in community metabolism to heterotrophic mode due to reduced photosynthesis.

Habitats chronically contaminated with oil

Some information is available on the ecology of algal communities in waters chronically contaminated with oils. Crosby *et. al.* (1954) found abundance of blue-green algae, particularly one *Oscillatoria* sp., in oil refinery effluent. Phytoplankton populations of oil refinery effluent holding ponds were studied by Minter (1964). Cooper and Wilhm (1975) employed plexiglass slides to study the colonization, productivity, pigment and species diversity of periphyton communities in a stream receiving petroleum refinery and domestic effluents. Effluent disposal caused detrimental effects on periphytic algae, but species diversity, productivity and pigment concentration increased downstream indicating regular improvement of water quality. Their study suffers from the serious drawback for not measuring the concentration of oil in the habitat. Effects of oil pollution on phytoplankton of Muddy river, near Boston, were evaluated by McCauley (1966). The study showed that reduced competition after the elimination of susceptible species had probably enabled the tolerant algae to maximize their reproductive potential. Composition of phytoplankton in an inshore environment chronically polluted by petroleum hydrocarbons from refineries was studied by Ignatiades and Mimicos (1977). A majority of phytoplankton species were able to withstand high concentrations of hydrocarbons. Le Petit *et. al.* (1977) did not observe harmful effects of refinery effluents on the chlorophyll of marine phytoplankton. Gaur and Kumar (1985a) conducted gradient analysis of algal communities in the effluents of oil refineries at the successive stages of treatment. They found abundance of *Oscillatoria* and other blue-green algal species in the effluents. Existence of a definite relation between algae and the amount of pollutants in the medium was demonstrated by them. Singh *et. al.* (1987) and Singh and Gaur (1989) studied epilithic algal communities in waters polluted with oil refinery effluents. They have related algal community characteristics, such as, species richness, evenness, diversity and similarity, with the level of oils and phenols in water. Algal assay of refinery effluents by Gaur and Kumar (1986) suggested a definite negative relationship between algal growth potential and the concentration of oil and phenol in wastewater.

Effects of petroleum on algal growth and physiology

Studies have been performed in laboratories using aqueous oil extracts, oil-in-water dispersions and whole oils. Aqueous oil extract has been preferred due to its uniform distribution throughout the volume of water, easy entry into cells, and the fact that it is more enriched in aromatic hydrocarbons than the parent oil (Boylan and Tripp, 1971; Anderson *et. al.*, 1974). Some researchers have reservations about using oil extract for toxicity testing (Østgaard, 1994). There is too much controversy regarding the appropriate method for the preparation of aqueous oil extract. A majority of workers have stirred one part of oil with 20 parts of sterilized medium for about 12 hours (Boylan and Tripp, 1971). The period of stirring as also the proportion of oil in the medium have, however, greatly varied in other studies (Anderson *et. al.*, 1974; Winters *et. al.*, 1976; Armstrong and Calder, 1978). Østgaard *et. al.*, (1984a) observed increased toxicity of Ekofisk crude extract following extended period of stirring. Sterilized aqueous extracts were used by some workers for oil toxicity bioassays. But autoclaving of oil extract decreases the amount of dissolved oil and changes the composition of aromatics (Vandermeulen and Ahern, 1976). The use of unsterilized extract is also not appropriate as it contains bacteria and other contaminating organisms. Vandermeulen and Ahern (1976) have demonstrated that autoclaving of whole oil before the preparation of aqueous extract does not change the chemical composition. Temperature and light conditions at the time of preparation of aqueous extract also matter a lot (Østgaard and Jensen, 1983; Maher, 1986). Another problem is that most of the studies do not indicate the exact concentration of oil in the aqueous extract; it has been rather expressed in terms of per cent saturation level (Kauss and Hutchinson, 1975a; Soto *et. al.*, 1975a; Gaur and Kumar, 1985b).

Some studies have been carried out to evaluate the toxicity of whole oils because algae may be affected by fractions of very low solubility in water (Pulich, *et. al.*, 1974; Batterton *et. al.*, 1978a; Gaur and Kumar, 1981). These workers have employed absorbent pads for administering oil into the culture suspension.

Growth

Petroleum oils and hydrocarbons seriously affect growth performance of algal cultures, though some workers report stimulatory effects at low levels of oils and hydrocarbons (Dunstan *et. al.*, 1975; Gaur and Kumar, 1981; Fabregas *et. al.*, 1984; Chan and Chiu, 1985; Al-Hasan *et. al.*, 1994). In some cases oils induced short lag prior to the onset of exponential phase, and the duration of the lag phase was dependent on the concentration of oil (Pulich *et. al.*, 1974; Kauss and Hutchinson, 1975a; Hsiao, 1978). Gaur and Kumar (1981), however, could not observe any lag and concluded that their test algae were capable of quickly adapting to petroleum oils.

Nuzzi (1973) found highly toxic effects of No. 2 fuel oil extract on *Phaeodactylum tricornutum*, *Skeletonema costatum* and *Chlorella* sp.; however, *Chlamydomonas* sp. remained unaffected. Mironov (1968) noticed delayed cell division in algae exposed to 10^{-5}-10^{-1} ml l^{-1} of various oils and oil products. Mironov (1972) reported mortality of several species of diatoms at 100 ml l^{-1} of kerosene and fuel oil; their growth rates, however, remained depressed at far lower concentrations. A 5 min exposure of fuel oil (1 ml l^{-1}) delayed the development of *Ditylum brightwelli* (Mironov, 1972). Depression of growth rate of *Phaeodactylum tricornutum* was observed 2 hours after exposure to crude oil extract and extension of incubation period increased the amount of inhibition (Kusk, 1981). Mommaerts-Billiet (1973) also observed reduction in the growth rate of the marine nanoplankton *Platymonas tetrathele* in a medium containing Libyan crude oil. Supplementation of oils to algal cultures also lowers the final yield (Chan and Chiu, 1985; Gaur and Kumar, 1985b). The studies conducted so far reveal concentration-dependent decrease in growth rate and yield of test algae in oil-enriched medium.

Although a majority of petroleum oils have been found toxic to algae, some are not. For example, Dogadina (1970) did not observe inhibitory effect of some crude oils on the growth of *Chlorella vulgaris* and *Scenedesmus obliquus*. Coffey *et. al.* (1977), Pulich *et. al.* (1974) and Winters *et. al.* (1976) were also unable to observe inhibitory effects of several petroleum oils. Variability in effects of oils on algae is undoubtedly related to their chemical composition (O'Brien and Dixon, 1976). Algal species too display variability in their response to a particular oil. Pulich *et. al.* (1974) noticed greater oil sensitivity of *Chlorella autotrophica* than *Dunaliella tertiolecta*. Kauss and Hutchinson (1975b) found that different species of *Ankistrodesmus* varied with regard to their sensitivity to oils. Similarly species-specific effects of crude oils and their products have been reported

(Nuzzi, 1973; Dunstan *et. al.*,1975; Coffey *et. al.*, 1977; Winters *et. al.*, 1977b; Hsiao, 1978; Mahoney and Haskin, 1980; Gaur and Kumar, 1981; Østgaard *et. al.*,1984b and Morales-Loo and Goutx, 1990). Diatoms have been found to be extremely tolerant to oils and hydrocarbons presumably due to their remarkable capability to adsorb and retain these pollutants. Thus they are able to maintain low concentrations of oils within the cells (Karydis, 1979). Selective toxicity of anilines, common in petroleum oils, to blue-green algae has been demonstrated by Batterton *et. al.* (1978b). What confers upon algal species tolerance or sensitivity to oils remains to be established, although physiological and genetical adaptations have been speculated (Gaur and Kumar, 1981). Since a given level of contamination of petroleum products affect various algal species differently, this may eventually lead to alteration of natural algal communities.

Petroleum hydrocarbons may induce morphological abnormalities in algal cells. Soto *et. al.* (1979a,b) studied the effects of aqueous crude oil extract (ACOE) and naphthalene on the green alga *Chlamydomonas angulosa*. The exposure of *C. angulosa* to ACOE and naphthalene caused shortening or complete loss of flagella, increase in cell size, formation of a wide space between the cell wall and the plasmalemma, distortion and disorganization of chloroplast lamellae, and decreased density of osmiophilic granules were also observed. Parasher *et. al.* (1978) noticed damaged cell membranes, disintegrated cytoplasm and disorganized cell organelles in *Chlorella pyrenoidosa* treated with hexachlorobenzene.

Reproductive stages during the life cycle of algae may be particularly sensitive to oils. Steele (1977) demonstrated inhibition of growth of fucoid zygotes at 20 ppm of No. 2 fuel oil and marked sensitivity of spermatozoa to petroleum. Sexual reproduction and auxospore formation in diatoms are inhibited by oils (Kustenko, 1981). Detergents may exert harmful effects on reproduction of marine algae. Boney (1968) observed marked toxic effects of detergents on fucoid spermatozoids, reproductive branch formation in *Bryopsis*, and cell rejuvenation and gonad settlement in *Prasinocladus marinus*.

Photosynthesis

Besides growth performance, photosynthesis has often been taken as a parameter to assess the effects of oils and hydrocarbons on algae. Depending on the concentration of oil in the medium, observations of various workers range from stimulation to inhibition (Gordon and Prouse, 1973; Parsons *et. al.*, 1976; Karydis, 1979; Kusk, 1981; Chan and Chiu, 1985; Gaur and Singh, 1990; Singh and Gaur, 1990). Lower concentrations are some times stimulatory but higher concentrations are always inhibitory. The mechanisms of stimulatory and inhibitory effects of oils and hydrocarbons are not adequately known. Oil coating around algae and a thin film of oil on the water surface interfere with gaseous exchange and inhibit photosynthesis by reducing the supply of carbon dioxide (Schramm, 1972). Van Overbeek and Blondeau (1954) suggested that the stimulatory and inhibitory effects of oils are probably related to the permeability of cell membrane; lower concentrations increase permeability while higher concentrations disorganise or disrupt the membranes causing damage to cells. Gordon and Prouse (1973) and Gaur and Kumar (1981) maintain that the growth regulating compounds are probably responsible for stimulation, and inhibition may be the result of enzymatic blockage. Hufford (1971) ascribed stimulatory effect of oils to trace elements. Singh and Gaur (1990) traced the stimulatory effect of a crude oil to its paraffinic fractions. Karydis (1979) suggested that the inhibition of photosynthesis is due to excessive accumulation of oil constituents in chloroplasts. It has also been proposed that the destruction of chlorophyll is one of the main reasons for the inhibition of photosynthesis (Chan and Chiu, 1985). The effects of oils on photosynthesis have been evaluated by measuring oxygen evolution (Pulich *et. al.*, 1974; Kusk, 1981, Chan and Chiu, 1985), ^{14}C incorporation (Gordon and Prouse, 1973; Karydis, 1979; Hilmer and Bate, 1983) and light-induced *in vivo* fluorescence of chlorophyll in presence of DCMU (Østgaard *et. al.*, 1984a,b, 1987). Pulich *et. al.* (1974) remarked that it would be interesting to pin point the action of oil on light reaction and possibly, but less likely, on the carbon fixation side of photosynthesis. Direct action of petroleum on the energy yielding photochemical electron transport systems has been proposed (Armstrong and Calder, 1978). Singh and Gaur (1988) showed that the action of Assam crude oil is mainly on the PS II of the photosynthetic electron transport system.

Respiration

Petroleum hydrocarbons generally inhibit respiration of algae (Hopkins and Kain, 1971; Kusk, 1978; Karydis, 1979). Although the mechanism of inhibition remains unexplained, various suggestions have been put forth. It has been proposed that oils may interfere with gas diffusion, but the effect is more pronounced for carbon dioxide than for oxygen (Schramm, 1972). Petroleum may disrupt membrane, then those in mitochondria could also be broken up to impair the TCA cycle and oxidative phosphorylation. Stimulation of respiration has been observed in some cases (Kusk, 1981; Chan and Chiu, 1985; Singh and Gaur, 1990) due probably to uncoupling of electron transport from phosphorylation. In such a situation, the rate of electron transport is increased but the energy is lost as heat.

Nutrient uptake

Hellebust *et. al.* (1985) observed that the short-term incubation of *Chlamydomonas angulosa* in hydrocarbon-enriched medium was either stimulatory or inhibitory to nitrate and methylamine (an analogue of ammonium) uptake, while long-term incubation was always inhibitory. Though naphthalene did not affect K_m for methylamine uptake, V_{max} was considerably lowered. Naphthalene had probably inactivated a proportion of the transport carriers rather than decreasing the transport affinity for the substrate. In a controlled ecosystem enclosure containing oil-Corexit mixture, Harrison *et. al.* (1986) found stimulation of photosynthesis as well as nitrate and ammonium uptake for the first four days, followed by a sharp decline. Hutchinson *et. al.* (1985) found that in oil-treated *Chlamydomonas angulosa*, phosphate uptake was inhibited at its high external concentration. But, greater amount of phosphate was released from the cells when its concentration was low in the medium. They also observed stimulation of phosphate uptake in the dark incubated cells of *C. angulosa* by the oil extract. This probably resulted through diversion of metabolic energy to phosphate transport from other energy-requiring metabolic processes. Singh and Gaur (1991), however, noticed only inhibition of phosphate uptake at the tested saturation levels of a crude oil extract.

Synthesis of macromolecules

Petroleum hydrocarbons generally inhibit the synthesis of macromolecules. The observation of several workers on action of petroleum on chlorophyll are contradictory. In some cases, the concentration of chlorophyll increased following oil treatment (Soto *et. al.*, 1977; Chan and Chiu, 1985). Vandermeulen and Ahern, 1976;

Hilmer and Bate (1983) found no change in the level of chlorophylls by oils. Inhibitory effects of oils on pigment synthesis have also been observed (Corner, 1978; Fabregas *et. al.*, 1984; Chan and Chiu, 1986). Karydis (1979) studied the influence of oils on the distribution of end products of photosynthesis. A decrease in biosynthesis of metabolic intermediates and lipids, and an increase in biosynthesis of polysaccharides and proteins were observed. In *Chlamydomonas angulosa* oil treatment did not change the carbohydrate content although total proteins decreased and lipids increased to significant levels (Soto *et. al.*, 1977). The increase in lipid and carbohydrate fractions of *C. angulosa* were probably contributed by carbon derived from the breakdown of proteins.

Davanin *et. al.* (1975) investigated the influence of oil pollution on biologically active compounds, particularly the nucleic acids. Although variable effects were observed, there was generally simultaneous inhibition of RNA and DNA synthesis in all the tested species.

Other physiological processes

Shindler *et. al.* (1975) speculated increased nitrogen fixation in freshwater systems polluted with petroleum, but Marowitch *et. al.* (1988) discounts any such possibility. Gaur and Singh (1989) observed inhibition of nitrogen fixation in *Anabaena doliolum* by Assam crude oil. The inhibitory effect was related with lowering of heterocyst frequency, nitrogenase activity and carbon fixation. Polynuclear aromatic hydrocarbons also hamper nitrogen fixing ability of cyanobacteria (Bastian and Toetz, 1985). Another physiological process greatly affected by petroleum oils and naphthalene is heterotrophic growth (Hellebust *et. al.*, 1982). These workers observed concentration-dependent stimulation or inhibition of acetate uptake in *Chlamydomonas angulosa*. At relatively low concentration of petroleum hydrocarbons increased acetate uptake and the level of isocitrate lyase were found which is essential for acetate assimilation.

High concentrations of oils have been found to cause leakage of photosynthetically fixed carbon compounds (Hutchinson *et. al.*, 1979). Further evidence of oil-induced changes in membrane permeability is provided by studies demonstrating leakage of cations from algal cells. Kauss and Hutchinson (1978) observed efflux of manganese and potassium from *Ankistrodesmus falcatus* subjected to oil treatment. In another communication, Hutchinson *et. al.* (1981) report leakage of manganese and calcium from *Chlamydomonas angulosa* in response to naphthalene or oil treatment.

Factors affecting petroleum toxicity

Composition of petroleum

Petroleum crude oils obtained from different sources vary greatly in colour, viscosity, specific gravity and chemical composition (details could be found in any text on organic chemistry). These variations are undoubtedly responsible for differences in action of different petroleum oils. The available reports point towards volatile hydrocarbons as the most toxic fraction of oils. Kauss and Hutchinson (1975a) found mitigation of oil toxicity with time and attributed it to escape of volatile toxic constituents. Hutchinson *et. al.* (see Atkinson *et. al.*, 1977) could not find toxic effects of benzene (1000 ppm) on algal cultures in cotton-stoppered flasks, although Atkinson *et. al.* (1977) noticed toxicity at its 20 ppm concentration in rubber stoppered vessels. The use of cotton-stoppered flasks facilitated evaporation and loss of volatile fractions which were apparently the most toxic. The toxicity became more pronounced due to retention of these constituents in rubber-stoppered flasks. Soto *et. al.* (1975a) similarly observed marked inhibition of growth of *Chlamydomonas angulosa* in glass-stoppered culture flasks. Improvement in growth occurred when the flasks were opened. The aged extracts did not exert toxic effects on the test alga. The volatile and water-soluble fractions of oils include low boiling paraffinics and aromatics, the latter considered to be more toxic (Blumer, 1969). Batterton *et. al.* (1978b) have shown that the toxic compounds in the fuel oils are environmentally persistent aromatics of high boiling point. Østgaard *et. al.* (1987) concluded that hydrocarbons, including aromatics, should not be considered as the dominant class of toxic constituents. Hutchinson *et. al.* (1979) tried to establish the relationship amongst the structure, solubility and toxicity of various hydrocarbons, including alkanes, aromatics, polynuclear aromatics and chlorinated hydrocarbons. They found highly significant negative relationship between solubility and toxicity. On the basis of their study it is possible to predict toxicity of a hydrocarbon by merely considering its solubility. In general, the solubility of hydrocarbons decreases with increase in methylation.

A majority of toxicological studies on petroleum oils neither define their chemical composition, nor pin point a particular fraction for toxic effects. Therefore some toxicologists have switched over to standard oils of known composition (Anderson *et. al.*, 1974), or distilled fractions, or to a particular oil constituent, e.g., benzene, naphthalene, aniline, etc. (Kauss *et. al.*, 1973; Soto *et. al.*, 1975a). Pulich *et. al.* (1974) separated different fractions of No. 2 fuel oil and found greater toxicity of those with low boiling point. But some species, for instance *Agmenellum quadruplicatum* and *Thalassiosira psudonana*, were most susceptible to high boiling point fractions. Batterton *et. al.* (1978a) did not find toxic effects of the paraffinic and asphaltic fractions of fuel oils. Some aromatic compounds which were not responsible for toxicity in their experiments include naphthalene, methylnaphthalene, dibenzothiophenes, phenanthrenes and others with volatilities greater than methylnaphthalene. Parker (1974) nevertheless observed paraffinics to be more toxic than the aromatics.

Light and temperature

Definite evidence are now available concerning the role of light on the toxicity of petroleum oils. Lacaze and Villedon de Naide (1976) observed that extraction of Kuwait crude oil in the light caused 2-3 fold increase in the toxicity of extract to *Phaeodactylum tricornutum* and marine phytoplankton communities. The toxicity of Algerian crude was also increased by light (Lacaze *et. al.*, 1982). Similarly, Bott and Rogenmuser (1978) observed greater toxicity to attached algal communities of three oil extracts prepared in light. These authors attribute light-induced accentuation of oil toxicity to formation of photooxidation products. Interestingly, even the normal indoor illumination has been found to increase the toxic effects of oils on *Skeletonema costatum* (Østgaard and Jensen, 1983). In a subsequent study, Østgaard *et. al.*,(1984b) found accentuation of toxicity of Ekofisk crude oil extract prepared under illumination. Karydis (1982) also showed light-induced toxicity of petroleum. It should, however, be kept in view that several peroxidation reactions can be induced even in the dark in the presence of oxygen (Østgaard, 1994).

It has been suggested that incorporation of oxygen into hydrocarbons greatly increases their solubility in water (Larson *et. al.*, 1979). For example, the solubility of naphthalene in water is approximately 35 mg l^{-1}, whereas the solubility of its oxidation product 2-naphthol is 1 g l^{-1}. The mechanism responsible for photooxidation of petroleum havae been discussed at length by Payne and Phillips (1985). Using GC-MS, Larson *et. al.* (1979) identified hydroperoxides (alkylated derivatives of tetralin hydroperoxide), phenolics (highly alkylated phenols and tetrahydronaphthols) and carboxylic acids as the main photooxidation products of No. 2 fuel oil. Of these, phenolics and carboxylic acids were the least inhibitory to ^{14}C incorporation by *Vaucheria* sp. and associated epiphytes. Tetralin hydroperoxide exerted pronounced toxic effect on *Vaucheria* sp. and also increased the excretion of carbon. Like synthetic detergent molecules, photooxidized products may become concentrated in organisms eventually leading to disruption of plasmamembrane and permeability. Heldal *et. al.* (1984) tested the toxic effects of different classes of organic compounds which are likely to be formed during photodecomposition of crude oils. Hydroperoxides and quinones were the most toxic, but they could not find any relationship between their chemical structure and the mode of toxic action. Winters *et. al.* (1977a) demonstrated wavelength dependent toxicity of phenlen-1-one to green algae.

Temperature may also influence the toxicity of oils (Graham and Hutchinson, 1975). It has been suggested that elevation of temperature increases the solubility of oils, and may promote the formation of toxic acids and peroxides (Baker, 1971). Baker (1971) further suggested that hot sunny weather might increase the toxicity of an oil by allowing it to penetrate more easily into the cells. LePemp *et. al.* (1976) similarly noticed rise in toxicity of Kuwait crude to phytoplankton with rise in temperature. However, when algae were exposed for a relatively short duration (4 h), the toxicity was greatly reduced. Gaur and Singh (1991) noticed enhancement of toxic effects on an oil extract with increase in light and temperature and this observation led them to speculate greater toxic effects of petroleum in tropical than in temperate waters. *Nitzschia* sp. and *Chaetoceros* sp. were much more sensitive to crude or fuel oil when grown at 0°C rather than 10°C (Van Baalen and O'Donnel, 1984). This study suggests that psychrophilic algae are more sensitive to oil pollution.

Dispersants

Several chemical dispersants are commercially available for the treatment of oil slicks. Dispersants break oil mass into fine droplets for homogenous distribution throughout the water column. Thus the surface area is increased thereby promoting the degradation of oil. The use of dispersants warrants proper toxicological evaluation because their misuse resulting in more severe damage has often been quoted. Most of the dispersants are toxic to algae at higher concentrations (Heldal *et. al.*, 1978; Fabregas *et. al.*, 1984; Chan and Chiu, 1985), although some are stimulatory at lower concentrations (Mommert-Billet, 1973; Hsiao *et. al.*, 1978; Bratbak *et. al.*, 1982; Fabregas *et. al.*, 1984; Chan and Chiu, 1985). The mixtures of oils and dispersants are generally more toxic than the oils or the dispersants alone (Chan and Chiu, 1985; Hsiao *et. al.*, 1978; Lee *et. al.*, 1985). In an enclosed marine ecosystem, Lindén *et. al.* (1987) found acute toxic effects of chemically dispersed North Sea oil. In this case, chemical dispersant brought about high initial concentration of oil which resulted in acute toxicity. If chemical dispersant was not used, toxic effects of oil would have lasted for a much longer time because of persistence of oil contaminants in the system. Wolfe *et. al.* (1998) showed that the use of dispersants enhanced the uptake of naphthalene by *Isochrysis galbana*.

Other factors

In addition to factors discussed before, population density of algae, availability of nutrients and the rate of dilution may have profound influence on oil toxicity. Østgaard and Jensen (1983) observed low toxicity of Ekofisk crude oil to *Skeletonema costatum* at higher cell density, whilst the reverse trend was obtained for cultures having low cell density. Phosphorus deficiency increases the sensitivity of diatoms to oils (Karydis, 1981). Dilution is another important factor mitigating the toxicity of oils. If an area polluted with oil has conditions congenial for dilution, severe toxic effects should not occur.

Uptake and metabolization of oil constituents

Hydrocarbons can be taken up and accumulated by algae from aquatic habitats. Thompson and Eglinton (1976) studied the distribution of aliphatic hydrocarbons in estuarine epipelic diatom populations by GC-MS. Similarity of aliphatic hydrocarbons in algal cells with those present in the Severn estuary sediment led them to point out the ability of algae to take up hydrocarbons from the milieu.

The uptake of hydrocarbons depend on the age of the algal culture and the composition of the medium. O'Brien and Dixon (1976) also cite a few papers reporting similar findings. More evidence have been provided by Boutry *et. al.* (1977) in *Chaetoceros simplex*. This diatom took up 37-90% of hydrocarbons from the medium, and the uptake depended on the age of culture and composition of the medium. Another marine diatom *Cyclotella cryptica* accumulated n-alkanes having C_{13} to C_{16} carbon atoms (Karydis, 1980). The uptake of alkanes with carbon atoms more than 12 was lower, while alkanes with C_{28}-C_{30} were not taken up by this diatom species. Thompson and Eglinton (1976) found exclusion of polynuclear aromatic hydrocarbons by epipelic diatom populations. They went on expressing the possibility of introduction of hydrocarbons accumulated by algae into the food chain.

Several examples of stimulation of algal growth and photosynthesis by petroleum are available. Such cases suggest that some petroleum constituents could be utilized by algae for their metabolic machinery. Dogadina (1970) found increased growth of two green algae exposed to ligroin (a mixture of paraffinic hydrocarbons and a lubricating oil). Likewise, Masters and Zajic (1971) observed stimulation of growth yield of *Scenedesmus quadricauda* treated with n-heptadecane. Of late, a number of evidence have accumulated showing the potential of cyanobacteria in degrading n-alkanes (Al-Hasan *et. al.*, 1998). In a long-term experiment, Graham and Hutchinson (1975) observed stimulation of algal growth by oil in the presence of light. They suspected it to be a case of photoheterotrophism and suggested that either hydrocarbons were directly taken up from the medium, or they were first degraded by contaminating bacterial population and the degradation products were in turn utilized by the test alga. They also thought that due to degradation of hydrocarbons by bacterial contaminants, carbon dioxide supply was probably increased leading to algal growth stimulation. *Chlamydomonas angulosa* could take up naphthalene from the medium but was able to depurate upon transfer to naphthalene-free medium (Soto *et. al.*, 1975b). Cerniglia *et. al.* (1980a) revealed that *Oscillatoria* sp. strain JCM could grow photoautotrophically in the presence of naphthalene and oxidizing it to Cis-1, 2-dihydroxy-1,2-dihydronaphthalene, 4-hydroxy-1 tetralone and 1-naphthol; the last compound being the major metabolite. Cerniglia *et. al.* (1979, 1980b,1982) observed that the capability of naphthalene oxidation exists in diverse groups of algae, such as blue-green, green, red and brown algae, and also in diatoms. Cerniglia *et. al.* (1980c) have also shown the ability of *Oscillatoria* sp. strain JCM to oxidize biphenyl to 4-hydroxyphenyl under photoautotrophic conditions. Biotransformation of aromatic amine to fomanilide, acetanilide and p-aminophenol has been demonstrated in *Agmenellum quadruplicatum* strain PR-6 and *Oscillatoria* sp. strain JCM (Cerniglia *et. al.*, 1981). These observations suggest that the ability of algal species to degrade hydrocarbons may be widespread in nature. Algae are also able to degrade phenolic compounds (Ellis, 1977).

Cyanobacteria may also assist in indirect degradation of petroleum by immobilizing oil degrading bacteria in their mucilage (Radwan and Al-Hasan, 2000). Cyanobacteria are supposed to provide oxygen, and sometimes fixed nitrogen in the case of diazotrophic cyanobacteria, to oil degrading microorganisms.

Conclusions

Petroleum oils are toxic to algae at higher concentrations. This results through inhibition of growth, photosynthesis, respiration, nitrogen fixation, nutrient uptake and other physiological processes. These pollutants are therefore non-specific in action. The aromatics of low and high molecular weight represent the most toxic component of petroleum. The specific compounds associated with the toxicity of petroleum extracts are benzene, naphthalene, di- and trimethylnaphthalenes, anilines and toluidines. The concentration of these compounds may vary from oil to oil. The complex nature of petroleum has always presented serious obstacles in toxicological studies. One important aspect completely ignored so far pertains to the possibilities of synergistic and antagonistic interactions among various compounds present in oils. It is necessary to further elaborate the relationships between chemical composition and toxicity of an oil. This would require joint efforts of chemists and biologists. Such concerted efforts have yielded valuable information in the U.S.A. and western Europe only. Similar efforts could not be made in other parts of the world due to lack of trained manpower and/or sophisticated analytical instruments.

Stimulation of algal growth by low concentrations of oils and hydrocarbons has been frequently observed in field and laboratory conditions. In nature, selective action of oil on the grazer populations perhaps allowed the tolerant algae to multiply and increase their populations. Some physiological explanations have also been given, however, stimulation remains an intriguing phenomenon till date. Some workers have reported the ability of algae to take up and metabolize hydrocarbons. But they did not give details about culture conditions. It may be that their experiments were based on contaminated cultures of algae. Recent work has clearly shown that algae can partially degrade some aromatic hydrocarbons. This aspect needs to be studied further.

Ecological studies have generated some very useful data. It may now be possible for various agencies to use algal community composition for monitoring of oil pollution in refinery effluents or the areas exposed to oil spills. The algae growing in oil-contaminated areas tend to accumulate hydrocarbons. If they are studied in detail, it may be possible to relate the hydrocarbon concentration in algae with that in the milieu. Similar approach has been successfully used for the monitoring of heavy metals (De Filippis and Pallaghy, 1994). Another aspect requiring further attention is the probability of biomagnification of hydrocarbons in aquatic food chains *via* algae.

Acknowledgement

We gratefully acknowledge the Head, Department of Botany and the Coordinator, Centre of Advanced Study in Botany, Banaras Hindu University, for facilities.

References

Al-Hasan, R.H., Al-Bader, D., Sorkhoh, N.A. and Radwan, S.S. 1998. Evidence for n-alkane consumption and oxidation by filamentous cyanobacteria from oil-contaminated coasts of the Arabian Gulf. *Appl. Mar. Biol.* **130**: 521-527.

Al-Hasan, R.H., Sorkhoh, N.A., Al-Bader, D. and Radwan, S.S. 1994. Utilization of hydrocarbons by cyanobacteria from microbial mats on oily coasts of the Gulf. *Appl. Microbiol. Biotechnol.* **41**: 615-619.

Anderson, J.W., Neff, J.M., Cox, B.A., Tatem, H.E. and Hightower, G.M. (1974). Characteristics of dispersions and water-soluble extracts of crude and refined oils and their toxicity to estuarine crustaceans and fish. *Mar. Biol.* **27**: 75-88.

Armstrong, J.E. and Calder, J.A. 1978. Inhibition of light-induced pH increase and O_2 evolution of marine microalgae by water-soluble components of crude and refined oils. *Appl. Environ. Microbiol.* **35**: 858-862.

Atkinson, L.P., Dunstan, W.M. and Natoli, J.C. 1977. The analysis and control of volatile hydrocarbon concentrations (e.g. benzene) during bioassays. *Water, Air Soil Pollut.* **8**: 235-242.

Baker, J.M. 1970. The effects of oils on plants. *Environ. Pollut.* **1**: 27-44.

Baker, J.M. 1971. Successive spillages. In: *Ecological Effects of Oil Pollution on Littoral Communities.* Edited by Cowell., E.B. Applied Science Publ., Barking, Essex. pp. 21-32.

Baker, J.M., Crothers, J.H., Mullett, J.A.J. and Wilson, C.M. 1980. Ecological effects of dispersed and non-dispersed crude oil: a progress report. In: *Proc. Institute of Petroleum Conference on Petroleum Development and the Environment, London,* Heyden & Sons Ltd, London. pp. 85-100.

Barsdate, R.S., Alexander, V. and Benoit, R.E. 1973. Natural oil seeps at Cape Simpson, Alaska: Aquatic effects. In: *Proc. Symposium on the Impact of Oil Resource Development on Northern Plant Communities, Occassional Publications on Northern Life,* No. I, University of Alaska, Fairbanks. pp. 91-95.

Bastian, M.V. and Toetz, D.W. 1985. Effect of polynuclear hydrocarbons on algal nitrogen fixation (acetylene reduction). *Bull. Environ. Contam. Toxicol.* **35**: 250-265.

Batterton, J.C., Winters, K. and Van Baalen, C. 1978a. Sensitivity of three microalgae to crude oils and fuel oils. *Mar. Environ. Res.* **1**: 31-41.

Batterton, J.C., Winters, K. and Van Baalen, C. 1978b. Anilines: Selective toxicity to blue-green algae. *Science* **199**: 1068-1070.

Bellamy, D.J., Clarke, P.H., John, D.M., Jones, D., Whittick, A. and Darke, P. 1967. Effects of pollution from Torrey Canyon on littoral and sublittoral ecosystems. *Nature* **216**: 1170-1173.

Blumer, M. 1969. Oil pollution of the ocean. In: *Oil on the Sea.* Edited by Hoult, D.P., Plenum Press, New York. pp. 5-13.

Boney, A.D. 1968. Experiments with some detergents and certain intertidal algae. *Fld. Stud.* **2**: 55-72.

Bott, T.L., and Rogenmuser, K. 1978. Effects of No.2 fuel oil, Nigerian crude oil, and used crankcase oil on attached algal communities:

acute and chronic toxicity of water-soluble constituents. *Appl. Environ. Microbiol.* **36**: 673-682.

Bott, T.L., and Rogenmuser, K. and Thorne, P. 1978. Effects of No. 2 fuel oil, Nigerian crude oil, and used crankcase oil on benthic algal communities. *J. Environ. Sci. Hlth.* **A 13**: 751-779.

Boucher, G., Chamroux, S. and Riaux, C. 1984. Modifications des caracteristiques Physico-Chemiques et Biologiques d'un Sable Sublittoral Pollue par Hydrocarbures. *Mar. Environ. Res.* **12**: 1-23.

Boutry, J.L., Borders, M., Feyzier, A. and Barbien, M. 1977. La diatomee marine *Chaetoceros simplex* calcitrans (Paulsen) et son environment N. Relations auec le milieu de culture: etude des hydrocarbures. *J. Exp. Mar. Biol. Ecol.* **28**: 41.

Boylan, D.B. and Tripp, B.W. 1971. Determination of hydrocarbons in seawater extracts of crude oil and crude oil fractions. *Nature* **30**: 44-47.

Bratbak, G., Heldal, M., Knutsen, G., Lien, T. and Norland, S. 1982. Correlation of dispersant effectiveness and toxicity of oil dispersants towards the alga *Chlamydomonas reinhardtii*. *Mar. Pollut. Bull.* **13**: 351-353.

Cerniglia, C.E., Freeman, J.P. and Van Baalen, C. 1981. Biotransformation and toxicity of aniline and aniline derivatives in cyanobacteria. *Arch. Microbiol.* **130**: 272-275.

Cerniglia, C.E., Gibson, D.T. and Van Baalen, C 1980b. Oxidation of naphthalene by cyanobacteria and microalgae. *J. Gen. Microbiol.* **116**: 495-500.

Cerniglia, C.E., Gibson, D.T. and Van Baalen, C 1982. Naphthalene metabolism in diatoms isolated from the Kachemak Bay region of Alaska. *J. Gen. Microbiol.* **128**: 987-990

Cerniglia, C.E., Gibson, D.T. and Van Baalen, C. 1979. Algal oxidation of aromatic hydrocarbons: formation of 1-naphthol from naphthalene by *Agmenellum quadruplicatum*, strain PR-6. *Biochem. Biophys. Res. Commun.* **88**: 50-58.

Cerniglia, C.E., Van Baalen, C. and Gibson, D.T. 1980a. Metabolism of naphthalene by the cyanobacterium *Oscillatoria* sp., strain JCM. *J. Gen. Microbiol.* **116**: 485-494.

Cerniglia, C.E., Van Baalen, C. and Gibson, D.T. 1980c. Oxidation of biphenyl by the cyanobacterium *Oscillatoria* sp. strain JCM. *Arch. Microbiol.* **125**: 203-209.

Chan, K. and Chiu, S.Y. 1985. The effects of diesel oil and oil dispersants on growth, photosynthesis, and respiration of *Chlorella salina*. *Arch. Environ. Contam. Toxicol.* **14**: 325-331.

Chassé, C., L'Hardy Halos, M.T. and Perrot, Y. 1967. Esquisse d'un bilan des pertes biologiques provoquecs par le mazout du "Torrey Canyon" sur le littoral du Trégor. *Penn an Bea*. **6**: 107-112.

Coffey, J.C., Ward, C.H. and King, J.M. 1977. Effects of petroleum hydrocarbons on growth of freshwater algae. Dev. *Ind. Microbiol.* **18**: 661-672.

Cooper, J.M. and Wilhm, J. 1975. Spatial and temporal variation in productivity, species diversity of periphyton in a stream receiving domestic and oil refinery effluents. *Southwest Nat.* **19**: 413-428.

Corner, E.D.S. 1978. Pollution studies with marine plankton. Part-I. Petroleum hydrocarbons and related compounds. *Adv. Mar. Biol.* **15**: 289-380.

Crossby, E.S., Rudolfs, W. and Heukelekian, H. 1954. Biological growths in petroleum refinery wastewaters. *Ind. Eng. Chem.* **46**: 296-300.

Dahl, E., Laake, M., Tjessem, K., Eberlein, K. and Bohle, B. 1983. Effects of Ekofisk crude oil on an enclosed planktonic ecosystem. *Mar. Ecol. Prog. Ser.* **14**: 81-91.

Davavin, I.A., Mironov, O.G. and Tsimbal, I.M. 1975. Influence of oil on nucleic acids of algae. *Mar. Pollut. Bull.* **6**: 13-15.

De la Cruz, A.A. 1982. Effects of oil on phytoplankton metabolism in natural and experimental estuarine ponds. *Mar. Environ. Res.* **7**: 257-263.

Dogadina, T.D. 1970. On the utilization of algae in purification of wastes which contain oil products. *Ukr. Bot. Zh.* **27**: 310-313.

Dunstan, W.M., Atkinson, L.P. and Natoli, J. 1975. Stimulation and inhibition of phytoplankton growth by low molecular weight hydrocarbons. *Mar. Biol.* **31**: 305-310.

el-Dib, M.A., Abou-Waly, H.F. and el-Naby, A.M. 1997. Impact of fuel oil on the freshwater alga *Selenastrum capricornutum*. *Bull. Environ. Contam. Toxicol.* **59**: 438-444.

Ellis, B.E. 1977. Degradation of phenolic compounds by freshwater algae. *Plant Sci. Lett.* **8**: 213-216.

Fabregas, J. Herrero, C. and Veiga, M. 1984. Effect of oil and dispersant on growth and chlorophyll *a* content of the marine microalga *Tetraselmis suecica*. *Appl. Environ. Microbiol.* **47**: 445-447.

Federle, T.W., Vestal, J.R., Hater, G.R. and Miller, M.C. 1979. Effects of Prudhoe Bay crude oil on primary production and zooplankton in Arctic tundra thaw ponds. *Mar. Environ. Res.* **2**: 3-18.

Foster, M., Neushul, M. and Zingmark, R. 1971. The Santa Barbara oil spill. Part 2. Initial effects intertidal and kelp bed organisms. *Environ. Pollut.* **2**: 115-134.

Gaur, J.P. and Kumar, H.D. 1981. Growth response of four microalgae to three crude oils and a furnace oil. *Environ. Pollut. (Ser. A)* **25**: 77-85.

Gaur, J.P. and Kumar, H.D. 1985a. The influence of oil refinery effluents on the structure of algal communities. *Arch. Hydrobiol.* **103**: 305-323.

Gaur, J.P. and Kumar, H.D. 1985b. Effects of aqueous oil extracts on four microalgae. *Acta Hydrochim. Hydrobiol.* **13**: 115-119.

Gaur, J.P. and Kumar, H.D. 1986. Effects of oil refinery effluents on *Selenastrum capricornutum* Printz. *Int. Rev. Ges. Hydrobiol.* **71**: 271-281.

Gaur, J.P. and Singh, A.K. 1989. Growth, photosynthesis and nitrogen-fixation of *Anabaena doliolum* exposed to Assam crude extract. *Bull. Environ. Contam. Toxicol.* **44**: 494-500.

Gaur, J.P. and Singh, A.K. 1991. Regulatory influence of light and temperature on petroleum toxicity to *Anabaena doliolum*. *Environ. Toxicol. Water Qual.* **6**: 341-350.

Gordon, D.C. Jr. and Prouse, N.J. 1973. The effects of three oils on marine phytoplankton photosynthesis. *Mar. Biol.* **22**: 329-333.

Graham, J.E.S. and Hutchinson, T.C. 1975. The role of temperature and heterotrophism in determining crude oil toxicity. In: *Water Pollution Research Proc. Xth Canadian Symposium*, Canada. pp. 73-83.

Hanna, B.M., Hellebust, J.A. and Hutchinson, T.C. 1975. Field studies on the phytotoxicity of crude oil to subarctic aquatic vegetation. *Verh. Int. Verein. Limnol.* **19**: 2165-2171.

Harrison, P.J., Cochlan, W.P., Acreman, J.C., Parsons, T.R., Thompson, P.A., Dovey, H.M. and Xiaolin, C. 1986. The effects of crude oil and Corexit 9527 on marine phytoplankton in an experimental enclosure. *Mar. Environ. Res.* **18**: 93-109.

Heldal, M. Norland, S., Lien, T. and Knutsen, G. 1978. Acute toxicity of several oil dispersants towards the green alga *Chlamydomonas* and *Dunaliella*. *Chemosphere* **3**: 247-255.

Heldal, M. Norland, S., Lien, T. and Knutsen, G. 1984. toxic responses of the green alga *Dunaliella bisculata* (Chlorophyceae, Volvocales) to selected oxidised hydrocarbons. *Environ. Pollut. (Ser. A)* **34**: 119-132.

Hellebust, J.A., Soto, C. and Hutchinson, T.C. 1982. Effect of naphthalene and aqueous crude oil extracts on the green flagellate *Chlamydomonas angulosa*. V. Heterotrophic responses. *Can. J. Bot.* **60**: 1495-1502.

Hellebust, J.A., Soto, C. and Hutchinson, T.C. 1985. Effect of naphthalene and aqueous crude oil extracts on the green flagellate *Chlamydomonas angulosa*. VII. Nitrate and methylamine uptake and retention. *Can. J. Bot.* **63**: 834-840.

Hilmer, T., and Bate, G.C. 1983. Observations on the effect of outboard motor fuel oil on phytoplankton cultures. *Environ. Pollut. (Ser. A)* **32**: 307-316.

Holmes, R.W. (1969). The Santa Barbara oil spill. In: *Oil on the sea.* Edited by Hoult, D.P. Plenum Press, New York. pp. 15-27.

Hopkins, K. and Kain, J.M. 1971. The effect of marine pollutants on *Laminaria hyperborea*. *Mar. Pollut. Bull.* **2**: 75-77.

Hsiao, S.I.C. 1978. Effects of crude oils on the growth of arctic marine phytoplankton. *Environ. Pollut.* **17**: 93-107.

Hsiao, S.I.C., Kittle, K.W. and Foy, M.G. 1978. Effects of crude oils and the oil dispersant Corexit on primary production of arctic marine phytoplankton and seaweed. *Environ. Pollut.* **15**: 209-221.

Hufford, G.L. 1971. *The biological response of oil in marine environment, a review.* Background report, U.S. Coast Guard Oceanographic unit, Office of Research and Development, U.S. Coast Guard Headquarters, Washington, D.C.

Hutchinson, T.C., Hellebust, J.A., Mackay, D., Tam, D. and Kauss, P. 1979. Relationship of hydrocarbon solubility to toxicity in algae and cellular membrane effects. In: *Oil Spill Conference*, Los Angeles, pp. 541-547.

Hutchinson, T.C., Hellebust, J.A. and Soto, C. 1985. Effect of naphthalene and aqueous crude oil extracts on the green flagellate *Chlamydomonas angulosa*. IV. Decrease in cellular manganese and potassium. *Can. J. Bot.* **59**: 742-749.

Hutchinson, T.C., Soto, C. and Hellebust, J.A., 1981. Effect of naphthalene and aqueous crude oil extracts on the green flagellate *Chlamydomonas angulosa*. VI. Phosphate uptake and retention. *Can. J. Bot.* **63**: 829-833.

Ignatiades, L. and Mimicos, N. 1977. Ecological responses of phytoplankton to chronic oil pollution. *Environ. Pollut.* **13**: 109-118.

Johansson, S., Larsson, U. and Boehm, P. 1980. The Tsesis oil spill – Impact on the pelagic ecosystem. *Mar. Pollut. Bull.* **11**: 284-293.

Kaas, R. 1980. Effects of the stranding of the 'Amoco-Cadiz' on exploitable algae populations. *Rev. Inst. Peches. Marit.* **44**: 157-194.

Kaas, R. 1981. *Development of exploitable algal populations since the wreck of 'Amoco-Cadiz'*. Amoco-Cadiz, Actes Coll. Int. Centre Oceanol. Bret., Brest, CNEXO Paris. pp. 687-702.

Karydis, M. 1979. Short term effects of hydrocarbons on the photosynthesis and respiration of some phytoplankton species. *Bot. Mar.* **22:** 281-285.

Karydis, M. 1980. Uptake of hydrocarbons by marine diatom *Cyclotella cryptica*. *Microbial Ecol.* **5**: 287-293.

Karydis, M. 1981. The toxicity of crude oil for the marine alga *Skeletonema costatum* (Greville) Cleve in relation to nutrient limitation. *Hydrobiologia* **85**: 137-143.

Karydis, M. 1982. Toxicity of a photooxidised crude oil on two marine microalgae. *Bot. Mar.* **25**: 25-29.

Kauss, P., Hutchinson, T.C., Soto, C., Hellebust, J.A. and Griffiths, M.

1973. The toxicity of crude oil and its components to freshwater algae. In: *Proc. Joint Conference on Prevention and Control of Oil Spills*, Washington, D.C. pp. 703-704.

Kauss, P.B. and Hutchinson, T.C. 1975a. The effects of water-soluble petroleum components on the growth of *Chlorella vulgaris* (Beijerinck). *Environ. Pollut.* **9**: 159-174.

Kauss, P.B. and Hutchinson, T.C. 1975b. Studies on the susceptibility of *Ankistrodesmus* species to crude oil components. *Ver. Int. Verein. Limno.* **19**: 2155-2164.

Kauss, P.B. and Hutchinson, T.C. 1978. Effects of benzene, a water-soluble component of crude oils, on membrane integrity and ionic content of the green alga *Ankistrodesmus falcatus* var. *mirabilis*. *Water Pollut. Res. Canada* **13**: 85-95.

Kusk, K.O. 1978. Effects of crude oil and aromatic hydrocarbons on the photosynthesis of the diatom *Nitzschia palea*. *Physiol. Plant.* **43**: 1-6.

Kusk, K.O. 1981. Effects of hydrocarbons on respiration, photosynthesis and growth of the diatom *Phaeodactylum tricornutum*. *Bot. Mar.* **24**: 413-418.

Kustenko, N.G. 1981. Effect of low oil concentrations on spermatogonia and auxospores in three marine diatom species. *Oceanology* **21**: 263-265.

Lacaze, J.C. 1975. Ecotoxicity of crude oils and the use of experimental marine ecosystems. *Mar. Pollut. Bull.* **7**: 73-76.

Lacaze, J.C., Gutierrez, P. and Joseph, M. 1982. Sensibilites de deux algues marines *Phaeodactylum tricornutum* et *Acetabularia mediterranea* bis-a-vis d'un petrole brut soumis a differentes conditions de photo-oxydation. Influence de l'evaporation. *Rev. Int. Océanogr. Méd. Tomes* **66-67**: 61-70.

Lännergren, C. 1978. Net- and nanoplankton: effects of an oil spill in the north sea. *Bot. Mar.* **21**: 353-356.

Larson, R.A., Bott, T.L., Hunt, L.L. and Rogenmuser, K. 1979. Photo-oxidation products of a fuel oil and their antimicrobial activity. *Environ. Sci. Technol.* **13**: 965-969.

Le Pemp, X, Lacaze, J.C. and Villedon de Naïde, O. 1976. Roxicite d'un pétrole brut vis-à-vis de la production primaire du phytoplacton de l'estuaire de la Rance: influence de la temperature et du temps de Contact. *Bull. Mus. Nat. His. Paris Ecologie Générale* **32**: 107-110.

Le Petit, J., N'Guyen, M.M. and Tagger, S. 1977. Quelques donnees sur l'ecologie d'une zone marine littorale recevant les rejets d'une reffinerie de petrole. *Environ. Pollut.* **13**: 41-56.

Lee, K., Wong, C.S., Cretney, W.J., Whitney, F.A., Parsons, T.R., Lalli, C.M. and Wu, J. 1985. Microbial response to crude oil and Corexit 9527: SEAFLUXES enclosures study. *Microbial. Ecol.* **11**: 337-351.

Lee, R.F., Takahashi, M., Beers, J.R., Thomas, W.H., Seibert, D.L.R., Koeller, P. and Green, D.R. 1977. Controlled ecosystems: their use in the study of effects of petroleum hydrocarbons on plankton. In: *Physiological responses of marine biota to Pollutants.* Edited by Vernberg, F.J., Calabrese, F.P. , Thurberg, F.P. and Vernberg, W.B.. Academic Press, New York. pp. 323-341.

Lindén, O., Elmgren, R. and Boehm, P. 1979. The Tsesis oil spill: its impact on the coastal ecosystem of the Baltic sea. *Ambio* **8**: 244-253.

Lindén, O., Rosemarin, A., Lindskog, A., Hoglund, C. and Johansson, S. 1987. Effects of oil and oil dispersant on an enclosed marine ecosystem. *Environ. Sci. Technol.* **21**: 374-382.

Maher, W.A. 1986. Preparation of water soluble fractions of crude oils for toxicity studies. *Bull. Environ. Contam. Toxicol.* **36**: 226-229.

Mahoney, B.M. and Haskin, H.H. 1980. The effects of petroleum hydrocarbons on the growth of phytoplankton recognised as food forms for the eastern oyster *Crassostera virginica* Gmelin. *Environ. Pollut. (Ser. A)* **22**: 123-132.

Manwell, C. and Baker, C.M.A. 1967. A study of detergent pollution by molecular methods, starch gel electrophoresis of a variety of enzymes and other proteins. *J. Mar. Biol. Assoc. U.K.* **47**: 659.

Marowitch, J., Dale, M.R.T. and Hoodinott, J. 1988. The effect of crude oil and oil spill chemicals on nitrogen fixation in cyanobacteria *Nostoc* sp. *Environ. Pollut.* **51**: 75-83.

Masters, M.J. and Zajic, J.E. 1971. Myxotrophic growth of algae on hydrocarbon substrates. In: *Developments in Industrial Microbiology*. Society for Industrial Microbiology, American Institute of Biological Sciences, Washington, D.C. pp. 77-86.

Maurin, C. 1984. Accidental oil spills. Biological and ecological consequences of accidents in French water on commercially exploitable living marine resources. In: *Effects of pollutants at the ecosystem level*. Edited by Sheehan, P.J. , Miller, D.R., Butler, G.C. and Bourdeau, P. John Wiley & Sons, New York. pp. 311-364.

McCauley, R. 1966. The biological effects of oil pollution in a river. *Limnol. Oceanogr*. **11:** 475-486.

Meghraj, M., Singleton, I., McClure, N.C. and Naidu, R. 2000. Influence of petroleum hydrocarbon contamination on microalgae and microbial activities in a long-term contaminated soil. *Arch. Environ. Contam. Toxicol*. **38**: 439-445.

Miller, M.C., Alexander, V. and Barsdate, R.J. 1978. The effect of oil spills on phytoplankton in an Arctic lake and ponds. *Arctic* **31**: 192-218.

Minter, K.W. 1964. *Standing crop and community structure of plankton in oil refinery effluent holding ponds*. Ph.D. thesis, Oklahoma State University.

Mironov, O.G. 1968. Hydrocarbon pollution of the sea and its influence on marine organisms. *Helgolander Wiss. Meeresunters* **17**: 335-339.

Mironov, O.G. 1972. Effect of oil pollution on flora and fauna of the Black Sea. In: *Marine pollution and sea life*. Edited by Ruivo, M., Fishing News Books, London. pp. 222-224.

Mitchell, C.T., Anderson, E.K., Jones, L.G. and North, W.J. 1970. What oil does to ecology? *J. Water Pollut. Cont. Fed*. **42**: 812-818.

Mommaerts-Billiet, F. 1973. Growth and toxicity tests on the marine nanoplanktonic alga *Platymonas tetrathele* G.S. West in presence of crude oil and emulsifiers. *Environ. Pollut*. **4**: 261-282.

Morales-Loo, M.R. and Goutx, M. 1996. Effects of water-soluble fractions of the Mexican crude oil "Isthumus Cactus" on growth, cellular content of chlorophyll *a* and lipid composition of planktonic algae. *Mar. Biol*. **104**: 503-509.

Nelson-Smith, A. 1968. The effects of oil pollution and emulsifier cleaning on shore life in South-west Britain. *J. Appl. Ecol.* 5: 97-107.

Nicholson, N.L. and Climberg, R.D. 1971. The Santa Barbara oil spills of 1969: a post spill survey of rocky intertidal. In: *Biological and oceanographical survey of the Santa Barbara Channel oil spill 1969-1970, Vol. I.* Edited by Straughan, D., Allan Hancock Fds, University of South California. pp. 325-399.

North, W.J., Neushul, M. and Clendenning, K.A. 1965. Successive biological changes observed in a marine cove exposed to a large spillage of mineral oil. In: *Symposium Commission internaionale exploration Scientifique Mer Mediterraée, Monaco.* pp. 335-354.

Nuzzi, R. 1973. Effects of water soluble extracts of oil on phytoplankton. In: *Conference on Prevention and control of oil spills.* Washington, D.C. pp. 809-813.

O'Brien, P.Y. and Dixon, P.S. 1976. The effects of oils and oil components on algae: a review. *Br. Phycol. J.* **11**: 115-142.

Østgaard, K. 1994. The oil, the water and the phytoplankton. In: *Algae and water pollution.* Edited by Rai, L.C., Gaur, J.P. and Soeder , C.J., E. Schweizerbart'sche Verlagsbuchhandlung, Stuttgart. pp. 157-193.

Østgaard, K. and Jensen, A. 1983. Preparation of aqueous petroleum solutions for toxicity testing. *Environ. Sci. Technol.* **17**: 548-553.

Østgaard, K., Aaberg, A., Klungsøyr, J. and Jensen, A. 1987. Comparative studies of phytotoxicity and chemical composition of aqueous oil solution affected by evaporation, illumination and extraction. *Water Res.* **21**: 155-164.

Østgaard, K., Eide, I. and Jensen, A. 1984a. Exposure of phytoplankton to Ekofisk crude oil. *Mar. Environ. Res.* **11**: 183-200.

Østgaard, K., Hegeseth, E.N. and Jensen, A. 1984b. Species-dependent sensitivity of marine planktonic algae to Ekofisk crude oil under different light conditions. *Bot. Mar.* **27**: 309-318.

Parasher, C.D., Özel, M. and Geike, F. 1978. Effect of hexachlorobenzene and acetone on algal growth: Physiology and ultrastructure. *Chem. Biol. Interactions* **20**: 89-95.

Parker, P.L. 1974. *Effects of pollutants on marine organisms*. Deliberations and Recommendations of the NSF/IDOE Effects of pollutants on marine organisms workshop, Sidney, British Columbia, Canada.

Parsons, T.R., Li, W.K.W. and Waters, R. 1976. Some preliminary observations on the enhancement of phytoplankton growth by low levels of mineral hydrocarbons. *Hydrobiologia* **51**: 85-89.

Payne, J.R. and Phillips, C.R. 1985. Photochemistry of petroleum in water. *Environ. Sci. Technol.* **19**: 569-579.

Pérez, R. 1978. Initial results on the impact of the stranding of the Amoco-Cadiz on exploitable algal populations.

Pulich, W.M. Jr., Winters, K. and Van Baalen, C. 1974. The effects of No. 2 fuel oil and two crude oils on the growth and photosynthesis of microalgae. *Mar. Biol.* **28**: 87-94.

Radwan, S.S. and Al-Hasan, R.H. 2000. Oil pollution and cyanobacteria. In: *The ecology of cyanobacteria: their diversity in time and space*. Edited by Whitton, B.A. and Potts, M. Kluwer, Dordrecht. pp. 307-319.

Riaux-Gobin, C. 1985. Long-term changes in microphytobenthos in a Brittany estuary after the 'Amoco-Cadiz' oil spill. *Mar. Ecol. Prog. Ser.* **24**: 51-56.

Schramm, W. 1972. Investigations on the oil pollutants on marine algae. I. The effects of crude oil films on the CO_2 gas exchange outside the water. *Mar. Biol.* **14**: 189-198.

Shindler, B.B., Scott, B.F. and Carlisle, D.B. 1975. Effect of crude oil on populations of bacteria and algae in artificial ponds subject to winter weather and ice formation. *Verh. Internat. Verein. Limnol.* **19**: 2138-2144.

Singh, A.K. and Gaur, J.P. 1988. Effect of Assam crude on photosynthesis and associated electron transport system in *Anabaena doliolum*. *Bull. Environ. Contam. Toxicol.* **41**: 776-780.

Singh, A.K. and Gaur, J.P. 1989. Algal epilithon and water quality of a stream receiving oil refinery effluent. *Hydrobiologia* **184**: 193-199.

Singh, A.K. and Gaur, J.P. 1990. Effects of petroleum and their paraffinic, asphaltic and aromatic fractions on photosynthesis and respiration of microalgae. *Ecotoxicol. Environ. Safety* **19**: 8-16.

Singh, A.K. and Gaur, J.P. 1991. Inhibition of NO_3^-, NH_4^- and PO_4^{3-} uptake in *Anabaena doliolum* exposed to a petroleum oil. *Ecotoxicol. Environ. Safety* **21**: 103-108.

Singh, A.K., Singh, M.P. and Gaur, J.P. 1987. Seasonal variation in periphyton composition and diversity in oil refinery effluents. *Acta Hydrochim. Hydrobiol.* **15**: 401-408.

Snow, N.B. and Scott, B.F. 1975. The effect and fate of crude oil spill on two arctic lakes. In: *Proc. 1975 Conference on Prevention and Control of Oil Pollution*. American Petroleum Institute, Washington D.C. pp. 527-534.

Soto, C., Hellebust, J.A., Hutchinson, T.C. and Sawa, T. 1975a. Effect of naphthalene and aqueous oil extracts on the green flagellate *Chlamydomonas angulosa. I. Growth. Can. J. Bot.* **53**: 109-117.

Soto, C., Hellebust, J.A. and Hutchinson, T.C. 1975b. Effect of naphthalene and aqueous oil extracts on the green flagellate *Chlamydomonas angulosa*. II. Photosynthesis and the uptake and release of naphthalene. *Can. J. Bot.* **53**: 118-126.

Soto, C., Hellebust, J.A., and Hutchinson, T.C. 1977. Effect of naphthalene and aqueous oil extracts on the green flagellate *Chlamydomonas angulosa*. III. Changes in cellular composition. *Can. J. Bot.* **55**: 2765-2777.

Soto, C., Hutchinson, T.C., Hellebust, J.A. and Sheath, R.G. 1979a. Effect of crude oil on the morphology of the green flagellate *Chlamydomonas angulosa. Can. J. Bot.* **57**: 2717-2728.

Soto, C., Hellebust, J.A., Hutchinson, T.C. and Sheath, R.G. 1979b. The effect of hydrocarbon naphthalene on the morphology of green flagellate *Chlamydomonas angulosa. Can. J. Bot.* **57**: 2729-2739.

Steele, R.L. 1977. Effects of certain petroleum products on reproduction and growth of zygotes and juvenile stages of the alga *Fucus edentatus* de la Pilay. In: *Fate and effects of petroleum hydrocarbons in marine organisms*. Edited by Wolfe, D.A. Pergamon Press, New York. pp. 138-142.

Straughan, D. 1971. Breeding and larval settlement of certain intertidal invertebrates in the Santa Barbara channel following pollution by oil. In: *Biological and Oceanographical Survey of the*

Santa Barbara Channel Oil Spill 1969-1970. Allan Hancock Foundation, University of Southern California, Los Angeles. **1**: 223-244.

Thompson, S. and Eglinton, G. 1976. The presence of pollutant hydrocarbons in estuarine epipelic diatom population. *Estuarine Coastal Mar. Sci.* **4**: 417-425.

Throndsen, J. 1982. Oil pollution and plankton dynamics. III. Effects on flagellate communities in controlled ecosystem experiments in Lindaspollene, Norway, June 1980-1981. *Sarsia* **67**: 163-169.

Van Baalen, C., and O'Donnel, R. 1984. Sensitivity of two psychrophilic diatoms to crude oil and fuel oil. *Mar. Environ. Res.* **12**: 63-68.

Van Overbeek, J. and Blondeau, R. 1954. The mode of action of phytotoxic oils. *Weeds* **3**: 55-65.

Vandermeulen, J.H. and Ahern, T.P., 1976. Effect of petroleum hydrocarbons on algae physiology: review and progress report. In: *Effects of pollutants on aquatic organisms.* Edited by Lockwood, A.P.M., Cambridge University Press, Cambridge. pp. 107-125.

Vandermeulen, J.H. and Lee, R.W. 1986. Lack of mutagenic activity of crude and refined oils in the unicellular alga *Chlamymomonas reinhardtii. Bull. Environ. Contam. Toxicol.* **36**: 250-253.

Vargo, G.A., Hutchens, M. and Almquist, G. 1982. The effect of low chronic levels of No. 2 fuel oil on natural phytoplankton assemblages in microcosms: 1. Species composition and seasonal succession. *Mar. Environ. Res.* **6**: 245-264.

Winters, K., Batterton, J.C. and Van Baalen, C. 1976. Water-soluble components of fuel oils: chemical characterization and effects on growth of microalgae. *Mar. Biol.* **36**: 269-276.

Winters, K., Batterton, J.C. and Van Baalen, C. 1977a. Phenalen-1-one: occurrence in a fuel oil and toxicity to microalgae. *Environ. Sci. Technol.* **11**: 271-273.

Winters, K., Van Baalen, C. and Nicol, J.A.C. 1977b. Water soluble extractives from petroleum oils: chemical characterization and effects of microalgae and marine animals. *Rapp. P.-V. Reun. Cons. Int. Explor. Mer.* **171**: 166-174.

Wolfe, M.F., Schwartz, G.J.B., Singaram, S., Mielbrecht, E.E., Tjeerdema, R.S. and Sowby, M.L. 1998. Effects of salinity and temperature on the bioavailability of dispersed petroleum hydrocarbons to the golden-brown alga *Isochrysis galbana*. *Arch. Environ. Contam. Toxicol.* **35**: 268-273.

Chapter 9

ALGAL PROTEIN: FUNCTIONAL PROPERTIES AND POTENTIAL FOR FOOD APPLICATIONS

Manjit Kaur

Abstract

Some blue-green algae (Cyanobacteria) are well known for their high protein content. Such proteins have been studied for their physico-chemical characteristics (functional properties) which determine their behaviour in food during processing, storage, preparation and consumption. These properties comprise water and fat absorption, viscosity, emulsification, foaming capacity etc. The information is important for exploring suitability of such algae/proteins for human consumption.

Key words: Algae, cyanobacteria, emulsification, foaming, proteins, viscosity.

Introduction

The food demand in the coming years may not be covered by conventional agriculture. Therefore, intensive efforts are being made to find alternate new and unconventional protein sources. Single cell protein, which now–a–days has attracted a great attention, holds a major potential for increasing protein supply (Lipinski and Litchfield, 1974; Becker and Venkataraman, 1984). In this context, algae as a source of single cell protein (SCP) holds a major potential for increasing protein supplies. The term single cell protein includes protein sources from unicellular or multicellular microbes like bacteria, yeast, fungi and algae. The role of single cell protein which includes algae as a newer protein source, is receiving attention for its technological feasibility and economy of production (Dabbah, 1970; Waslien, 1975).

Cultivation of algae is comparable to agricultural cropping with the difference of cultivation in liquid media and quicker period for harvesting. Tropical countries like India are well suited for algal cultivation the year round except for one or two monsoon months. Algae can be cultured on marginal land in arid region of the world utilizing waters unsuitable for conventional agriculture (Thomas *et. al.*, 1983; Ciferri and Tiboni, 1985; Richmond, 1986). The yield of protein from algae is much higher compared to other agricultural crops. Under favourable conditions, microbial cultures can produce upto 20–35 times more protein than soybean and about 50 times more than rice, wheat or maize for the same area (Ciferri, 1983). Chemical composition of the algae gives basic information on the nutritive potential of the algal biomass. The proportion of different macromolecules can be modified by specific cultivation measures (Aaronson *et. al.*, 1980). The biomass and protein content of algae increase with the increasing concentration of nitrate in the medium (Baasch *et. al.*, 1985). Among microalgae *Spirulina* spp. have come out as the most favourable ones for Biomass protein (BMP) followed by species of *Chlorella, Scenedesmus, Phormidium* and *Dunaliella*.

Not all the species of algae have characteristics desired of an algal SCP. The genes of algae under consideration should have the ability to produce high levels of protein. Their amino acid pattern should be comparable to that of conventional protein foods and the organism should present either no or minimum problem of digestibility. Algal proteins are generally deficient in sulphur containing amino acids (cystine and methionine) while they are fairly rich in lysine. Since cereals like rice, ragi and wheat are rich in sulphur amino acids, it is appropriate to make a balanced protein blend of cereal-algal combination.

Proteins not only provide a nutritional component in a food system but also perform a number of other functions. The most important attribute of a protein is its various functional properties. Functionality has been defined as any property of a food ingredient except its nutritional values that affects its utilization (Kinsella, 1976). Functional properties are, however, often used to denote any property of protein that affects their use, either as a processing aid or as a direct contributor of product attributes (Wilding *et. al.*, 1984). Due to increasing cost of food grade proteins and a trend towards complete food formation from refined ingredients and unconventional sources, there is a growing need for less expensive proteins with good functional properties. Various forms of protein additives are added to foods to achieve functional, nutritional or economic goals. Potential functional benefits include emulsification, foam formation, viscosity, solubility, binding of fat or water. To be useful and successful in food application, protein product should possess desirable functional properties.

The potential for incorporation of a protein in food is directly related to their water retention. The storage quality of food is related to their moisture content as well. Water absorption capacity of algae is twice or more than that of the protein concentrate on material basis and nearly three times on protein basis (Devi and Venkataraman, 1984) and they are comparable with soybean meal, sunflower and cotton seed protein.

Several factors such as amino acid composition, protein conformation, surface hydrophobicity etc, affect the water binding by proteins (Kinsella, 1982). According to Breuer and Kennerley (1971) water retention of protein is related to polar groups of the proteins and high value of water absorption capacity of the algal protein shows that they contain good amount of polar residues.

Fat absorption capacity is the binding of fat by nonpolar side chains of protein (Sathe *et. al.*, 1982). The ability of protein to bind fat is likely due to nonpolar side chains that bind hydrocarbon chains, thereby contributing to increased oil adsorption (Lin and Zayas, 1987). Algal proteins have higher ability to bind fat. The ability of protein to bind fat is very important for applications such as meat replacement and extenders mainly because it enhances flavour retention and improve mouth feel. The ability of algal protein to absorb and retain water and fat would perform well in foods such as low fat bakery products and dough nuts, as well as in emulsion type foods.

The solubility of a protein is the manifestation of the equilibrium between protein–solvent and protein–protein interactions. Conditions which favour protein–solvent interactions, increase the solubility, and those having a positive impact on protein–protein interactions, disfavour the solubility.

A highly soluble protein is desirable to obtain optimum functionality in foods. The extent of solubility depends upon the intensity and duration of the process, treatment, alongwith some other factors such as pH, temperature, nature and ionic strength of salt etc. Protein solubility at various pH values may serve as a useful indicator of how well a protein isolate will perform its application in food systems under heat or chemical treatment. Solubility of *Spirulina, Scenedesmus* and *Klebsormidium* protein is high at extreme pH values and the least at isoelectric point (Kaur *et. al.*, 1992; Dua *et. al.*, 1993). Solubility profile of algal proteins with respect to pH is similar to cereals (Hettiarachchy *et. al.*, 1996; Ahmedna *et. al.*, 1999). Solubility of algal protein decreases with increasing concentration of sodium chloride. Major part of algal protein usually exists in the form of complexes with the pigments. Chlorophyll protein complex has been isolated from algae (Thornber, 1971). As the proteins are already in the form of complexes, the force of attraction between the protein ion and salt ion is probably less which accounts for its low solubility. Algal proteins differ from legumes where solubility increases in the presence of sodium chloride (Shen, 1976). Solubility is affected by a wide range of pH, temperature and ionic strength (Wagner and Anon, 1990; Zayas, 1997).

Viscosity reflects the flow behaviour of a substance in liquid state; it is a manifestation of the attractive forces between the molecules in the liquid or fluid state (Tanford, 1962). Proteins in solution do not behave as rigid particles but rather exist in a thermodynamic equilibrium between various conformational states. The hydrodynamic volume and molecular shape of a protein which profoundly influences the flow behaviour of a protein solution, is dependent on physico–chemical state of the protein under the given conditions. When protein is dispersed in water, it imbibes water and swells. This increases the hydrodynamic volume and induces long range effects on the flow behaviour of solvent molecules (Kuntz, 1971).

The relative viscosity of algal proteins is minimum at their isoelectric point (Dua *et. al.*, 1993). Apparent viscosity of phosphorylated yeast protein enhanced with increasing protein concentration and pH especially above 6.5 (Huang and Kinsella, 1986). A good correlation between viscosity, solubility and pH was observed for soy protein, mustard and rapeseed (Gururaj Rao and Narasinga Rao, 1981), peanut (Cater and Rhee, 1975) and sunflower (Fleming *et. al.*, 1974).

Particle shape and axial ratio also affect the viscosity measurements of macromolecules (Jirgenson and Strawmanis, 1952). Axial ratio of *Scenedesmus* and *Klebsormidium* proteins are 5.43 and 5.32 respectively showing thereby that these proteins are either rod shaped or ellipsoidal in form. Relative viscosity increases with increase in concentration of protein and sodium lauryl sulphate but decreases with urea due to unfolding or swelling of protein molecules (Kaur, 1991).

Foaming or whipping is the capacity to form stable foam with air and it depends on several parameters such as types of protein and degree of denaturation, the presence or absence of calcium ions, pH, temperature, and whipping methods (Townsend and Nakai, 1983; Fennema, 1996). Foam contributes to smoothness, lightness, flavour dispersions, and palatability of foods. Proteins with good foam serve as a potential replacement of these proteins in food applications requiring high foamability and foam stability, such as in cakes, bread marshmallow, whipped topping, ice–cream, and desserts. The foam capacity of algal flour and protein concentrate are pH dependent and closely resemble with nitrogen *vs.* pH profile suggesting that it is also dependent on solubilised protein (Devi *et. al.*, 1981; Devi and Venkataraman, 1984; Kaur *et. al.*, 1992). Similar observations have been reported in soy protein isolate alfaalfa leaf protein (Hermansson, 1975; Wang and Kinsella, 1976). Protein solubility requires high net charges, which influence the adsorption of protein at the air–water interface (Cherry and McWatters, 1981). High protein solubility is believed to improve foaming capacity. Calcium chloride enhances both foam capacity and stability through ionic bridging between protein molecules (Fennema, 1996).

Foam capacity of algal proteins increases upto 0.4M of NaCl and it gets reduced above this concentration. Ionic strength of the medium affects foaming properties because ions may reduce the forces between the polypeptide chain and hence modify stability. Sodium chloride reduces surface viscosity and rigidity of protein films but increases spreading rate; i.e. it weakened the interpeptide attraction and increased foam volume for certain proteins (Buckingham, 1970).

Emulsifying capacity is the volume of oil that can be emulsified by proteins before phase inversion or collapse of emulsion occurs and it is inversely related to protein concentration. Efficiency of emulsification varies with the type of protein, its concentration, solubility, pH, ionic strength, temperature and method of preparation of the emulsion (Saffle, 1968). Emulsification properties of *Scenedesmus* and *Klebsormidium* proteins changed with pH, and these resembled a typical protein solubility curve with a single point of minimum around isoelectric point (Dua *et. al.*, 1993). Emulsifying capacity of *Spirulina* protein concentrate was much better than that of *Spirulina* flour and it closely resembles nitrogen solubility vs pH profile (Devi and Venkataraman, 1984). This pattern of algal protein is comparable with that of soybean, rapeseed, cotton seed, non–fat dry milk, sunflower and groundnut protein (Crenwelge *et. al.*, 1974; Ramanathan *et. al.*, 1978; Volkert and Klein, 1979; Dev and Mukherjee, 1986). Emulsifying capacity of *Spirulina* protein increased with increasing salt concentration and had its maximum at 0.6M NaCl (Devi and Venkataraman, 1984; Kaur *et. al.*, 1992). The studies concerning emulsifying properties of single cell protein have been limited. However, a good emulsification capacity has been reported for yeast protein isolates (Vananuvat and Kinsella, 1975).

Acknowledgement

Thanks are due to the Chairman, Department of Botany, Panjab University, Chandigarh for providing necessary facilities during the course of this work.

References

Aaronson, S., Berner, T. and Dubinsky, Z. 1980. Microalgae as a source of chemicals and natural products. In: *Algal Biomass*. Edited by Shelet , G. and Soeder, C.J. Elsevier/North–Holland Biomedical Press. pp.575–601.

Ahmedna, M., Prinyawiwatkul, W. and Rao, R.M. 1999. Solubilized wheat protein isolate: Functional properties and potential applications. *J. Agric. Food. Chem.* 47: 1340–1345.

Baasch, K.H., Kohlhase, M. and Pohl, P. 1985. Axenic mass culture of prokaryotic and eukaryotic microalgae and lipid production: Influence of the temperature and nitrate concentration in the nutrient medium. *Dev. Plant. Biol.* 9: 587–590.

Becker, E.W. and Venkataraman, L.V. 1984. Production and utilization of the blue-green alga *Spirulina* in India. *Biomass* 4: 105-125.

Breuer, M.M. and Kinnerley, M.G. 1971. Hydration of synthetic polypeptides. *J. Colloid Interfaces Sci.* **37**: 124-131.

Buckingham, J.H. 1970. Effect of pH, concentration and temperature on strength of cytoplasmic protein foams. *J. Sci. Food. Agric.* **21**: 441–445.

Catar, C.M. and Rhee, K.C. 1975. Protein concentrates and isolates. *Peanut Prod. Tex.* **3**: 115-125.

Cherry, J.P. and McWatters, K.H. 1981. Whippability and aeration. In: *Protein Functionality in Foods*. Edited by Cherry, J.P., Amer. Chem. Soc., Washington, D.C. pp. 149-176.

Ciferri, O. 1983. *Spirulina*, the edible microorganism. *Microbiol. Rev.* **47:** 551–578.

Ciferri, O. and Tiboni, O. 1985. The biochemistry and industrial potential of *Spirulina*. *Ann. Rev. Microbiol.* **39:** 503–526.

Crenwelge, D.D., Dili, C.W., Tybor, P.T. and Landmann, W.A. 1974. A comparison of the emulsification capacity of some protein concentrates. *J. Food. Sci.* **39:** 175–177.

Dabbah, R. 1970. Protein from microorganisms. *Food Technol.* **24:** 659–666.

Dev, D.K. and Mukherjee, K.D. 1986. Functional properties of rapeseed protein products with varying phytic acid contents. *J. Agric. Food. Chem.* **34**: 775-780.

Devi, A.M., Subbulakshmi, G., Madhavi Devi, K. and Venkataraman, L.V. 1981. Studies on the proteins of mass cultivated blue-green alga (*Spirulina platensis*). *J. Agric. Food Chem.* **29**: 522-525.

Devi, A.M. and Venkataraman, L.V. 1984. Functional properties of protein products of mass cultivated blue–green alga *Spirulina platensis*. *J. Food. Sci.* **49:** 24–27.

Dua, S., Kaur, M. and Ahluwalia, A.S. 1993. Functional properties of two pollutant grown green algae. *J. Food. Sci. Technol.* **30:** 25–28.

Fennema, O.R. 1996. *Food Chemistry*, 3rd ed., Dekker, New York, pp.365–424.

Fleming, S.E., Sosulski, F.W., Kilara, A. and Humbert, S.E. 1974. Viscosity and water absorption characteristics of slurries of sunflower and soyabean flours, concentrates and isolates. *J. Food Sci.* **39**: 188-191.

Gururaj Rao, A. and Narasinga Rao, M.S. 1981. Comparative study of the high molelcular weight protein fraction of mustard (*B. juncea*) and rapeseed (*B. campstris*). *Int. J. Peptide Protein Res.* **18:** 154-161.

Hermansson, A.M. 1979. Methods of studying functional characteristics of vegetable proteins. *J. Amer. Oil. Chem. Soc.* **56:** 272–279.

Hettiarachchy, N.S., Griffin, V.K. and Gnanasambandam, R. 1996. Preparation and functional properties of a protein isolate from defatted wheat germ. *Cereal Chem.* **73:** 364–367.

Huang, Y.T. and Kinsella, J.E. 1986. Functional properties of phophorylated yeast protein: Solubility, water holding capacity and viscosity. *J. Agri. Food Chem.* **34:** 670—674.

Jirgenson, B. and Strawmanis, M.F. 1952. *A Short Text Book of "Colloid Chemistry"*. John Wiley and Sons. Inc., New York.

Kaur, M. 1991. *Biochemical studies on some algae of polluted areas.* Ph.D Thesis, Panjab University, Chandigarh, India. pp. 277.

Kaur, M., Dua, S. and Ahluwalia, A.S. 1992. Biochemical studies on *Spirulina* proteins. In: *Spirulina ETTA*. Edited by Seshadri, C.V. and Jeeji Bai, N. pp. 78–84.

Kinsella, J.E. 1976. Functional properties of protein in food: A survey. *CRC Crit. Rev. Food Sci. Nutr.* **7:** 219–280.

Kinsella, J.E. 1982. Protein structure and functional properties–emulsification and flavour binding effects. *ACS Symp. Ser.* **206:** 301–326.

Kuntz, I.D. 1971. Hydration of macromolecules III. Hydration of polypeptides. *J. Amer. Chem. Soc.* **93:** 514–516.

Lin, C.S. and Zayas, J.F. 1987 . Functionality of defatted corn germ proteins in a model system: fat binding and water retention. *J. Food. Sci.* **52:** 1308–1311.

Lipinski, E. and Litchfield, J.H. 1974. Single cell protein in perspective. *Food Technol.* **28:** 16–22.

Ramanathan, G., Ran, L.H. and Urs, L.N. 1978. Emulsification properties of ground nut protein. *J. Food. Sci.* **43:** 1270–1273.

Richmond A. 1986. Microalgae of economic potential. In: *CRC Handbook of Microalgal Mass Culture.* Edited by Richmond , A. CRC Press Inc., Boca Raton, Fla. pp.199–243.

Saffle, R.L. 1968. Meat emulsion. *Adv. Food Res.* **16**: 105–110

Sathe, S.K., Deshpande, S.S. and Salunkhe, D.K. 1982. Functional properties of winged bean proteins. *J. Food. Sci.* **47**: 503–509.

Shen, L.J. 1976. Solubility profile, intrinsic viscosity and optical rotation studies of acid precipitated by protein and commercial soy isolate. *J. Agric. Food Chem.* **24**: 784–789.

Tanford, C. 1962. *Advances in Protein Chemistry* **17**: 69–74.

Thomas, W.H., Seibert, D.N.R., Alden, M., Eldridge, P., Neori, A. and Gaines, S. 1983. Microalgae from desert saline waters as potential biomass production. *Proc. Solar Energy*. **6**: 143–145.

Thornber, J.P. 1971. Chlorophyll *a* protein complex of blue–green algae. *Meth. Enzymol.* **23**: 682–687.

Townsend, A. and Nakai, S. 1983. Relationship between hydrophobicity and foaming characteristics of food proteins. *J. Food. Sci.* **47**: 503–509.

Vananuvat, P. and Kinsella J.E. 1975. Extraction of protein low in nucleic acid from *Saccharomyces fragilis* grown continuously on crude lactose. *J. Agric. Food. Chem.* **23**: 216–224.

Volkert, M.A. and Klein B.P. 1979. Protein dispersibility and emulsion characteristics of flour soy products. *J. Food Sci.* **44**: 93–97.

Wagner, J.R. and Anon, M.C. 1990. Influence of denaturation, hydrophobicity and sulfhydryl content on solubility and water absorbing capacity of soy protein isolate. *J. Food. Sci.* **55**: 765–770.

Wang, J.C. and Kinsella, J.E. 1976. Functional properties of alfaalfa leaf protein: foaming. *J. Food. Sci.* **41**: 498–501.

Waslien, C.J., 1975. Unusual sources of protein for man. *Crit. Rev. Food Sci. Nut.* **6**: 77–151.

Wilding, P., Lillford, P.J. and Regenstein, J.M., 1984. Functional properties of proteins in foods *J. Chem. Technol. Biotechnol.* **34B**: 182–189.

Zayas, J.F. 1997. *Functionality of proteins in Foods*. Springer Verlag, Berlin, Germany.

Chapter 10

ROLE OF MAGNESIUM AND PHOSPHATE LIMITATIONS AND LOW TEMPERATURE IN STIMULATING ALGICIDE PRODUCTION IN A CYANOBACTERIUM, *OSCILLATORIA LAETEVIRENS*

Soma Ray, Rahul Shrivastava and Suvendra N. Bagchi

Abstract

The effects of illumination, temperature and orthophosphate, magnesium and hydrogen ion concentrations on growth and algicide production by *Oscillatoria laetevirens* were studied under laboratory conditions. Growth was determined by measuring dry weight, while amount of algicide, henceforth called oscillatorin, within the cells and in culture medium was analyzed by high-performance liquid chromatography. Further, a bioassay was used to deduce the minimum inhibitory dose of oscillatorin present in the culture filtrate, using a sensitive strain, *Synechococcus*. High oscillatorin production correlated with low incubation temperatures, high pH and reduced magnesium and phosphate concentrations, and in several conditions this was not correlated with optimal growth. The magnitude of the combined effect of reduced temperature and phosphate limitation on oscillatorin production was two-fold greater than the sum of their individual effects. Fluence rate was, however, without much effect on production. The implication of tested conditions on allelopathic influence in nature is discussed.

Key words: Algicide, Allelopathic effect, Cyanobacteria, *Oscillatoria*, Oscillatorin.

Introduction

In phytoplankton communities several interactive environmental factors *viz.* light, temperature and nutrients, regulate dominance of cyanobacteria. Cyanobacteria can grow only in two dimensions as biofilms and therefore competitors for available surface are related cyanobacteria rather than other organisms. Cyanobacteria mainly belonging to the genera *Scytonema, Oscillatoria, Fischerella, Calothrix* and *Nostoc* produce algicidal substances (Schlegel *et. al.*, 1998), which are presumably responsible for species-succession in nature (Keating,1977).Frequently, these metabolites inhibit photosystem II-dependent electron transport in a large number of sensitive organisms (Gleason and Case, 1986; Bagchi and Marwah, 1994; Bagchi, 1995). Such algicides are expected to be synthesized and released at concentrations large enough to be effective in suppressing the predecessor and contemporary phytoplankton. The influence of environmental factors on algicidal production and growth of bioactive cyanobacteria has not been systematically investigated (Ray and Bagchi, 2001),though such information is required in order to understand the dynamics of phytoplankton assemblages in nature.

A planktonic filamentous cyanobacterium, *Oscillatoria laetevirens* produces and releases outside the cells an algicidal metabolite (Ray and Bagchi, 2001 and references therein). The effective substance called oscillatorin (OS) has been purified from culture filtrates and from broken cell-free extracts. Using the purified compound the concentration of OS was determined in both the preparations. Further, the amount of algicide in culture filtrate could also be measured by involving a bioassay in which a minimum inhibitory dose against a sensitive cyanobacterium *Synechococcus* was determined.

In the present work, we took the advantage of the quantification measures for OS and attempted to investigate the regulatory responses of nutrients, temperature and light intensity on growth and production of this secondary metabolite in *O. laetevirens*.

Materials and methods

Organisms and culture conditions

Oscillatoria laetevirens var. minimus, an isolate from local lake Dev Tal (Bagchi *et. al.*, 1990), and test cyanobacterium *Synechococcus* PCC7942 were grown axenically in modified BG 11 medium containing, apart from other nutrients, 0.17 mM K_2HPO_4 and 0.3 mM $MgSO_4 \cdot 7H_2O$(Rippka *et. al.*, 1979). Continuous illumination at fluence rates of 75 µ mole photons $m^{-2} s^{-1}$ with cool flourescent lamps and temperature of 28 ±2°C was provided. For experiments, cultures (fresh wt. of cells = 150-200 mg equivalent to approx. 20 mg dry wt.) were transferred to 100 ml of BG12 medium (BG11 medium supplemented with 5 mM Hepes/NaOH buffer, pH 8.0; Ferris and Hirsch, 1991).

For changing incubation temperatures during growth, the cultures were placed in a BOD incubator at 15, 21, 28 and 35°C and were illuminated at above stated fluence rates. The cultures were processed for various determinations after 20 d.

The distance of flasks from light source was adjusted to achieve desired variations in fluence rates. Additional light source was provided for high intensity. Light intensity was measured using a Lux meter. In these experiments the ambient temperature was maintained.

The pH of the medium was raised to 9.0 with 5 mM Hepes buffer. Mg^{2+} and phosphate concentrations were reduced by reducing $MgSO_4$ and K_2HPO_4 concentrations and by supplementing equivalent-balancing counter-ions with appropriate concentrations of Na_2SO_4 and KCI respectively.

Growth estimation

Growth of *Oscillatoria* was recorded by filtering the cultures on glass microfibre filters previously dried to constant weight. After filtration, the filter discs (4.25 cm Whatman GF/C) plus residue were freeze-dried and re-weighed. Growth of *Synechococcus* was determined by measuring the change in absorbance of the cultures at 665 nm. The initial absorbance after inoculation was adjusted in 25 ml medium at 0.05.

Determination of the amount of oscillatorin

For estimating the minimum inhibitory dose (MID) of oscillatorin in culture filtrates of *O.laetevirens* cultures, a bioassay procedure was established (Ray and Bagchi, 2001). In triplicate sets, decreasing volumes of filtrates (50-500µl) were added to 25ml of *Synechococcus* cultures. If growth was inhibited by more than 80% of control (kept without addition), the dose was considered as inhibitory. For precision, filtrates at lower increment (10µl) were added to judge the accurate minimum volume, which inhibited the growth. Minimum inhibitory dose is that least volume of filtrate, which inhibited growth of *Synechococcus* in 1 litre of culture.

Oscillatorin was also quantified by ODS C-18 high-performance liquid chromatography using water methanol gradient according to Ray and Bagchi (2001). For this, 1.6-2 g fresh wt. of *O.laetevirens*, equivalent to 220-260 mg dry wt., was used to inoculate one litre medium. A cell-free extract was prepared according to Bagchi *et. al.* (1990) and crude OS was extracted in ether. After evaporation of solvent, the residue was dissolved in 1 ml methanol and an aliquot was applied to the column. Since OS in culture filtrate is bound to a protein (Bagchi and Marwah, 1997), it was released by mild proteolysis. For this, the filtrate was concentrated in a freeze drier and proteins precipitated by using saturated ammonium sulfate. After dialysis, proteolysis was achieved by trypsin-digestion. Subsequently, an ether extract was prepared and residue was dissolved in methanol for application to column. Oscillatorin concentration was determined based on peak area at retention time of 4.5 min. using a calibration curve prepared with purified OS.

All experiments were done at least in triplicate and the average values are shown. The deviation from mean did not exceed ±10%. Correlation analysis was done using Karl Pearson's method (Agarwal, 1988).

All chemicals were purchased at their purest grades from Sigma Chemical Co., U.S.A. and Qualigens, India.

Results

Upon growth of O. *laetevirens* in BG12 medium, OS started accumulating in the medium from mid-exponential phase and paralleled the biomass yield. The highest OS concentrations in the medium and within the cells were 51 mg l^{-1} and 9.3 mg g^{-1} dry weight of biomass respectively, and these levels were attained after 20 d of growth. The MID of the culture filtrate corresponded to 100± 10µl for one litre of *Synechococcus* culture (Ray and Bagchi, 2001).

In the present work we have mimicked the conditions prevailing in the lake coinciding with the period of *Oscillatoria* dominance. The P value in the lake was found to be 0.4 mg l^{-1}, magnesium hardness 40-50 ppm, pH 8.5-8.8 and temperature 10-15 °C.

Effect of independent parameters

The effect of temperature on growth and OS accumulation is shown in Fig. 10.1. *O. laetevirens* exhibited optimal growth in a broad range of temperature (21-35°C) , whereas growth was suppressed at lower temperatures. Increase in temperature from ambient (28°C) did not affect algicide accumulation or MID of the filtrate. With rise in temperature, growth was positively ($r = 0.8$) and production was negatively ($r = -0.9$) correlated.

Maximum biomass yield was recovered at 75 μ mol. photons m^{-2} s^{-1}. Illumination at higher or lower intensities suppressed the growth. By and large, variation in light intensities had little effect on OS production and related MID values.

The effect of magnesium and phosphate was examined in BG12 medium by lowering Mg^{2+} and phosphate concentrations to one-tenths of original and then compensating the counter-ion concentrations (Fig. 10.1). Lowering of these nutrients significantly enhanced OS production and its inhibitory activity ($P < 0.001$; Student's t-test). In such deprived medium growth was only marginally inhibited.

O.laetevirens also exhibited a broad range of pH choice for growth especially towards alkaline side. H^+ ion concentration however imposed a remarkable influence on OS production. As shown in Fig. 10.1, with one unit of increase in pH, the amount of OS was enhanced by nearly 400%.

Effect of combination of factors

The above parameters independently stimulated or suppressed OS accumulation,or were without any effect. In order to examine the combined effect, above parameters were imposed in a pair-wise manner. The magnitude of suppression/stimulation by variation in temperature, pH or by depletion of Mg^{2+} or phosphate was unchanged at different intensities of light, although there was a general reduction of biomass yield at low or high light intensity.

Effect of Mg^{2+} deprivation was discernable only at 28°C. At temperatures above or below, this parameter did not impose any particular additive effect on OS production and MID values (Fig. 10.1).

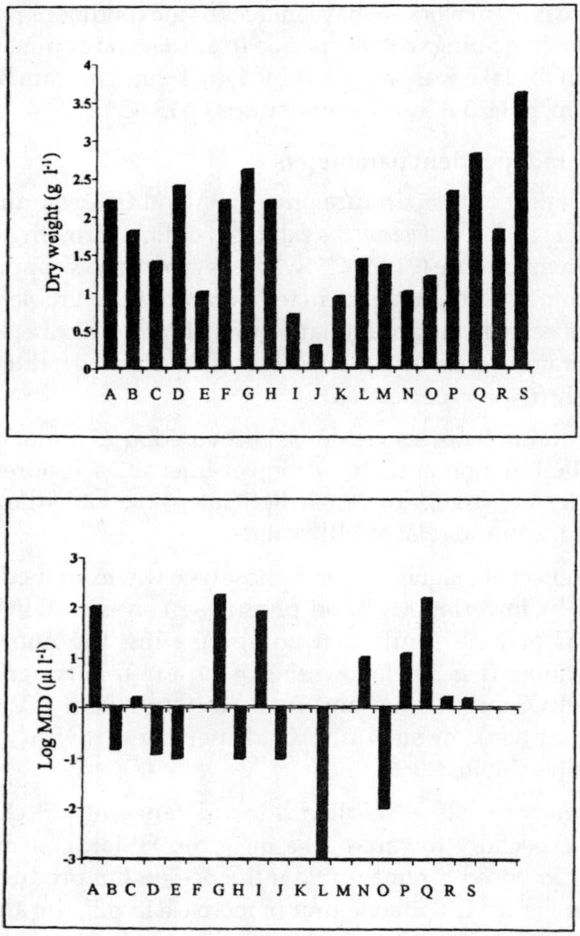

Fig. 10.1: Effect of incubation temperatures, nutrients and pH on growth, cellular and extracellular oscillatorin accumulation and its algicidal activity (MID values) in culture filtrates of *O. laetevirens*.

Cultures were grown for 20 d in (A) BG12 medium (pH 8.0, 28°C controls), (B) with reduced Mg^{2+} (0.03 mM) or (C) phosphate (0.017mM), or at (D) pH 9, or at (E) 15°C, (F) 21°C or (G) 35°C, or in a combination of reduced nutrients, pH and temperatures such as (H) reduced Mg^{2+} and pH 9.0, (I) reduced P and pH 9.0, (J) reduced Mg^{2+} and P, (K) 15°C and reduced Mg^{2+} and reduced P, (M) 15°C and pH 9.0, (N) 21°Cand reduced Mg^{2+}, (O) 21°C and reduced P, (P) 21°C and pH 9.0, (Q) 35°C and reduced Mg^{2+}, (R) 35°C and reduced P or (S) 35°C and pH 9.0.

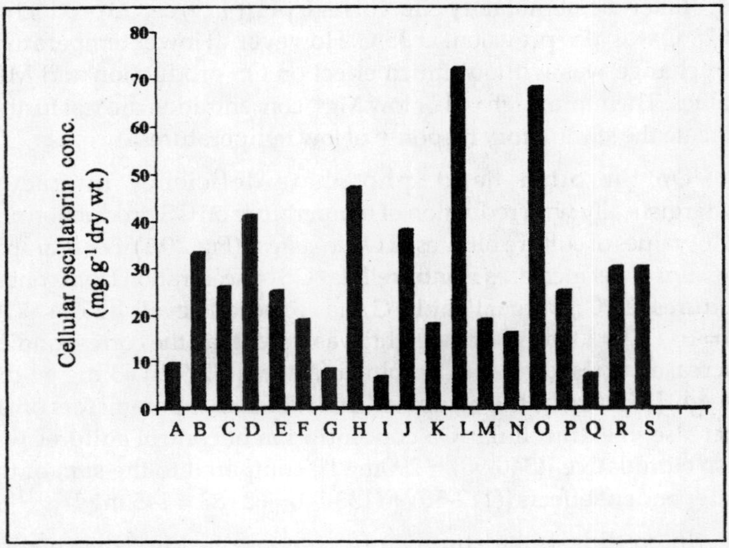

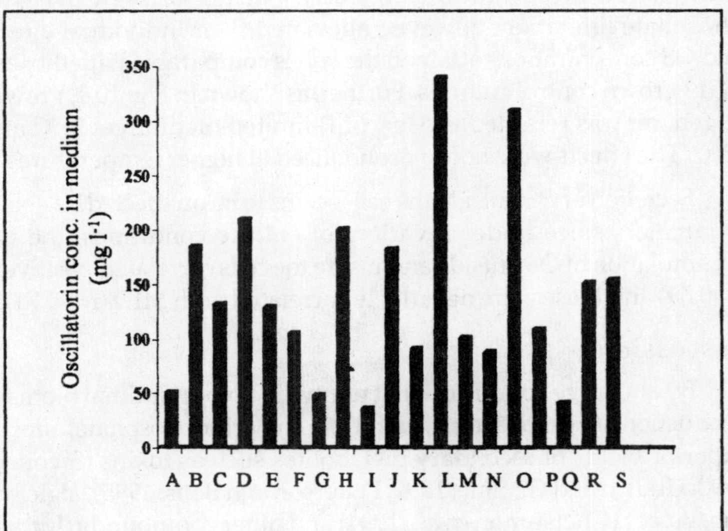

Fig. 10.1 (contd...)

Likewise, stimulatory effect of high pH (pH 9) on OS production at 28°C was also prevalent at 35°C. However, at lower temperature, pH change was without much effect on OS production and MID values. Therefore, high pH or low Mg^{2+} concentration did not further elevate the stimulatory response of low temperatures.

On the other hand, phosphate deficiency interacted synergistically with reduction of temperature on OS production and MID values of culture filtrates of *O. laetevirens* (Fig. 10.1). For example, the sum of net increases in intracellular OS concentration from control cultures (28°C, normal P) at 15°C and reduced P levels [(25-9) + (26-9) = 16 + 17 = 33 mg g^{-1} dry weight] was lesser than the corresponding increase imposed by the combined factors [(72-9) = 63 mg g^{-1} dry weight]. Similarly, the combined effect of above parameters on an increase in extracellular OS concentration of control cultures was also cumulative {(340-50) = 290mg l^{-1}} compared to the sum of the independent effects [(112-50) + (133-50)= 62+83 = 145 mg l^{-1}] .

Interestingly, the stimulatory response fo Mg^{2+} deprivation persisted when combined with conditions such as high pH or low phosphate concentration (Fig. 10.1). Pairing of high pH with phosphate limitation, however, alleviated their individual effects and OS concentrations attained the levels comparable with those of BG12 grown control cultures. Further, as shown in Fig. 10.1, growth in general was retarded in Mg^{2+} or P-limited medium at 21°C and 15°C. The effects were not so pronounced at higher temperatures.

Secretion of OS outside the cells seems to be unaffected by above parameters since under a variety of culture conditions, the net accumulation of OS outside and inside the cells correlated positively (r=0.89) and these were negatively correlated with MID (r= -0.71).

Discussion

By and large conditions that normally favour optimal biomass production of cyanobacteria, *viz* nitrate and orthophosphate, affords superior yields of secondary metabolites such as toxins (Sivonen, 1990; Utkilen and Gijolme, 1992; Patterson and Bolis, 1995), algicidal substances (Chetsumon *et. al.*, 1994) and other compounds (Juttner *et. al.*, 1983).

In contrast, the results indicate that low temperature, reduced concentrations of magnesium and phosphate and high pH stimulate OS production in *O. laetevirens* and such conditions did not necessarily support optimal biomass production. The parameters tested in laboratory were the ones which prevailed in field during the bloom period of *Oscillatoria* (September-January every year; Bagchi *et. al.*, 1990). Thus one may expect that despite the fact that conditions may not be appropriate for optimal growth, the concentration of algicide (OS) in nature is high enough to sustain *Oscillatoria* bloom.

Loading of nutrients in lakes, especially N and P, and elevated temperatures, encourage bloom formation by cyanobacteria (Pearson, 1990) . When nutrients are depleted and temperature falls, cyanobacteria face tough competition with other phytophankton, particularly the green algae. Over production and secretion of allelochemicals may be advantageous for sustaining the bloom regime until growth-favourable conditions are resumed.

If the results obtained in the laboratory are extrapolated in the field we may get some interesting information. Roughly, one litre lake water contains 5-10 g of cyanobacterial bloom. If we apply the data obtained from BG 12 grown cultures, this amount of cells, which is equivalent to 0.6-1.2 g dry wt., will accumulate within the cells 6-12 mg of OS. A similar estimated value for OS in one litre of lake water will be 15-30 mg.

Once the data of production under conditions of limited phosphate and low temperatures, which prevails coinciding with bloom period of *Oscillatoria* in nature, is considered, the quantity of OS in bloom material and in lake water of one litre volume would be 48-96 mg and 160-320 mg respectively. These estimated values in water are about 350-700 times greater than what is required for the algicidal response (minimum inhibitory concentration of pure OS = 0.45 mg l^{-1}, Bagchi and Ray, 2001).

In conclusion, we suggest that the quantity of allelochemicals during growth-sub-optimal conditions in nature is large enough to be effective in controlling species succession and dominance. More insight on signals that regulate algicide production will help in optimization of product yield for biotechnological processes.

Acknowledgement

The authors wish to thank the Head, Department of Biological Sciences, R.D. University, Jabalpur for laboratory facilities, and the Department of Science & Technology, Govt. of India, New Delhi for financial help (project no. SP/SO/AO2/95).

References

Agarwal, B.L. 1988. *Basic Statistics.* Wiley Eastern Limited, New Delhi, India.

Bagchi, S.N. 1995. Structure and site of action of an algicide from a cyanobacterium. *Oscillatoria laetevirens. J. Plant Physiol.* **146**: 372-374.

Bagchi, S.N. and Marwah, J.B. 1994. Production of an algicide from cyanobacterium *Fischerella* sp. which inhibits photosynthetic electron transport. *Microbios* **79**: 187-193.

Bagchi, S.N. and Marwah, J.B. 1997. Selection by means of acridine orange renders defect in algicide production and carrier-mediated release in a cyanobacterium, *Oscillatoria laetevirens. Cytobios* **87**: 181-188.

Bagchi, S.N., Palod, A. and Chauhan, V.S. 1990. Algicidal properties of a bloom-forming blue-green alga, *Oscillatoria* sp. *J. Basic Microbiol.* **30**: 21-29.

Bagchi, S.N. and Ray, S. 2001. Extraction and purification of an algicidal metabolite from a cyanobacterium, *Oscillatoria laetevirens. Indian J. Microbiol.* **41**: 163-167.

Chetsumon, A., Maeda, I., Umeda, F., Yagi, K.,Miura, Y. and Mizoguchi,T.1994. Antibiotic production by the immobilized cyanobacterium, *Scytonema* sp. TISTR 8208, in a saeweed-type photobioreactor. *J. Appl. Phycol.* **6**: 539-543.

Ferris, M.J. and Hirsch, C.F. 1991. Method for isolation and purification of cyanobacteria. *Appl. Environ. Microbiol.* **57**: 1448-1452.

Gleason, F.K. and Case, D.E. 1986. Activity of the natural algicide, cyanobacterin on angiosperms. *Plant Physiol.* **80**: 834-837.

Juttner, F., Leonhardt, J. and Mohren, S. 1983. Environmental factors affecting the formation of mesityloxide, dimethylallylic alcohol and other volatile compounds excreted by *Anabaena cylindrica. J. Gen. Microbiol.* **129**: 407-412.

Keating, K.I. 1977. Allelopathic influence on blue-green bloom sequence in a eutrophic lake. *Science* **196**: 885-887.

Patterson, G.M.L. and Bollis, C.M. 1995.Regulation of scytophycin accumulation in cultures of *Scytonema ocellatum* II Nutrient requirements. *Appl. Microbiol. Biotech.* **43**: 692-700.

Pearson, M.J. 1990.*Toxic blue green Algae*. Report of the National Rivers Authority. Water quality series No.2, United Kingdom.

Ray, S. and Bagchi, S.N. 2001. Nutrients and pH regulate algicide accumulation in cultures of the cyanobacterium *Oscillatoria laetevirens*. *New Phytol.* **149**: 455-460.

Rippka, R., Deruelles, J., Waterbury, J., Herdman, M. and Stanier, R.Y. 1979. Generic assignments, strain histories and properties of pure cultures of cyanobacteria . *J. Gen. Microbiol.* **111**: 1-61.

Schlegel, I., Doan, N.T., Chazal, N.D. and Smith, G.D. 1998. Antibiotic activity of new cyanobacterial isolates from Australia and Asia against green algae and cyanobacteria. *J. Appl. Phycol.* **1**: 1-9.

Sivonen, K. 1990. Effects of light, temperature,nitrate, orthophosphate and bacteria on growth and hepatotoxin production by *Oscillatoria agardhii* strains. *Appl. Environ. Microbiol.* **56**: 2658-2666.

Utkilen,H. and Gijolme, N. 1992. Toxin production by *Microcystis aeruginosa* as a function of light in continuous cultures and its ecological significance. *Appl. Environ. Microbiol.* **58**: 1321-1325.

Chapter 11

CONTRIBUTIONS TO THE UNDERSTANDING OF NITROGEN FIXATION AND NITROGENASE REGULATION IN CYANOBACTERIA BY INDIAN SCIENTISTS

N. Anand and Selvi Gnanasekaran

Abstract

Nitrogenase regulation has been one of the important steps involved in the process of biological nitrogen fixation in cyanobacteria. Various aspects of this group of organisms have been studied by many Indian scientists. Cyanobacterial organisms have been systematically documented from paddy field soils, mass multiplied in the fields to be used as inoculum for algalization and improvement in the fertility of soils. The work done on these aspects by Indian scientists has been reviewed.

Key words: Agrochemicals, biofertilisers, cyanobacteria, nitrogen fixation, nitrogenase.

Introduction

Biological nitrogen fixation assumes great significance in sustainable agriculture as an alternative and renewable source of fertiliser nitrogen. The recognition of the capacity of cyanobacteria to fix atmospheric nitrogen and the importance of this process in sustaining the nitrogen levels of the rice field soils in the tropics (De, 1939) established their value as biofertilisers. Cyanobacteria are of ubiquitous distribution in soils and reach their greatest abundance in paddy field soils of the tropical regions. Heterocystous filaments belonging to the genera *Aulosira, Anabaena, Anabaenopsis, Calothrix, Cylindrospermum, Fischerella, Hapalosiphon, Microchaete, Nostoc, Tolypothrix* and *Westiellopsis* are dominant in the rice fields of India and are promoted as suitable strains for biofertilisers (Singh 1961; Subrahamanyan and Sahay, 1964; Patnaik, 1966; Anand, 1990; Ahluwalia *et. al.*, 1993).

Many Indian scientists have systematically surveyed and documented the diversity of cyanobacterial population from paddy field soils with a view to select efficient nitrogen fixers (Pandey, 1965; Kolte and Goyal, 1986; Anand, 1989; Singh and Bisoyi, 1989; Anand and Shanthakumar Hopper, 1995; Suseela and Goyal, 1995; Nayar *et. al.*, 1996). Application of cyanobacterial biofertiliser developed as a technology through the efforts of Venkataraman (1972, 1981) and mass cultivation of selected heterocystous forms were undertaken at agricultural research institutes (Kannaiyan, 1995; Goyal, 2000). Improvements in the biofertiliser production technology include the use of polypacks as low cost bioreactors and immobilisation of selected isolates in artificial and natural matrices (Kannaiyan, 1995; Kannaiyan *et. al.*, 1997).

Influence of environmental factors and agrochemicals

Anand (1990) discussed the importance of the choice of strains for biofertiliser programmes and suggested the use of pH tolerant, halotolerant and ammonia releasing strains. One such strain, *Nostoc calcicola* isolated from the rice field of Tamil Nadu was able to physiologically adapt to varying pH levels (Anand and Revathy, 1992) and exhibited best growth at pH 9 and 10. Thomas and Apte (1984) studied the requirement of sodium for cyanobacterial nitrogen fixation and observed that a brackish water form, *Anabaena torulosa* is capable of profuse growth in saline/alkaline soils. They also advocated the use of such salt tolerant forms in the reclamation of coastal soils. Certain strains of nitrogen fixing cyanobacteria are known to release varying quantities of ammonia into the extracellular medium during their active growth phase (Subramanian and Shanmugasundaram, 1986; Anand and Parameshwaran, 1992; Anand and Gayathri, 1999). Several workers have sought to utilise such strains for the continuous photoproduction of ammonia by immobilising them in various matrices (Kannaiyan *et. al.*, 1997).

Cyanobacteria in the aquatic environment are exposed to heavy metal pollution and UV- B irradiation. As a result., primary photochemical reactions and metabolic processes dependent on these are affected. Rai *et. al.* (1991, 1996b) have studied the response of *Anabaena doliolum* to Cu stress and reported that exposure to 1 μg/ml Cu causes significant oxidative damage. Also heavy metal uptake is facilitated by prior exposure to UV-B thereby resulting in increased toxicity.

As modern agricultural practices demand the use of agrochemicals, their effect on cyanobacterial growth and nitrogen fixation was investigated in many laboratories across the country. Anand (1992) reported that heterocystous filaments of *Anabaena torulosa, Cylindrospermum muscicola* and *Nostoc calcicola* had the capacity to fix nitrogen even in the presence of commercial fertilisers like urea, potash, NPK, superphosphate and DAP. Low concentrations of chemical fertilisers promoted good growth of nitrogen fixing as well as non-nitrogen fixing strains (Anand and Karuppasamy, 1987). Singh (1975) showed that high concentrations of chemical fertilisers reduced the growth of many blue-greens . But certain algal species such as *Aulosira fertilissima* not only adapted to the fungicide Difolatan at 300 ppm, it also exhibited stimulation of algal growth (Gangawane and Saler, 1979). They also observed that among the blue-green algal strains screened for tolerance to MBC (Methyl carbamoyl benzimidazole) and Hexacap, *Calothrix* was less affected than the others. Similarly the response of cyanobacteria to pesticides have been studied extensively , so that pesticide resistant and pesticide tolerant strains may be considered for the algal biofertiliser programme (Nagpal and Goyal, 1995). Anand and Veerappan (1980) reported that higher concentrations of insecticides were lethal to cyanobacteria whereas fungicides such as Bavistin and Dithane did not completely limit nitrogenase activity (Anand and Subramanian, 1997). Organisms with a well defined sheath tolerated higher concentrations of pesticide (Rath and Adhikary, 1994) while some cyanobacteria were even capable of using organophosphate pesticides as their phosphate source (Subramanian *et. al.*, 1994). Herbicide tolerance has been extensively investigated (Roychoudary and Kaushik, 1986; Kolte and Goyal, 1992; Kaur *et. al.*, 1997). At the recommended field doses, carbaryl and BHC did not show any deleterious effect on the biological activities of *Anabaena variabilis* whereas higher concentrations of these chemicals were lethal (Pabbi and Vaishya, 1992).

Nitrogen fixation by non-heterocystous forms

Much of the work has been done with strains isolated from paddy field soil and maintained *in vitro* under well defined cultural conditions. Realising the limitations of extrapolating results obtained from such experiments, a few researchers have attempted to study the process of nitrogen fixation under field conditions. Anand and Murugesan (1996) reported that algal mats developing in waterlogged paddy field soils, consist of both heterocystous and non-heterocystous filaments which co-exist and carry on the process of nitrogen fixation round the clock- the heterocystous filaments during the day and the non-heterocystous forms at night. Aerobic diazotrophy has been demonstrated earlier in unicellular cyanobacteria belonging to the genus *Aphanothece* (Singh, 1973).

Light dark modulation of the nitrogen fixation process by non-heterocystous cyanobacteria has now been recorded in several paddy field isolates (Murugesan, 1998). In the nitrogen fixing unicellular and non-heterocystous filaments, nitrogenase activity reaches a maximum during the dark phase and the energy for nitrogen fixation is derived from the respiratory pathway. Efficient oxygen scavenging mechanism in these forms probably aids the temporal separation of the process. Rajalakshmi and Subramanian (1992) reported the simultaneous functioning of both nitrogen fixation and carbondioxide fixation in *Gloeocapsa* (IUCC 795) under altered CO_2/O_2 regimes indicating the protection of nitrogenase by RuBP carboxylase.

Enzymes regulating nitrogen fixation

Attempts to understand nitrogenase regulation in cyanobacteria are still continuing. Certain environmental signals such as nitrogen deficiency, trigger the differentiation process, leading to the formation of heterocysts and commencement of nitrogen fixation in strains of *Anabaena* (Apte *et. al.*, 1998). Intracellular oxygen levels initiate phycocyanin degradation and heterocyst formation. Carbon/nitrogen starvation in *Anabaena* sp. (Strain PCC 7120) regulates rearrangement and expression of nif genes and thereby nitrogenase activity. Kumar and Kumar (1989) studied the role of sodium, molybdenum and phosphorus on nitrogen fixation in *Nostoc linckia*. Though both the components of nitrogenase were synthesised in metal deficient cultures, all three elements were required for the activation of nitrogenase activity. In *Cylindrospermum muscicola*, both nitrogenase and glutamine synthetase are affected by various combined nitrogen sources (Anand and Radha, 1990). In the presence of glutamine, both enzymes showed an inverse relationship. As GS activity increased, nitrogenase was suppressed suggesting that either glutamine or products of ammonia assimilation may be involved in the regulation of nitrogenase.

Genetics of nitrogen fixation

Singh *et. al.* (1977) developed several mutants of *Nostoc muscorum* to study heterocyst differentiation and nitrogenase regulation. They reported the operation of two distinct glutamine sensitive regulatory units, one specific for nitrogenase and the other specific for heterocyst formation. Further mutational evidence for the involvement of three distinct categories of genes in functional organisation of nitrogen fixing heterocysts- the nif genes, het genes and the ntr genes were also provided by Singh and Singh (1983). Isolation of a GS deficient glutamine auxotrophic mutant of *Nostoc muscorum* and *Anabaena cycadeae* helped in establishing GS as the key enzyme involved in ammonia assimilation and glutamine production in heterocystous cyanobacteria (Singh, 1989). Transformation as a method of gene transfer in *Nostoc muscorum* was recognised by Singh (1989) and paved the way for manipulation of cyanobacterial genetic systems to understand the molecular basis of nitrogen fixation, heterocyst differentiation and related physiological processes.

Nitrogen fixation by cyanobacteria in associations

Though cyanobacteria are capable of independent existence, they form symbiotic associations with a wide variety of plants and animals. Such cyanobionts show higher heterocyst frequency and increased nitrogenase activity. Moreover, the fixed nitrogen is transferred to the partner either as nitrogen derived ammonia or as amino acids. One such association that has been extensively used for paddy cultivation in Asia is the *Azolla-Anabaena* system (Kannaiyan, 1995). The cyanobiont *Anabaena azollae* has reduced levels of glutamine synthetase and releases more than 50% of the nitrogen derived ammonia (Rai, 1994; Pabby *et. al.*, 2001). Such strains can be immobilized and grown in large air-lift fermentors for photobiological production of ammonia (Rai *et. al*., 1996a). Alternatively, cyanobacterial strains with efficient diazotrophic growth may be selected for herbicide resistance and combined with GS defective strains. Such mutants can survive in the presence of herbicides, as well as excrete ammonia (Tiwari *et. al*., 1991). Tiwari and Singh (1997) also employed transposon mutagenesis to develop strains for the photoproduction of nitrogen fixed ammonia for use in biofertiliser programmes.

One of the recent approaches aims at improving the nitrogen economy of crops by establishing artificial symbiosis between the roots of crop plants and nitrogen fixing cyanobacterial strains. In associations involving whole plants, cyanobacteria have been isolated from stomatal pores, xylem vessels and primary cortical parenchyma. Colonisation of roots of rice plants by *Nostoc* isolated from *Anthoceros* has been attempted (Rai *et. al*., 1996a). Cyanobacterial association with the pneumatophores of mangrove plants *Avicennia sp*. has been investigated. Non-heterocystous, filamentous cyanobacteria colonising the pneumatophores contribute considerably to the nitrogen status of mangrove ecosystem (Anand *et. al*., 1996).

Concluding remarks

Indian scientists have been providing major insights on cyanobacterial nitrogen fixation but we are still dependent on chemical fertilisers to attain the target yield for many of our crops. Inspite of extensive research on biological nitrogen fixation, we are yet to realise our dream of fertilising our crops solely with biological manures. The biofertiliser programme largely promotes rhizobia for legumes and BGA for paddy. A decade back Desikachary (1992) urged the scientists to identify natural wild types with a distinct advantage over existing species and to develop new strategies for rainfed crops and even plantation crops. We must renew our efforts to popularise indigenous technologies and native strains that will help sustain biodiversity at the macro and microlevel.

Acknowledgement

The authors thank, the Director, CAS in Botany, University of Madras, for the facilities provided and one of us (SG) thanks the UGC for the award of FIP fellowship.

References

Ahluwalia,A.S., Reena and Dahuja, S. 1993. Studies on the role of algalization on growth and yield of rice (*Oryza sativa* Linn.) *Res. Bull. Panjab Univ.* **43**: 55-60.

Anand, N. 1989. *Handbook of Blue-green Algae.* BMPS, Dehradun. pp. 74.

Anand, N. 1990. Selection of nitrogen fixing blue-green algae tolerant to pesticides. In: *Biotecnology of Biofertilisers for Rice Crop.* Edited by S. Kannaiyan, TNAU, Coimbatore. pp. 206-211.

Anand, N. 1992. Physiological responses of nitrogen fixing blue-green algae (cyanobacteria) to commercial fertilisers. In: *Natl. Symp. Cyanobacterial Nitrogen Fixation.* Edited by Kaushik, B.D., IARI, New Delhi, pp. 83-91.

Anand, N. and Gayathri, V. 1999. Ammonia release by a nonheterocystous filamentous cyanobacterium *Lyngbya spiralis* Geitler. *Phykos* **38**: 27-30.

Anand, N. and Karuppasamy, A. 1987. Responses of nitrogen fixing and non-nitrogen fixing BGA (cyanobacteria) in the presence of certain common fertilisers. *Phykos.* **26**: 22-26.

Anand, N. and Murugesan, V.S.R. 1996. Nitrogen fixation at night in nature. *Curr. Sci.* **20**: 119-120.

Anand, N., Murugesan, C., Murugesan, V.S.R., Geetha, S. and Mourugasamy, N. 1996. Nitrogen fixation by cyanobacteria in mangrove ecosysterms in South India. Abst. In: *Internat. Symp. Cyanobacterial Biotechnol.* Bharathidasan University, Tiruchirapalli. pp. 59.

Anand, N. and Parameshwaran, P. 1992. Nitrogen fixation and ammonia excretion by blue-green algae. *Seaweed Res. Utiln.* **15**: 15-19.

Anand, N. and Radha, L. 1990. Glutamine synthetase-Nitrogenase relationship in *Cylindrospermum muscicola* Kutz. ex. Born. et. Flah. In: *Perspectives in Phycology.* Edited by Raja Rao, V.N. Today and Tomorrow Printers and Publ, New Delhi. pp. 439-442.

Anand,N. and Revathy, G. 1992. Physiological responses of heterocystous filamentous blue-green alga *Nostoc calcicola* Breb. ex. Born.et. Flah. to varying pH. In. *Natl. Symp. Cyanobacterial Nitrogen Fixation.* Edited by Kaushik, B.D. IARI, New Delhi, pp. 151-161.

Anand, N. and Shanthakumar Hopper, R.S. 1995. Distribution of blue-green algae in rice fields of Kerala state, India. *Phykos*. **34**: 55-64.

Anand, N. and Subramanian, T.D. 1997. Effect of certain pesticides on the physiology of *Nostoc calcicola*. *Phykos*. **36**: 15-20.

Anand, N. and Veerappan, B. 1980. Effect of pesticides and fungicides on blue-green algae. *Phykos*. **19**: 210-212.

Apte, S.K., Fernandes, T., Iyer, V., Alahari, A. and Badran, H. 1998. Stress regulated gene expression in N_2 fixing cyanobacteria. In: *Cyanobacterial Biotechnology*. Edited by Subramanian, G., Kaushik , B.D. and Venkataraman, G.S. Oxford & IBH Publ. Co. Pvt. Ltd., New Delhi, pp. 197-200.

De, P.K. 1939. The role of BGA in N_2 fixation in rice fields. *Proc. R.. Soc. Lond.* B 127: 121-139.

Desikachary, T.V. 1992. Biological Nitrogen fixation – What next? In: *Natl. Symp. Cyanobacterial Nitrogen Fixation*. Edited by Kaushik, B.D. IARI, New Delhi. pp. 5-10.

Gangawane, L.V. and Saler, R.S. 1979. Tolerance of certain fungicides by nitrogen fixing blue-green algae. *Curr. Sci*. 48: 306-308.

Goyal, S.K. 2000. Preparation of soil based algal inoculum. In: *Biofertilizer-Blue green Algae and Azolla*. Edited by Singh, P.K.Dhar, D.W., Pabbi, S., Prasanna, R. and Arora, A. Venus Printers and Publ. , New Delhi. pp. 100-106.

Kannaiyan, S. 1995. Algal biofertilizer technology for rice crop. In: *Rice Management Biotechnology*. Edited by Kannaiyan, S. Assoc. Publ. Co., New Delhi. pp. 299-307.

Kannaiyan, S., Aruna, S.J., Merrina, S., PremKumari and Hall, D.O. 1997. Immobilised cyanobacteria as a biofertilizer for rice crops. *J. Appl. Phycol*. **9**: 167-174.

Kaur, M., Dahuja, S. and Ahluwalia, A.S. 1997. Response of diazotrophic cyanobacteria to butachlor. *Phykos* **36**: 93-101.

Kolte, S.O. and Goyal, S.K. 1986. Natural variation in the N_2 fixing potential of cyanobacteria. *Phykos* **25**: 166-172.

Kolte, S.O. and Goyal, S.K. 1992. Inhibition of growth and nitrogen fixation in *Calothrix marchica* by herbicides *in vitro*. In: *Nat. Symp. Cyanobacterial Nitrogen Fixation*. Edited by Kaushik, B.D. IARI, New Delhi. pp.507-510.

Kumar, D. and Kumar, H.D. 1989. Effect of sodium, molybdenum and phosphorus on growth and nitrogen fixation in *Nostoc linckia*. *Phykos* **28**: 254-259.

Murugesan, V.S.R. 1998. *Nitrogen fixation by non-heterocystous Cyanobacteria (blue-green algae)*. Ph.D. thesis, University of Madras, Chennai.

Nagpal, V. and Goyal, S.K. 1995. Induction of weedicide resistance in a diazotrophic cyanobacterium *Anabaena variabilis*. *Phykos* **34**: 113-122.

Nayar, A., Sahu, J.K. and Adhikary, S.P. 1996. Blue green algae of rice fields of Orissa state. II Growth & N_2 fixing potential. *Phykos* **35**: 111-118.

Pabbi, S. and Vaishya, A.K. 1992. Effect of insecticides on cyanobacterial growth and nitrogen fixation. In: *Natl. Symp. Cyanobacterial Nitrogen Fixation*. Edited by Kaushik, B.D. IARI, New Delhi. pp. 489-495.

Pabby, A., Dua,S. and Ahluwalia, A.S. 2001. Changes in ammonia assimilation enzymes in response to different nitrate levels in *Azolla pinnata and Azolla microphylla*. *J. Plant Physiol.* **158**: 889-903.

Pandey, D.C. 1965. A study of the algae from paddy soils of Ballia and Ghazipur districts of Uttar Pradesh, India 1. Cultural and ecological considerations. *Nova Hedw*. **9**: 299-334.

Patnaik, H. 1966. Growth and nitrogen fixation by *Westiellopsis prolifica* Janet . *Ann. Bot*. **39**: 231-238.

Rai, A.N.1994. Cyanobacteria in symbiosis- a Review. In: *Recent Advances in Phycology*. Edited by Kashyap, A.K. and Kumar, H.D. Rastogi & Co. Meerut. pp. 31-38.

Rai, A.N., Borthakur, M. and Paul, D. 1996a. Symbiotic Cyanobacteria: Biotechnological applications. *J. Sci. Ind. Res*. **55**: 742-752.

Rai, L.C., Mallick, N., Singh, J.B. and Kumar, H.D. 1991. Physiological and biochemical characterization of a copper tolerant and a wild type strain of *Anabaena doliolum* under copper stress. *J. Plant Physiol*. **138**: 68-78.

Rai, L.C., Tyagi, B. and Mallick, N. 1996b. Alternation in photosynthetic characteristics of *Anabaena doliolum* following exposures to UV-B and Pb. *Photochem. Photobiol*. **64**: 658-663.

Rajalakshmi, N. and Subramanian, D. 1992. Regulation of N_2 fixation by ribulose bisphosphate carboxylase. In: *Natl. Symp.Cyanobacterial Nitrogen Fixation.* Edited by Kaushik., B.D. IARI, New Delhi. pp. 35-39.

Rath, B. and Adhikary, S.P. 1994. Relative tolerance of several nitrogen fixing cyanobacteria to commercial grade furadan (carbofuran 3%). *Indian J. Expt. Biol.* **32**: 213-215.

Roychoudary, P. and Kaushik, B.D. 1986. Response of cyanobacterial growth and nitrogen fixation to herbicides. *Phykos* **25**: 36-43.

Singh, H.N. 1989. Genetic analysis of inorganic nitrogen assimilation in diazotrophic cyanobacteria. *Phykos* **28**: 278-285.

Singh, H.N., Ladha, J.K. and Kumar, H.D. 1977. Genetic control of heterocyst formation in the blue green algae *Nostoc muscorum* and *Nostoc linckia. Arch. Mircobiol.* **114**: 155-159.

Singh, H.N. and Singh, H.N. 1983. Loss of aerobic nitrogen fixation associated with kasugamycin resistance in the heterocystous nitrogen fixing streptomycin resistant blue green alga *Nostoc muscorum. Pl. Cell Repor.* **2**: 310-313.

Singh, P.K. 1973. Nitrogen fixation by the unicellular blue-green alga *Aphanothece. Arch. Mocrobiol.* **103**: 297-302.

Singh, P.K. 1975. Fertiliser tolerance of blue-green algae and their effect on heterocyst differentiation. *Phykos* **14**: 81-88.

Singh, P.K. and Bisoyi, R.N. 1989. Blue green algae in rice fields. *Phykos* **28**: 181-188.

Singh, R.N. 1961. *Role of Blue green Algae in N_2 economy of Indian Agriculture.* Indian Council of Agricultural Research New Delhi. pp. 175.

Subrahamanyan, R. and Sahay, M.N. 1964. Observations on nitrogen fixation by some blue green algae and remarks on its potentialities in rice culture. *Proc. Indian Acad. Sci. Sect.* B **60**: 145-154.

Subramanian, G., Sekar, S. and Sampoornam, S. 1994. Biodegradation and utilization of organophosphorus pesticides by cyanobacteria. *Int. Biodet. Biodeg.* **33**: 129-143.

Subramanian, G. and Shanmugasundaram, S. 1986. Uninduced ammonia release by the nitrogen fixing cyanobacteriun *Anabaena. FEMS Microbiol. Lett.* **37**: 151-154.

Suseela, M.R. and Goyal, S.K. 1995. Growth and nitrogen fixing potential of cyanobacteria. *Phykos.* **34**: 131-134.

Thomas, J. and Apte, S.K. 1984. Sodium requirement and metabolism in N_2 fixing cyanobacterium. *J. Biosci.* **6**: 771-794.

Tiwari, D.N., Kumar, A. and Mishra, A.K. 1991. Use of cyanobacterial diazotrophic technology in rice agriculture. *Appl. Biochem. Biotech.* **29**: 387-396.

Tiwari, D.N. and Singh, A.P. 1997. Physiological characterisation of transposon generated methionine sulfoximine and ethylenediamine resistant mutants of *Anabaena sp*. PCC 7120. *J. Plant Physiol.* **150**: 9-15.

Venkataraman, G.S. 1972. *Algal Biofertilisers and Rice Cultivation.* Today and Tomorrow Publ., New Delhi. pp. 75.

Venkataraman, G.S. 1981. Blue green algae for rice production: a manual for its promotion. *FAO Soils. Bull.* **46**: 102.

Chapter 12

BIOCHEMICAL AND ULTRASTRUCTURAL STUDIES ON THE EFFECT OF DIFFERENT LIGHT INTENSITIES ON *HYPNEA MUSCIFORMIS* AND *H. VALENTIAE* FROM RAMESWARAM COAST

K. Sivakumar and Rengasamy, R.

Abstract

The carrageenophytes of *Hypnea musciformis* (Wulf.) Lamour. and *H. valentiae* (Turn.) Mont. were investigated under different light intensities from 2.0 to 30.0 $\mu Em^{-2}s^{-1}$ under laboratory conditions. At all light intensities, the biomass of *H. valentiae* was compared to *H. musciformis*. Both species showed maximum growth under 20 $\mu Em^{-2}s^{-1}$ light intensity. In contrast to biomass, concentration of pigments were high in *H. musciformis* than *H. valentiae*. The concentration of PE was high followed by APC, PC and chl. a in both algae. Maximum concentration of chl. a and accessory pigments were recorded when both the species exposed to 2 and 10 $\mu Em^{-2}s^{-1}$, of light intensities. The yield of kappa-carrageenan of *H. musciformis* was more than the *H. valentiae*. An inverse correlation was observed between carrageenan yield and light intensity. Ultrastructural studies made on *H. valentiae* under 2-3 $\mu Em^{-2}s^{-1}$ light intensities revealed that the thylakoidal lamellae were poorly developed in the former and well developed in the latter conditions respectively.

Key words: Growth, pigments, kappa-carrageenan, light intensity, ultrastructure.

Introduction

Light intensity is one of the most important factors governing growth and productivity of macrophyte communities. The growth of *Gracilaria folifera* (Laponite, 1981), *Gracilaria spp.* (Rotem *et. al.*, 1986) and *Porphyridium purpureum* (Levy and Gantt, 1988) increased with increasing irradiance. An inverse relationship between pigment concentration and irradiance has been reported in many red algae (Waaland *et. al.*, 1974; Ramus *et. al.*, 1976, Levy and Friedlander, 1990). Increasing photon flux densities decreased the photosynthetic pigments in *Porphyridium* sp. (Levy and Gantt, 1988) and *Griffithsia* sp. (Walland *et. al.*, 1974). The interaction of irradiance with phycocolloid yield was critically investigated in many red algae (Morgan and Simpson, 1981; Rotem *et. al.*, 1986; Ekman and Pederson, 1990). Species of *Hypnea* are well-known sources of their phycocolloid, kappa-carrageenan and other compounds of pharmaceutical interest (Rama Rao, 1982; Davies *et. al.*, 1984). Increased shading usually leads to slow growth (Philips and Mitsui, 1982). The light intensity affects not only the growth but also photosynthesis, pigment composition and carrageenan yield and quality (Mollion, 1978). For an efficient management of seaweed mariculture, the growth potential of algae should be understood under the physico-chemical conditions for good productivity. In the present study, a comparative account was made on the effect of different light intensities on *Hypnea musciformis* (Wulf.) Lamour. and *H. valentiae* (Turn.) Mont. (Gigartinales, Rhodophyta) in relation to growth, pigments, kappa-carrageenan content and ultrastructure.

Materials and methods

Vegetative thalli of *Hypnea musciformis* and *H. valentiae* collected along the coast of Rameswaram, Tamil Nadu, India during December, 1995 were brought to the laboratory within 48 hours, in between cotton pads soaked in seawater. They were removed carefully from the cotton pads and washed thoroughly with filtered seawater to remove sand particles and epiphytes. Samples of *H. musciformis* and *H. valentiae* were inoculated in sterilized PES/2 medium (Provasoli, 1968; McLachlan, 1973) and kept under 20 $\mu Em^{-2}s^{-1}$ light intensity 24 ± 1°C and 12/12 h light/dark regime. The contaminants such as diatoms and blue-green algae were eliminated (Lewin, 1966; Rengasamy *et. al.*, 1987). Specimens acclimatized for a period of one month under laboratory conditions were taken for further experimental studies. Algal samples were kept under the various light intensities *viz.*, 2, 10, 20 (control) and 30 $\mu Em^{-2}s^{-1}$. For each experiment 40 mg of 5 apical portions was used as initial inoculum in 50 ml of the medium. The medium was replenished every three days interval (Mshigeni, 1976). The data recorded at the end of 15th day included (a) total fresh weight (mg); (b) photosynthetic pigments such as chlorophyll *a* (chl. *a*), phycoerythrin (PE), phycocyanin (PC) and allophycocyanin (APC) (Jeffery and Humphery, 1975; Kursar *et. al.*, 1983; Ramus, 1983) and (c) kappa- carrageenan fractions such as total carbohydrate using galactose as standard (Dubois *et. al.*, 1965; Hellebust and Craigie, 1978), 3,6 anhydrogalactose (3,6-AG) (Yaphe and Arsenault, 1965) and sulphate (Doshi *et. al.*, 1987). The values of pigments and kappa-carrageenan content were expressed as µg/mg fresh weight and per cent fresh weight respectively. Statistical analysis was carried out using the statistical software SPSSPC 4.01 (Statistical package for Social Sciences). Apical portions of the thalli measuring 1-2 mm in length collected under 2 (low) and 30 $\mu Em^{-2}s^{-1}$ (high) were fixed in 3% glutaraldehyde, post fixed in 1% OsO_4, dehydrated through upgraded series of acetone and embedded in spur resin for transmission electron microscopy. Ultrathin (60 – 90 nm) sections were taken in Richert Jung-Ultracut-ultramicrotome with glass knives. These sections were poststained with uranyl acetate and lead citrate (Reynolds, 1963) and examined under CM 10 Philips transmission electron microscope.

Results

Maximum fresh wt. of 225 mg and 188 mg was recorded in *H. valentiae* and *H. musciformis* respectively at 20 $\mu Em^{-2}s^{-1}$ light intensity. At 2 $\mu Em^{-2}s^{-1}$ light intensity, minimum fresh weight of 82 mg and 45 mg was recorded in *H. valentiae* and *H. musciformis* respectively (Figs. 12.1a, 12.2a). Analysis of variance for two way showed a significant ($p<0.01$) difference in fresh weight within *H. musciformis* and *H. valentiae* and between two species at different light intensities. T-test showed a significant ($p<0.01$) difference in fresh weight at 0.01% level between two species at each light intensity (Table 12.1). In contrast to fresh weight the concentration of pigments were high in *H. musciformis* than *H. valentiae* (Figs. 12.1b, 12.2b). The concentration of PE was high followed by APC, PC and chl. *a* in both algae. Maximum concentrations of 0.25, 1.46, 0.71 and 0.96

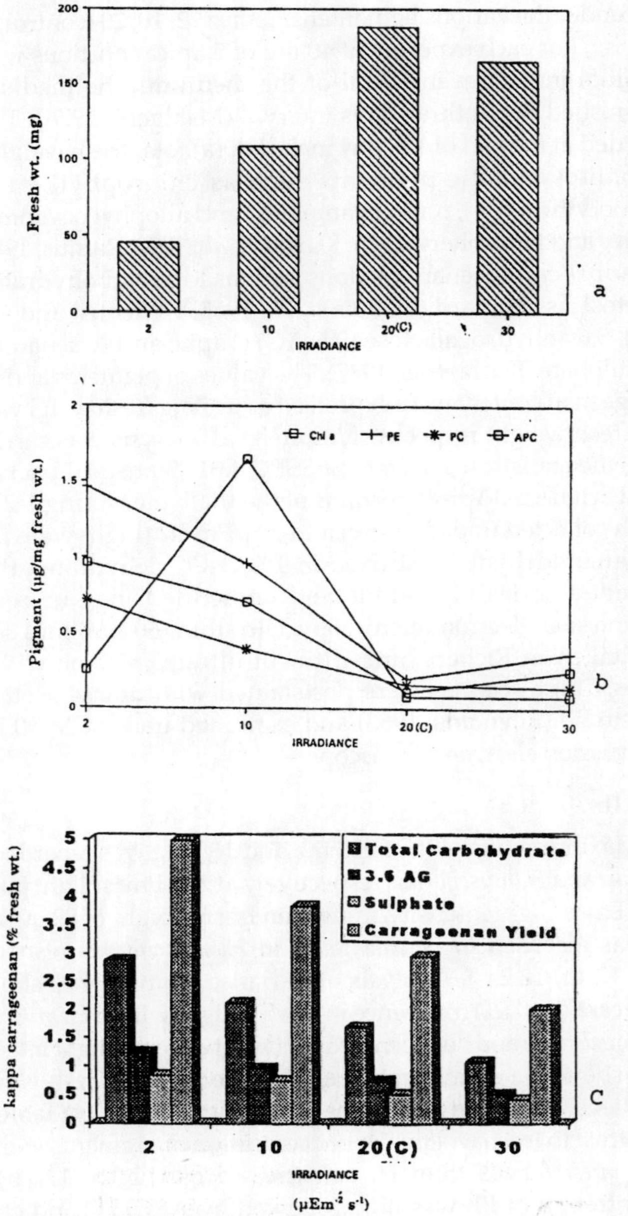

Fig. 12.1: Effect of different irradiances on *H. musciformis*.

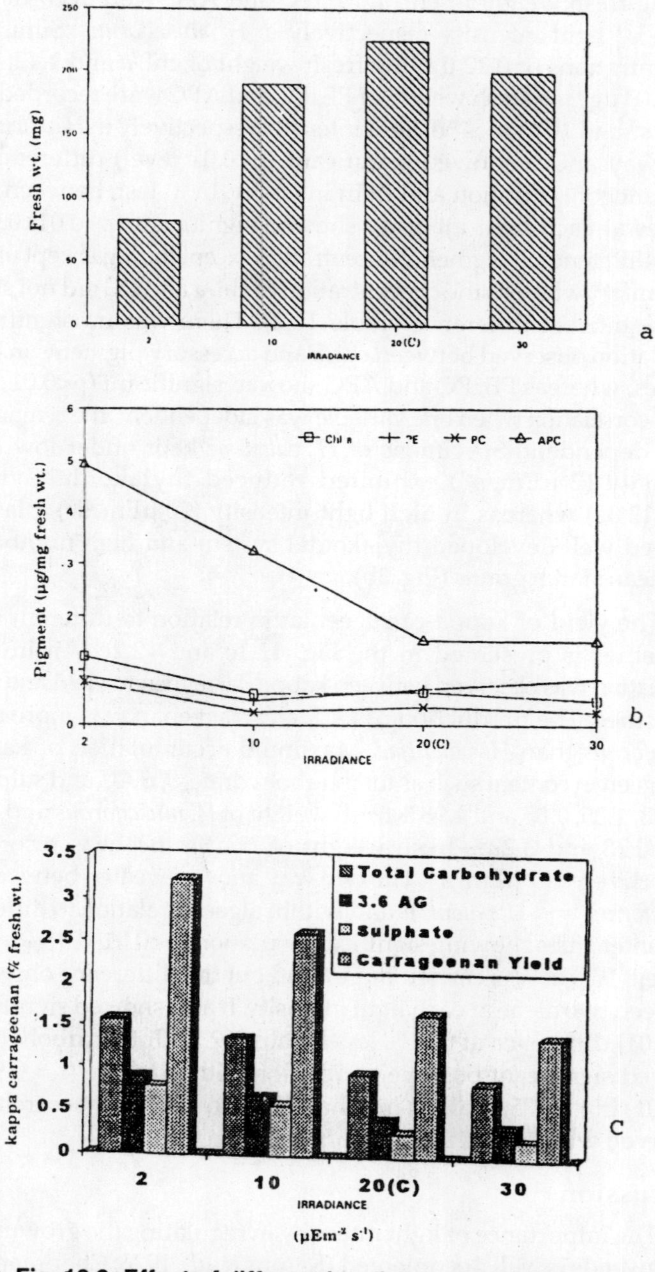

Fig. 12.2: Effect of different irradiances on *H. valentiae*.

μg/mg fresh weight of chl. *a*, PE, PC and APC were recorded at 2μEm^{-2}s^{-1} light intensity respectively in *H. musciformis*. Similarly, concentrations of 0.22 μg/mg fresh weight of chl. *a* and 1.42, 0.48 and 1.03 μg/mg fresh weight of PE, PC and APC were recorded at 2 μEm^{-2}s^{-1} and 10 μEm^{-2}s^{-1} of light intensity respectively in *H. valentiae*. Two way anova showed significant ($p<0.01$ level) difference in pigment concentration and light intensity. In 't' test, between two species at each light intensity showed significant ($p<0.01$, 0.01% level) difference in pigment concentration except the algae kept under 20 μEm^{-2}s^{-1}, whereas the concentration of chl. *a* and PC did not show any significant difference (Table 12.2). There was no significant correlation observed between chl. *a* and accessory pigments in both species, whereas PE, PC and APC showed significant ($p<0.01$, 0.01 level) correlation when one variable was independent, the remaining were dependent. Specimens of *H. valentiae* kept under low light intensity (2 μEm^{-2}s^{-1}) exhibited reduced thylakoidal system (Fig. 12.3a) whereas in high light intensity (20 μEm^{-2}s^{-1}), plastids showed well developed thylakoidal system and high number of floridean starch grains (Fig. 3b).

The yield of kappa-carrageenan in relation to different light intensities is presented in the Fig. 12.1c and 12.2c. An inverse correlation was observed between kappa-carrageenan yield and light intensities. The production of kappa-carrageenan was more in *H. musciformis* than *H. valentiae*. Maximum accumulation of kappa-carrageenan content such as total carbohydrate, 3, 6 AG and sulphate of 2.88, 1.25, 0.85 and 4.98% fresh weight of *H. musciformis* and 1.56, 0.88, 0.78 and 3.22% fresh weight of *H. valentiae* was recorded, respectively at 2 μEm^{-2}s^{-1}. The two way anova revealed between *H. musciformis* and *H. valentiae* and within algae in relation to different light intensities showing significant variation ($p<0.01$, 0.01% level). Student 't' test was carried out to find out the difference observed between two algae at each light intensity. It also showed significant ($p<0.01$) difference at 0.01% level (Table 12.3). Total carbohydrate showed significant positive correlation with sulphate ($r = 0.9751$, $p<0.01$) ($r = 0.875$, $p<0.01$) but there was no significant correlation observed with 3,6 AG in *H. valentiae*.

Discussion

The importance of light intensity in regulating the growth rate of seaweeds is well documented (Ramus *et. al.*, 1976; Chapman and

Table 12.1: Student 't' test for fresh weight (mg) of *H. musciformis* (H.m.) and *H. valentiae* (H.v.)

		Mean		*Standard deviation*		*Standard error*		*t. value*	*Sig.t*
		H.m.	*H.v.*	*H.m.*	*H.v.*	*H.m.*	*H.v.*		
Irradiance ($\mu Em^{-2}s^{-1}$)	2	45.00	82.00	4.35	2.64	2.51	1.52	-12.57**	0.01
	10	108.00	190.00	2.64	5.00	1.52	2.88	-25.11**	0.01
	20©	188.00	225.00	2.64	4.35	1.52	2.51	-12.57**	0.01
	30	165.00	200.00	5.00	4.35	2.88	2.51	-9.14**	0.01

** Significant at 0.01% level, $p<0.01$

Table 12.2: Effect of different irradiances on pigments of *Hypnea musciformis* (H.m.) and *H. valentiae* (H.v.) 't' test

	Irradiance $mEm^{-2}s^{-1}$	*Mean*		*Standard deviation*		*Standard error*		*t. value*	*Sig. t*
		H.m.	*H.v.*	*H.m.*	*H.v.*	*H.m.*	*H.v.*		
Chlorophyll a	2	0.25	0.22	0.03	0.01	0.01	0.006	171	N.S
	10	1.65	0.13	0.04	0.01	0.02	0.006	58.87**	0.01
	20©	0.08	0.08	0.01	0.01	0.006	0.006	0.00	N.S.
	30	0.08	0.19	0.01	0.01	0.006	0.006	-13.47**	0.01
Phycoerythrin	2	1.46	1.05	0.10	0.02	0.05	0.01	6.91**	0.01
	10	0.96	1.42	0.05	0.01	0.03	0.01	-14.31**	0.01
	20©	0.20	0.56	0.02	0.01	0.01	0.006	-22.05**	0.01
	30	0.37	1.03	0.01	0.01	0.006	0.006	-80.83**	0.01
Phycocyanin	2	0.71	0.37	0.05	0.01	0.03	0.006	10.94**	0.01
	10	0.39	0.48	0.02	0.02	0.01	0.01	-4.17**	0.01
	20©	0.16	0.20	0.02	0.01	0.01	0.006	-2.45	N.S.
	30	0.14	0.35	0.01	0.01	0.006	0.006	-25.72**	0.01
Allophycocyanin	2	0.96	0.52	0.05	0.01	0.03	0.006	14.15**	0.01
	10	0.70	1.03	0.02	0.01	0.01	0.006	-20.21**	0.01
	20©	0.11	0.41	0.01	0.01	0.006	0.006	-36.74**	0.01
	30	0.25	0.69	0.01	0.01	0.006	0.006	-53.89**	0.01

** Significant at 0.01% level. $P<0.01$; N.S. Not significant

Table 3: Effect of different irradiances on kappa-carrageenan composition and yield of *Hypnea musciformis* (H.m.) and *H. valentiae* (H.v.) 't' test

	Irradiance $\mu Em^{-2}s^{-1}$	*Mean*		*Standard deviation*		*Standard error*		*t. value*	*Sig. t*
		H.m.	*H.v.*	*H.m.*	*H.v.*	*H.m.*	*H.v.*		
Total carbohydrate	2	2.88	1.26	0.33	0.04	0.19	0.02	8.28**	0.01
	10	1.31	0.57	0.03	0.06	0.02	0.03	18.13**	0.01
	20©	1.78	0.60	0.02	0.04	0.01	0.02	40.08**	0.01
	30	1.63	0.46	0.07	0.07	0.04	0.04	19.73**	0.01
3,6 – Anhydrogalactose	2	1.11	0.88	0.03	0.02	0.02	0.01	8.91**	0.01
	10	0.62	0.45	0.03	0.04	0.02	0.02	5.21**	0.01
	20©	0.78	0.66	0.02	0.01	0.01	0.00	6.61**	0.01
	30	0.45	0.85	0.04	0.05	0.02	0.02	-10.44**	0.01
Sulphate	2	0.50	0.78	0.17	0.02	0.10	0.01	-2.77	0.05
	10	0.10	0.24	0.01	0.03	0.01	0.02	-6.06**	0.01
	20©	0.21	0.34	0.06	0.03	0.03	0.02	-3.01*	0.05
	30	0.13	0.25	0.02	0.03	0.01	0.01	-5.76**	0.01
Kappa-carrageenan	2	4.49	4.92	0.20	0.04	0.11	0.02	-3.55*	0.05
	10	2.03	3.26	0.01	0.05	0.01	0.03	-36.52**	0.01
	20©	2.77	1.57	0.08	0.05	0.04	0.03	21.44**	0.01
	30	2.21	1.56	0.13	0.10	0.07	0.06	6.69**	0.01

** Significant at 0.01% level. P<0.01; * Significant at 0.05% level, p<0.05

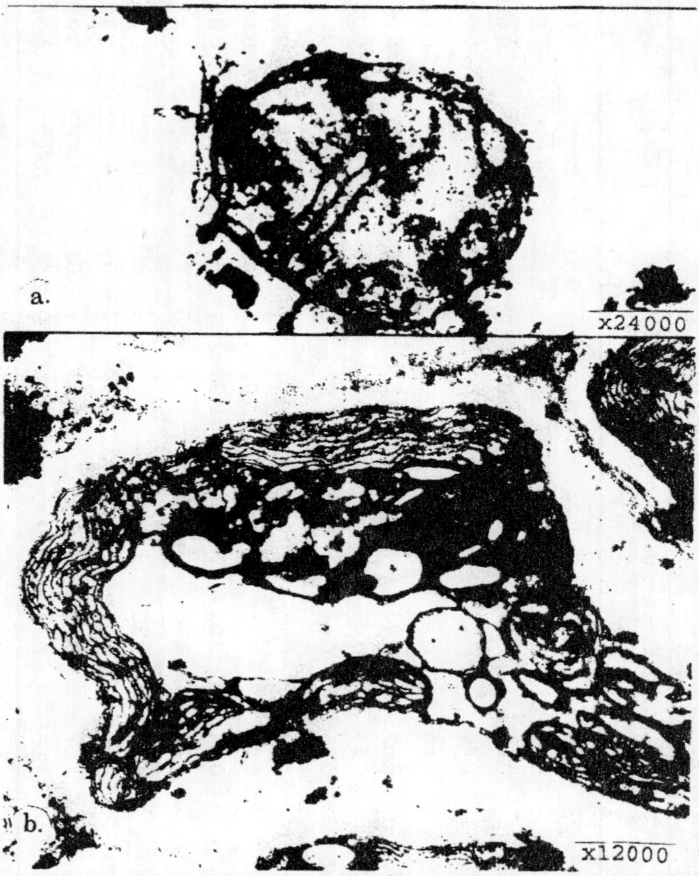

Fig. 12.3 (a): ***Hypnea valentiae*** **at 2 μEm^{-2}s^{-1} shows plastids with a peripheral thylakoids and three thylakoids at the centre. (x 24000).**

Fig. 12.3 (b): ***Hypnea valentiae*** **grown at 20 μEm^{-2}s^{-1} shows well developed thylakoidal system and high number of floridean starch grain. (x 12000).**

Craigie, 1977; Laponite, 1981; Beer and Levy, 1983). Species of *Gracilaria* grown under 100 μEm^{-2}s^{-1} showed a maximum increase of 150% in fresh weight per week and species of *Falkenbergia rufolanosa* showed good growth under 1000 – 1200 lux light intensities compared to the least growth observed under 100 – 400 lux (Oza, 1989). In the present study, at all light intensities, the fresh weight of *H. valentiae* was higher than *H. musciformis*. Analysis of variance for two way anova showed significant ($p<0.01$) difference in fresh weight within *H. musciformis* and *H. valentiae* and between two species at different light intensities. Concentration of pigments in seaweeds are known to decrease with increasing light. *Gracilaria tikvahiae* cultured under different levels of light showed more chl. *a* (2.29 mg/g d.wt^{-}) and PE (12.25mg/g d.wt^{-}) under low light compared to chl. *a* (1.36mg/g d. wt) and PE (3.9 mg/g d. wt) under high light intensity (Laponite and Duke, 1984). In the present study, concentrations of chl. *a*, PE, PC and APC were high in *H. musciformis* at low light intensity ($2\mu Em^{-2}s^{-1}$). The anova for two way and one way showed a significant ($p<0.01$ level) difference in pigment concentration when both algae were grown at different light intensities.

Red algae are known to concentrate high photosynthetic pigments under low light intensity but this did not correlate with their growth capacity. In *Gracilaria folifera* var. *angustissima*, though there were two fold changes in PE and chl. *a*, the growth was not affected (Laponite 1987). A similar change in chl. *a* did affect the growth of *Ulva* (Ramus *et. al.*, 1976). These results support the hypothesis that photosynthesis and growth increase with increasing pigment content in optically thin but not in optically thick algae (Ramus *et. al.*, 1976). In the present study, 't' test between two species at each light irradiance showed significant ($p<0.01$, 0.01% level) difference in pigment concentration except the algae exposed at $20\mu Em^{-2}s^{-1}$. Negative significant correlation was observed between chl. *a* and accessory pigments in both the species. Increase in total carbohydrate with high photon flux density was reported in *Palmaria palmata*. In the present study, an inverse relationship was observed between carrageenan yield and light intensity. The yield of kappa-carrageenan of *H. musciformis* was higher than *H. valentiae*. Low light intensity enhanced the accumulation of kappa-carrageenan content such as total carbohydrate, 3,6-AG. and SO_4 (Neish and Shacklock, 1977). A significant ($p<0.01$) difference at 0.01% level was recorded among total carbohydrate, 3,6 – AG., sulphate and yield of kappa-carrageenan. While one parameter was independent variable, the rest were dependent. In *H. valentiae*, total carbohydrate showed significant positive correlation with sulphate ($r=0.9751$, $p<0.01$) and yield of kappa-carrageenan ($r=0.875$, $p<0.01$) but there was no significant correlation observed with 3,6 –AG. It was not correlated significantly with sulphate and kappa-carrageenan yield. However, sulphate showed significant correlation with carrageenan yield (r=o.8004, $p<0.01$) whereas in *Hypnea musciformis* all the parameters showed significant positive correlation where one value was independent and the rest were dependent.

Sheath *et. al.* (1977) found that the plastids fragmented and thylakoids dilated in *Porphyra leucostica* under low light intensity. Similar observations were also made on the fresh water red alga *Batrachospermum moniliform* (Sheath *et. al.*,1979). In the present study under low light intensity, the plastids of *H. valentiae* showed reduced thylakoidal system either by having a single peripheral thylakoid or one to three thylakoids at the central region and in contrast, well developed thylakoids, higher number of floridean starch grains, distinct cell wall and vacuoles were observed under conditions of high light intensity ($20\mu Em^{-2}s^{-1}$).

References

Beer, S. and Levy, I. 1983. Effects of photon fluence rate and light spectrum composition on growth, photosynthesis and pigment relation in *Gracilaria* sp. *J. Phycol.* **19**: 516-522.

Chapman, A.R.O. and Cragie, J.S. 1977. Seasonal growth in *Laminaria longicruris* ; Relations with dissolved inorganic nutrients and internal reserves of nitrogen. *Mar. Biol.* **40**: 147-205.

Christiaen, D., Stadler, T., Ondarza, M. and Verdus, M.C. 1987. Structures and functions of the polysaccharides from the cell wall of *Gracilaria verrucosa* (Rhodophyceae, Gigartinales) Proc. *Int. Seaweed Symp.* **12**: 139-146.

Davies, L.P., Jamieson, D.D., Baird-Lambert, J.A. and Kazlauskas, R. 1984. Halogenated pyrrolopyrimidine analogs of adenosine from marine organisms. Pharmacological activities and potent inhibition of adenosine kinase. *Biochem. Pharmacol.* **33**: 347-356.

Doshi, Y.A., Parekh, R.G. and Chauhan, B. 1987. Studies on marine alga *Laurentia papillosa* for carrageenan. *Phykos.* **26**: 113-118.

Dubois, M., Gilles, K.A., Hamilton, J.K., Kebers, P.A. and Smith, F. 1956. Calorimetric method for determination of sugars and related substances. *Anal. Chem.* **28**: 350-356.

Ekman, P. and Pederson, M. 1990. The influence of photon irradiance, day length, dark treatment, temperature and growth rate on the agar composition on *Gracilaria sordida* Nelson and *Gracilaria verrucosa* (Hudson) Papenfuss (Gigartinales, Rhodophyta). *Bot. Mar.* 33: 483-495.

Hellebust, J.A. and Craigie, J. S. 1978. *Hand book of Phycological Methods. Physiological and Biochemical Methods.* Cambridge Univ. Press, Cambridge.pp. 512.

Jeffrey, S.W. and Humphrey, G. F. 1975. New spectrophotometric equations determining chl. a, b, c, and c_2 in higher plants, algae and natural phytoplankton. *Biochem. Physiol Pflanzen.* **167**: 191-194.

Kursar, T. A., Van Deermeer, J. and Alberte, R. S. 1983. Light harvesting system of the red alga *Gracilaria tikvahiae. Plant Physiol.* **73**: 353-360.

Laponite, B.E. 1981. The effects of light and nitrogen on growth, pigment content and biochemical composition of *Gracilaria folifera* var. *angustissima* (Gigartinales, Rhodophyta). *J. Phycol.* **17**: 90-95.

Laponite, B.E.1987. Phosphorus and nitrogen–limited photosynthesis and growth of *Gracilaria tikvahiae* (Rhodophyceae) in the Florida Keys. An experimental fields study. *Mar. Biol.* **13**: 561-568.

Laponite, B.E. and Duke, C.S. 1984. Biochemical strategies for growth of *Gracilaria tikvahiae* (Rhodophyta) in relation to light intensity and nitrogen availability. *J. Phycol.* **20**: 488-495.

Levy, I. and Gantt, E. 1988. Light acclimatization in *Porphyridium purpureum* (Rhodophyta): growth, photosynthesis and phycobilisomes. *J. Phycol.* **24**: 452-458.

Levy, I. and Friedlander, M. 1990. Strain selection in *Gracilaria* sp. I. Growth, pigment and carbohydrate characterization on strains of *G. conferta* and *G. verrucosa* (Rhodophyta, Gigartinales). *Bot. Mar.* **33**: 339-345.

Lewin, J. 1966. Silicon metabolism in diatoms V. Germanium dioxide-a specific inhibitor of diatom growth. *Phycologia.* **6**: 1-12.

McLachlan, J. 1973. Growth media In: *Handbook of Phycological Methods*. Edited by Stein, J.R., Cambridge Univ. Press, Cambridge. pp. 25-51

Mollion, J. 1978. Seasonal variations in carrageenan from *Hypnea musciformis* (Florideophyceae) in Senegal. *Proc. Int. Seaweed Symp.* **9**: 233-240.

Morgan, K.C. and Simpson, F.J. 1981. The cultivation of *Palmaria palmata*. Effect of light intensity and temperature on growth and chemical composition. *Bot. Mar.* **24**: 547-552.

Mshigeni, E.K. 1976. Studies on the reproduction of selected species of *Hypnea* (Rhodophyta, Gigartinales) from Hawaii. *Bot. Mar.* **19**: 341-346.

Neish, A.C., Shacklock, P.F., Fox, C.H. and Simpson, F.J. 1977. The cultivation of *Chondrus crispus* factors affecting growth under greenhouse conditions. *Can. J. Bot.* **55**: 2263-2271.

Oza, R. M. 1989. Growth of red alga *Falkenbergia rufolanosa* (Harvey) Schmitz in response to temperature, irradiance and photoperiod. *Indian J. Mar. Sci.* **18**: 210-211.

Philips, E.J. and Mitsui, A. 1982. Light intensity preference and tolerance of aquatic photosynthetic macroalgae In: *CRC Hand Book of Biosolar Resources* Vol – I *Basic Principles Part 2*. Edited by Mitsui, A. and C. S. Blade. pp. 309 – 333.

Provasoli, L. 1968. Media and prospects for the cultivation of marine algae: culture and collection of algae In: *Japan Soc. P1. Physiol. USA – Japan Conference*. Edited by Watanabe, A. and Hattori, I. pp. 63 – 75.

Ramus, J. 1983. A physiological test of the theory of complementary chromatic adaptation in brown, green and red seaweeds. *J. Phycol.* **19**: 173 – 178.

Ramus, J., Beale, S. I. and Mauzerall, D. 1976. Correlation of changes in pigment content with photosynthetic capacity of seaweeds as a function of water depth. *Mar. Biol.* **37**: 231 – 238.

Rama Rao, K. 1982. Biology of *Hypnea spp.* and their possible utilization in India. *Seaweed Res.Utiln.* **5**: 11 – 19.

Rengasamy, R., Prema, M., Govindarajan, S. and Ilanchelian, K. 1987. Effect of antibiotics on growth of *Hypnea valentiae* (Turn.) Mont. (Gigartinales, Rhodophyta). *Curr. Sci.* **56**: 676 – 677.

Reynolds, E. S. 1963. The use of lead citrate at high pH as an electron opaque stain in electron microscopy. *J. Cell. Biol.* **17**: 208.

Rotem, A., Bejerano, N.R. and Arad, S. 1986. Effect of controlled environmental conditions on starch and agar contents of *Gracilaria* sp. (Rhodophyceae). *J. Phycol* **22**: 117 – 121.

Sheath, R. G., Hellebust, J. A. and Sawa, T. 1977. Changes in plastid structure, pigmentation and photosynthesis of the conchocelis stage *Porphyra leucostica* (Rhodophyta Bangiophyceae) in response to low light and darkness. *Phycologia.* **16**: 265 – 276.

Sheath, R.G., Hellebust, J.A. and Sawa, T. 1979. Effect of low light and darkness on structural transformation in plastids of the Rhodophyta. *Phycologia.* **18**: 1-125.

Waaland, J.R., Waaland, S.D. and Bates, G. 1974. Chloroplast structure and pigment composition in the red alga *Griffithsia pacifica*: regulation by light intensity. *J. Phycol.* **10**: 193 – 198.

Yaphe, W. and Arsenault, G.P. 1965. Improved resorcinol reagent for the determination of fructose and of 3,6 anhydrogalactose in polysaccharides. *Anal.Biochem.* **13**: 143–148.

Chapter 13

SYMBIOTIC ASSOCIATION OF N_2 FIXING CYANOBACTERIUM *ANABAENA AZOLLAE* IN AQUATIC WATER FERN *AZOLLA*

S. Kannaiyan and K. Kumar

Abstract

Azolla is a small sized free-floating water fern with an endosymbiont *Anabaena azollae* lying in its leaf cavities. This association has a great potential as a biofertilizer in paddy fields. Ultrastracture, morphology and growth pattern of *Azolla*, and also the differentiation of the cyanobiont have been discussed. Like free living *Anabaena* 7120, the genes *nif*H, *nif*D and *nif*K are transcribed from a single *nif* HDK operon in the endosymbiont. Immobilization of the algal symbiont in solid matrices has shown certain advantages. Cyanobacterial cells (*A. azollae*) immobilized in polymer foam exhibited an increase in heterocyst frequency. Efforts have been made to review various aspects of this association (*Azolla – Anabaena azollae*).

Key words: *Anabaena azollae, Azolla,* cyanobacterium, symbiosis, nitrogen fixation.

Introduction

Nitrogen is an important element which is highly essential for crop growth and crop productivity. Atmosphere contains about 80 per cent nitrogen by volume in a free state but unfortunately plants cannot assimilate this and they essentially require a combined form such as nitrate or ammonia. However, tremendous increase in the cost of nitrogenous fertilizers generally limit the use of fertilizer nitrogen by the farmers.

In India, rice crop contributes around 45 per cent of the total cereal production which holds the key to sustain food sufficiency. The rice production is to be increased by more than 60 per cent of the current level by 2020 AD to feed the millions of people (Lampe, 1998). The present production and productivity is reported to be 82.1 million metric tonnes and 1851 kg/ha (1997-98) and the estimate for 2006-07 is 100 million metric tonnes and 2,454 kg/ha respectively (Siddiq, 1999). The widespread use of chemical nitrogen fertilizer had led to large increase in rice production. However, chemical N fertilizers cause pollution hazards, land degradation and also from one of the most expensive inputs for rice production (Groswell and Godwin, 1984). Therefore, a new strategy for increasing rice production particularly in developing countries should take into account more realistic, inexpensive and pragmatic programmes to utilise the biological fertilizers which will not only increase the rice crop productivity but also improve the long term soil fertility in sustainable rice farming.

The biological origin of nitrogen is estimated to be around 175 million metric tonnes per year with approximately 90 million metric tonnes fixed in soil. *Azolla - Anabaena* symbiotic complex, which is capable of growing under the canopy of rice, has been identified as prospective N biofertilizer for wetland rice culture both in the tropics and in temperate regions, for increasing rice yields (Kannaiyan, 1990). The nitrogen fixing cyanobacterium *Anabaena azollae* Stras. as a symbiont inhabits the leaf cavities of the aquatic fern *Azolla* Lam. and these two partners maintain a stable relationship based on exchange of nitrogen (fixed by *Anabaena*) and carbohydrates (provided by *Azolla*).

Azolla is one of the most important green manures in China, where currently more than 1.34 million ha of rice fields are cultivated with *Azolla*. Maximum estimated inputs during exponential growth of *Azolla* range from 335-670 kg N/ha/year and the increase in grain yield due to *Azolla* nitrogen incorporation is around 14-20%.

History of *Azolla* cultivation

The aquatic fern *Azolla* is native of Asia, Africa and America. The genus was established by Lamarck in 1983 (Svenson, 1944) and the name *Azolla* is derived from the two greek words, *azo* means 'to dry' and *allyo* means 'to kill' implying that the plant dies under dry conditions (Moore, 1969). Rice growing areas in South East Asia and other third world countries have recently been evincing increasing interest in the use of the symbiotic N_2 fixing water fern *Azolla* either as an alternate nitrogen source or as a supplement to commercial nitrogen fertilizer (Watanabe, 1982). *Azolla* is used as a biofertilizer for wetland rice and it is known to contribute 40-60 kg N/ha per year for rice crop (Kannaiyan, 1989).

The utilization of *Azolla* as dual crop with wetland rice is gaining importance in The Philippines (Watanabe, 1990), Srilanka (Kulasooriya, 1985) and India (Kannaiyan, 1992). The quick decomposition of *Azolla* in soil and efficient availability of its nitrogen to rice have upgraded *Azolla* as a biofertilizer for rice. The applied *Azolla* in tropical rice soils mineralizes very rapidly and its nitrogen is available to rice crop in a short period (Kannaiyan, 1990).

The spontaneous occurrence of *Azolla* in paddy fields probably contributed to its eventual recognition and cultivation as green manure in Asia. The exact period when cultivation began is lost in hisotry, but is reported to extend back to the 11th century in Vietnam (Dao and Tran, 1979) and at least to the Ming dynasty (1368-1644 AD) in China.

Taxonomy of *Azolla*

Historically, *Azolla* has been placed in the family Salviniaceae with *Salvinia*, another heterosporous fern. However, major differences between the two, particularly with respect to reproductive structures have resulted in the recategorization of *Azolla* into the monotypic family Azollaceae (Moore, 1969; Konar and Kapoor, 1974; Martin, 1976; Fowler and Stennet – Willson, 1978; Lumpkin and Plucknett, 1982).

On the basis of reproductive and morphological features, seven distinct species of *Azolla* have been recognized. They are grouped into two sections, *Azolla* and *Rhizosperma*. The primary basis for this division is sporocarp morphology. Species that belong to section *Azolla* exhibit three floats on the megaspore apparatus, whereas species in section *Rhizosperma* have nine (Moore, 1969). The section *Azolla* comprises five species which include *A. caroliniana*, *A. filiculoides*, *A. mexicana*, *A. microphylla* and *A. rubra*. The section

Rhizosperma includes *A. pinnata* and *A. nilotica*. Based on differences in leaf branching morphology, *A. pinnata* is further divided into the varieties *A. pinnata* var. *pinnata* and *A. pinnata* var. *imbricata* (Sweet and Hills, 1971). Reynaud and Franche (1986) reported a new variety of *A. pinnata viz., A. pinnata* var *africana* from Africa.

However, based on their chromosome numbers, Stergianou and Fowler (1990) grouped *A. pinnata* and *A. nilotica* under *Rhizosperma* as they had n=26. This finding was supported by the further evidence from the genomic DNA product studies which indicated that the dominant symbionts in the section *Azolla* and those in *A. pinnata* belong to divergent evolution lines (Plazinski *et. al.*, 1990).

Morphology of *Azolla*

The *Azolla* sporophyte, which normally floats freely on the water surface and multiply vegetatively, are usually 1-3 cm in diameter. Some species, however, can become appreciably larger. It mainly consists of pinnately branched floating stem in the simple roots which occur at some nodes and hangs down into the water. The stem and branches are covered with small, alternate, overlapping leaves. Each leaf is divided into a thick chlorophyllous dorsal lobe and a thin achlorophyllous ventral lobe. The heterocyst forming blue-green alga (cyanobacterium) *Anabaena azollae* is found enclosed within a special cavity in the ventral side of the chlorophyllous dorsal lobe (Moore, 1969). The specialized cavity is extracellularly located in the proximal half of the ventral side of the dorsal leaf lobe. Numerous heterocystous cyanobacterial filaments line the periphery of the cavity wall. In younger leaves, filaments are usually dispersed throughout the cavity. The development of leaf cavities and hair cells proceeds normally in the absence of the endophyte (Walmsley *et. al.*, 1973; Duckett *et. al.*, 1975; Ashton and Walmsley, 1976; Uheda, 1980). Algal packets are limited by an envelope (10 to 20 nm thick) that is rich in fucose and insensitive to digestion with cellulases, proteases, pectinases or lipases (Peters *et. al.*, 1978; Uheda, 1980). When algal packets are ruptured, cyanobacteria and their associated microflora are released in a mucilagenous matrix (Peters *et. al.*, 1978) that is rich in amino acids and ammonia (Newton and Cavins, 1976).

Growth of *Azolla*

Azolla has rapid growth rate and it doubles its weight in a mere 2-3 days. Due to the fast relative growth rate, the biomass production is heavy in a given time and hence the biomass produced is exploited as biofertilizer for rice. The environmental conditions largely influence the growth and multiplication of *Azolla.* Growth rate of *Azolla* is expressed in terms of daily relative growth rate.

Under optimum conditions, doubling time of *Azolla* falls between 2 and 3 days. Watanabe *et. al.*(1977) reported that doubling time for *A. pinnata* is 3 days under the most favourable conditions. Talley *et. al.* (1977) observed a shoot doubling time of 2.8 days in *A. mexicana* and about twice in *A. filiculoides*. Kannaiyan (1988) observed a doubling time of 3.1 days in *A. maxicana* and *A. filiculoides*. The doubling time of *A. pinnata* GP-BT, *A. pinnata* SK-1 and *A. pinnata* BR-M was recorded as 3.71, 3.85 and 3.68 days respectively (Kannaiyan, 1990). Plant density markedly influences the growth rate of *Azolla*. When undistributed exponential growth occurred, a growth rate of 0.230 g/g/ day was measured and decreased to 0.032 g/g/day under severe crowding (Becking, 1979). Kulasooriya (1985) recorded 8 fold increase in biomass in 2 weeks and 50 fold increase in 22 days giving a doubling time of 4.8 and 3.9 days respectively for *A. pinnata*.

Nutrient contents of *Azolla*

The dry weight of *Azolla* ranges from 4.8 to 7.7 per cent (Peters and Calvert, 1983). The dry weight nitrogen content of *Azolla* has been reported to be as high as 6.5 per cent for aseptic cultures (Peters *et. al.*, 1980) but will probably be closer to 3.5 to 4.0 per cent. The C: N ratios of 7:1 to 18:1 have been reported for *Azolla* (Peters, 1977; Peters *et. al.*, 1980). *Azolla* recorded 13-30 per cent of dry weight of protein (Tran and Dao, 1973 ; Buckingham *et. al.*, 1978). Other constituents are given in Table 13.1 (Buckingham *et. al.*, 1998). Gopalaswamy and Kannaiyan (1998) recorded increase in total nitrogen, phosphorus and potassium content due to incorporation of *Azolla* hybrids in sodic soil. Kumar and Kannaiyan (2000) analyzed the major and micronutrient contents of the mutants and wild parents of *Azolla* and observed that the maximum nitrogen, potassium, manganese and zinc contents were recorded in *A. filiculoides* KK-SK-C5, whereas *A. microphylla* KK-SK-E20 accumulated higher phosphorus, iron and copper content compared to other mutants and wild parents.

Performance of different species of *Azolla*

Wide variation in the performance of different strains in *A. pinnata* was noticed and a strain collected from Anamalai area was found to be superior in terms of N_2 fixation and biomass production and also adapted well under rice fields conditions (Kannaiyan, 1986). Kannaiyan and Somporn (1988) found that *A. pinnata* GP-B7 and *A. pinnata* BR-M have shown tolerance to high temperature at 38 ±1°C / 25±1°C day/night regime. Higher nitrogen fixing, fast growing and better biomass producing strains of *A. microphylla* were located through field screening of different *Azolla* species for adaptability (Kannaiyan, 1988). *A. microphylla* is known to adapt well under rice field conditions and also produces sporocarps abundantly which forms an important trait for its survival during rice fallow summer season and subsequent continuation of its life in rice growing seasons (Kannaiyan, 1989). *A caroliniana* is also fixing higher amount of nitrogen coupled with pests and disease resistance under field condition (Kannaiyan, 1985).

Table 13.1. Chemical composition of *Azolla*

Constituents	*Composition on dry wt basis (%)*
Ash	10.5
Crude fibre	9.1
Crude fat	3.0-3.3
Starch	6.5
Soluble sugar	3.5
Crude protein	24.0-30.0
Nitrogen	4.0-5.0
Phosphorus	0.5-0.9
Potassium	2.0-4.5
Calcium	0.4-1.0
Magnesium	0.5-0.9
Manganese	0.11-0.16
Iron	0.06-0.26
Chlorophyll	0.34-0.55

Algal symbiont of *Azolla*

The water fern *Azolla* contains a symbiotic heterocystous cyanobacterium *Anabaena azollae,* within specialized leaf cavities. The symbiotic alga was named as *Anabaena azollae* by Strasburger (1973) and it is the only species found in symbiotic association with

Azolla. The alga occurs as an unbranched filament consisting of vegetative cells, heterocysts and akinetes or sproes (Moore, 1969). The algal symbiont belongs to the phylum Cyanophyta, order Nostocales and the family Nostocaceae. It has been found that the *Anabaena* filaments in young *Azolla* seedlings originated from the algal cells/akinetes in meagasporocarps (Watanabe, 1978). The species has trichomes (threads) composed of bead like or barrel shaped cells without a sheath (Lumpkin and Plucknett, 1980).

The actively dividing filaments of algal symbionts occur among the apical meristem and the developing leaf primordia of *Azolla* - *Anabaena azollae* has three types of cells *viz*., vegetative cells – primary site of photosynthesis, heterocysts – site of nitrogen fixation, and akinetes – thick walled resting spores formed from vegetative cells (Lumpkin and Plucknett, 1980). The frequency of heterocysts is higher in the symbiotic alga than in the free-living heterocystous cyanobacteria. The average heterocyst frequency has been reported as 15-20% (Becking, 1976), 23-30% (Singh, 1977) and 12-16% (Kannaiyan, 1991).

The size of the vegetative cells are 6-8μ broad (Singh, 1979). The growth and development of *Azolla* and *Anabaena* are synchronous (Hill, 1977). The growth of the algal symbiont (*A. azollae*) under artificial medium is poor (Kannaiyan, 1986). The difficulty of culturing *A. azollae* in isolation has been a major obstacle in elucidating the symbiosis.

Azolla-Anabaena symbiosis

The *Azolla* – *Anabaena* symbiosis is a mutualistic association between the eukaryotic water fern *Azolla* and the prokaryotic nitrogen fixing cyanobacterium *Anabaena*. This association has gained attention in recent years because it may represent a viable alternative to chemical nitrogen fertilizers. The *Azolla* sporophytes possess leaves comprising of dorsal and a ventral lobe. On the ventral side of the dorsal lobes, ellipsiodal cavities occur which harbour the cyanobiont. The cyanobiont is embedded in mucilage and enclosed by an envelope in these cavities. The cavities do not have direct contact with the *Azolla* vascular system and the metabolic exchange between the symbionts takes place through simple and branched hair in these cavities (Calvert *et. al.*, 1985). Special structures in cavities, referred as simple hair, have been suggested to be involved in exchange of photosynthates.

Growth and development of the partners is highly coordinated; an ontogenetic sequence of the fern and cyanobacterium is expressed along every stem axis (Hill, 1975; Calvert and Peters, 1981; Peters and Calvert, 1983). Partitioning of filaments of the apical *Anabaena* colony into the forming leaf cavities is facilitated by a special epidermal trichome, called the primary branched hair (PBH), which originates in the axil of each leaf primordia and rapidly differentiates a stalk cell and a palmate array of elongated terminal cells which protrude into the apical *Anabaena* colony (Calvert and Peters, 1981; Peters and Calvert, 1983; Calvert *et. al.*, 1985). When *Azolla* sporulates, *Anabaena* filaments becomes associated with the developing sporocarps as well as with the developing leaf cavities. Information pertaining to the specific manner in which the *Anabaena* become associated with the fern's reproductive structures is limited. *Anabaena* has been stated to creep (Campbell, 1983), or slip (Becking, 1987) into developing megasporocarp through a small pore in the tip of the indusium, occupying the cavity formed by the envelopment of the megasporangia by the indusia. According to Becking (1987), the *Anabaena* within this sporocarp cavity multiplies after the indusium closes at the tip. It was likely that the *Anabaena* cells changed into a resting stage, or akinetes, in the later stages of megasporocarp development. In spite of the shared ontogeny of micro and megasporocarps, Becking (1987) concluded that the *Anabaena* is not enclosed within the microsporocarp, but it seemed to involve a specific and distinct group of epidermal trichomes. The ontogeny of the four sporangial trichomes associated with each sporocarp pair appears to be even more rapid than that of leaf cavity trichomes (Calvert *et. al.*, 1985). There is also the possibility that the sporangial hairs release other specific, non-nutrient compound(s), which effectively 'attract' growth of the *Anabaena* filaments towards the developing sporocarps in order to accomplish inoculation. The PBH physically displaces filaments from the apical *Anabaena* colony directly into the developing cavity of the dorsal leaf lobe (Peters and Calvert, 1983). As with the other cavity trichomes (Calvert and Peters, 1981), the PBH has been suggested to facilitate post-symbiont metabolite interchange and remains associated with the *Anabaena* filaments throughout the functional life of the leaf (Calvert *et. al.*, 1985).

The *Anabaena* within the microsporocarps is lost to the environment and play no role in maintaining the continuity of the symbiosis (Calvert *et. al.*, 1983; Peters and Calvert, 1983). In contrast, the *Anabaena* cells within the megasporocarps are essential to the continuity of the symbiosis. They are precisely packaged, occurring inside the indusium at its tip and above the megaspore (Rao, 1935; Konar and Kapoor, 1974; Ashton and Walmsley, 1976; Calvert *et. al.*, 1983; Herd *et. al.*, 1986; Dunham and Fowler, 1987). Due to the involvement of epidermal trichomes, the events leading to the re-establishment of the symbiosis are appreciably more controlled, complex and sophisticated than noted in the account of the process provided by either Becking (1987) or Lin and Watanabe (1988) and somewhat more so than in the account provided by Dunham and Fowler (1987). In *Azolla*, the age of the megagametophyte, and hence the position of archegonia, at the time of fertilization is critical for attaining successful reinoculation. The first archegonia to arise are positioned directly below the inoculation chamber. If one of these is fertilized, the megaspore apparatus not only protects the embryonic sporophyte as it develops but also confers a path of least resistance leading to the inoculation chamber. The cotyledonary leaf of the developing embryo encompass the *Anabaena*, as the developing sporophyte penetrates into the inoculation chamber. In this respect, it functions in a manner somewhat analogous to that of the indusium during sporocarp development (Perkins and Peters, 1992). Due to the inversion of the apical membrane during megagametophyte development, the cotyledonary hairs and shoot apex quickly contact and penetrate (or rupture) the apical membrane as the vase – like cotyledonary leaf develops around them to engulf the *Anabaena* cells in the inoculation chamber. This brings the cotyledonary leaf hairs and shoot apex into close proximity to, if not direct contact with, the *Anabaena* cells as the modified true leaves develop. These unlobed true leaves, all of whih contain a cavity – like area and at least one multibranched hair, appear to serve in a manner somewhat analogous to the involucre during sporocarp development (Perkins and Peters, 1992) and the forming leaves at the apices of the mature sporophyte. They arch over the meristematic shoot tip, protecting it while furthering the degree of intimacy between it and the *Anabaena*. The germinating akinetes establish the apical *Anabaena* colony from which generative filaments are partitioned into the cavities which form in the dorsal lobe of the developing true leaves. Germination of the akinetes in the inoculation chamber is clearly dependent upon and controlled by, the developing sporophyte. Dunham and Fowler

(1987) simply noted that all leaves except the cotyledonary leaf contain apically derived *Anabaena* and that first two true leaves are entire. Peters and Perkins (1993) postulated that the entrapment of the *Anabaena* cells in the modified leaf cavities disrupts their interaction with the cotyledonary hairs, arresting their development such that they may be cannibolized as a source of nitrogen and other nutrients. These would be assimilated by the cells of the branched hairs and used to support apical growth and differentiation of subsequent true leaves until the *Anabaena* in the true leaf cavities differentiates heterocysts and provide the shoot apex with fixed nitrogen. The latter has been shown to occur in the sporophyte symbiosis (Kaplan and Peters, 1981, Peters *et. al.*, 1985).

Differentiation of cyanobiont heterocysts is similar to that observed in other free-living and symbiotic cyanobacteria. However, in the *Azolla pinnata* cyanobiont, an elaborate honeycomb configuration of unknown function is formed (Braun - Howland *et. al.*, 1988). Nitrogenase protein is detectable in the intermediate stage of heterocyst development, but is not present in akinetes or vegetative cells. Further, development of polar nodules and cyanophycin granules in heterocysts indicate that the cyanobionts are not nitrogen starved. Subcellular localization of nitrogenase was difficult to determine, with nitrogenase being uniformly distributed within the heterocysts. Observations of older heterocysts, which are highly vesiculated, or heterocysts with tightly packed membranes, both contain nitrogenase protein which appears to be associated with membranes (Braun - Howland *et. al.*, 1988).

It has been suggested that one method for protecting nitrogenase from oxygen damage is the inactivation of PS II (Tel-Or and Stewart, 1977). However, molecular studies have indicated that *psba* gene transcripts, which encode the 32 - kDa protein of PS II, are seven times higher in *Azolla caroliniana* cyanobionts than in free-living *Anabaena azollae* isolated from *A. caroliniana* (Nierzwicki - Bauer and Haselkorn, 1986). This results cannot be explained based on difference in cell types and numbers. Furthermore, while the reported function of 32 - kDa protein involves PS II quinone - mediated electron transfer, the high transcript levels have suggested that there may be additional functions for this protein in the endosymbiont.

Immunolocalization studies were performed to determine whether these transcripts were being actively translated into protein (Braun - Howland and Nierzwicki - Bauer, 1990). Immunologically recognizable 32 - kDa protein was localizaed on membranes of *Azolla* chloroplasts, vegetative cyanobacterial cells, akinetes and heterocysts of all ages. Using serial sections, it was shown that the 32- kDa protein was actually present in the identical heterocysts that also contained nitrogenase. This suggests the possibility that the 32-kDa protein is serving some other, as yet to be elucidated role, in heterocysts of the cyanobiont.

In free - living filamentous cyanobacteria, hetercysts comprise approximately 8-10% of the total cells, however, in *Azolla*, heterocyst frequencies increase to approximately 20% in all but the youngest leaf cavities. However, nitrogenance activity, as measured by acetylene reduction, increases to a peak around leaves 6 or 7 (from the apical meristem), and then steadily decreases and becomes undetectable in leaves older than 15. Immunolocalization studies have demonstrated the presence of nitrogenase in the heterocysts present within older leaf cavities (Lindblad *et. al.*, 1991). The reason that heterocysts present in older leaf cavities are not nitrogen fixing remains unknown.

Anabaena fixes N_2 and meets the total nitrogen requirement of the symbiotic system. In contrast to the free-living cyanobacteria where combined nitrogen represses nitrogenase, the cyanobiont in *Azolla* continues to fix N_2 at 50-60% of the normal rate, in the presence of combined N_2 (Brarun - Howland and Nierwizcki - Bauer, 1990). *Azolla* - *Anabaena* complex fixes 3-7 kgN/ha/day in flooded fields (Singh, 1994) which is equivalent to the forage leguminous crops. So it can be asserted that with its doubling time of 2-3 days, *Azolla* exhibits a biological potential as a green manure.

Arrangement of structural nitrogenase (*nif*) genes in *Anabaena azollae*

Franche and Cohen - Bazire (1985, 1987) have investigated on the physical arrangement of structural *nif* genes of nine symbiotic *Anabaena azollae* spp. It was determined that the arrangement of *Anabaena azollae* and *Anabaena* 7120 *nif*HDK genes was quite similar; *nif*H and *nif*D were linked, and *nif*K was some distance away from *nif*D. It was also demonstrated that the arrangement of *nif*HDK genes was strongly conserved among *Azolla* species, regardless of their geographical origin. It is important to note that slight differences in hybridization banding patterns make it possible to use DNA hybridization techniques for differentiation among cyanobacterial isolates from different *Azolla* species, even through the general arrangement of the *nif* genes was conserved. Meeks *et. al.* (1988) have discussed the organization of *nif* genes in symbiotic cyanobacteria having association with *Azolla* and *Anthoceros*.

Transcript levels of key enzymes in *Anabaena azollae*

Nierzwicki – Bauer and Haselkorn (1986) carried out a molecular study on *Anabaena azollae* that was concerned with the quantification of transcript levels of enzymes involved either in N_2 fixation or in the heterocyst differentation processes. More specifically, using cloned genes from *Anabaena* 7120 for glutamine synthetase (GS), ribulose – 1, 5 – bisphosphate carboxylase/ oxygenase (RUBISCO), nitrogenase, and the 32 – kDa protein of photosystem II, mRNA levels of the corresponding genes in the *Anabaena* endosymbiont were determined by northern blot and dot blot hybridization.

Glutamine synthetase (GS)

The *Anabaena gln*A gene was of particular interest because (GS) is very important in the *Azolla – Anabaena* association. In symbiotically associated *Anabaena* GS activities are very low, as is the level of *Anabaena* GS antigen. While a vast amount of information on the regulation of GS activity in a number of different systems was available, comparatively little was known about the regulation of GS synthesis in endosymbiotic *Anabaena*. It was reported that the GS protein levels in the *Anabaena* symbiont were only 10% of those present in the free-living form. On the other hand, Peters *et. al.*(1980) suggested that regulation of GS in the association is at the level of activity rather than synthesis. In the study it was shown that the GS mRNA transcript levels were ten-fold lower in symbiotic versus free – living *Anabaena*. This result suggests that the host fern regulates transcript of the *Anabaena gln*A. This, of course, makes sense because the fern is prepared to utilize the ammonia given up by *Anabaena*.

In enteric bacteria, *gln*A transcription is regulated by the products of two genes, *ntr*A and *ntr*C, that respond to the ratio of glutamine to alpha – ketoglutarate. A high ratio leads to a repression of *gln*A. A different mechanism would appear to be called for in the *Azolla – Anabaena* association. It would be futile for the fern to pump glutamine into the endosymbiont to shut down *gln*A because the glutamine would simply be metabolized.

RuBP carboxylase (RUBISCO)

Transcript levels for RUBISCO in endosymbiotic and free – living *Anabaena* were also determined. RUBISCO is the first enzyme required in the Calvin pathway for the fixation of atmospheric CO_2 Like carboxylase of higher plants, the enzyme from *Anabaena* consists of eight large and eight small subunits. Using a carboxylase – specific probe, Northern dot blot hybridization showed that caboxylase transcript were approximately five to seven times less abundant in the symbiont than in free – living *Anabaena*. This very low amount of carboxylase message in the symbiont RNA is consistent with the reduced CO_2 fixation reported for the symbiont *nif* gene transcripts.

In addition to the quantitative results, it was found that the *nif*H, *nif*D and *nif*K genes are transcribed from a single *nif*HDK operon in the endosymbiont, as originally determined for the free – living cyanobacterium *Anabaena* 7120.

Ultrastructural studies

In order to obtain a more complete understanding of structure function relationship in the *Azolla-Anabaena* symbiosis, ultrastructural studies have been conducted on the cyanobacterial and bacterial endosymbionts, within leaf cavities of different ages in the *Azolla caroliniana* association. In the *A. caroliniana* association, three different cyanobacterial cell types are formed; vegetative cells, heterocysts and akinetes. The percentages of these cell types varied with age of cavity. With growth conditions of 26°C under a 16:8 (light:dark) photoperiod with light from fluorescent and incandescent lamps at a photon fluence rate of 60-100 mol. m^{-2} s^{-1}, vegetative cells made up approximately 91% of the total cell population in the youngest leaves (1-3). The percentage of vegetative cells steadily declined in older leaves, to only 50% of the total population by leaf 15. In contrast, akinetes, which did not appear until leaf 4 where they made up 4% of the cell population, continually increased in numbers so that by leaf 15 they made up approximately 30% of the cell population. Heterocyst frequency also increased by 3% and by leaf 15 they had increased the frequency to 33%.

Vegetative cells

Cyanobacterial filaments associated with the stem apex of *Azolla* were actively dividing, undifferentiated and lacked nitrogenase activity. Many of the vegetative cells present within the apical meristem constained large cyanophycin granules. The ultrastructural characteristics of vegetative cells within very young leaf cavities (leaves 1-3) were relatively consistent. These vegetative cells were very uniform in size and shape and usually contained numerous caboxysomes. The thylakoid membranes within these cells were often, peripherally located. With increasing leaf age, vegetative cells varied with respect to cell size and shape, thylakoid membrane abundance and arrangement, numbers of inclusions, and glycogen content.

Heterocysts

As leaf cavity closure commensed, vegetative cells, become enlarged and divided more slowly with many cells differentiating into heterocysts. The process of heterocyst differentiation from vegetative cells, involved a sequence of ultrastructural changes. The first observable changes were external. The cell become enlarged, a loosely organized layer of envelope material was layed down outside of the cell wall, and a "neck" began to form. As heterocyst development continued, the carboxysomes and cyanophycin granules gradually disappeared. Membranes become more diffuse in the center of cells, and "honeycomb" membrane lattuces were formed adjacent to the polar plugs, In addition to cyanophycin plugs, mature heterocysts of *A. caroliniana* contained small, spherical granules of varying electron density.

Regardless of leaf age, the sequence of ultrastructural changes during heterocyst differentiation was the same. However, the size and shape of heterocysts varied considerably, even within a single cavity. Degree of membrane vesiculation within heterocysts was dependent upon age, with increased membrane vesiculation in older heterocysts.

Akinentes

Of particular interest in the *A. caroliniana* association, was the differentiation of cyanobacterial akinetes. The ultrastructural characteristics that distinguish akinetes from vegetative cells in a number of free-living cyanobacterial species, were also observed in the symbiotic cyanobionts, a thick multilayered external envelope, large glycogen accumulation and large amounts of cyanophycin. Germinating akinetes were occasionally observed in leaves 13 and older. Ultrastructural characteristics of these akinetes indicate that endosymbiotic *Nostoc* is contained within leaf cavities of *A.*

caroliniana. However, this does not exclude the possibility that endosymbiotic *Anabaena* is also present in this association.

The function of akinetes in *Azolla* leaf cavities remains unknown. However, cyanobacterial cells resembling akinetes constitute the cyanobacterial inoculum in sporocarps of at least some *Azolla* species.

Physiological studies

In an attempt to elucidate the contributions made by either partner of the symbiosis, results from studies on the intact association have often been compared to those obtained using endophyte-free plants and freshly isolated cyanobacteria. It is unlikely that the physiological characteristics of isolated *Anabaena azollae* accurately reflect the *in situ* physiological characteristics of the organism. The physiological capabilities of the cyanobacteria have also been shown to change with leaf age. Thus, it is apparent that many basic questions concerning the ontogenic physiological changes which are imposed on *Anabaena* by the symbiotic state remain unanswered.

Separation methods

Because *Anabaena* – free *Azolla* occur infrequently in nature (Moore, 1969; Hill, 1977), several methods have been employed to generate *Anabaena* –free cultures for use in laboratory studies. *Azolla* fronds have reportedly been freed of cyanobacteria using methods such as antibiotic treatment (Ashton and Walmsley, 1976; Beacking, 1979; Peters and Mayne, 1974) changes in light intensity (Hill, 1975, 1977), high temperature (Tung and Watanabe, 1983), cold treatment (Moore, 1969), dilution of culture medium (Moore, 1969), nitrogen starvation and treatment with hypochlorite (Moore, 1969). Endophyte free culture of both *A. caroliniana* and *A. filiculoides* have been obtained by regenerating new plants from apical tips which had been excised from surface sterilized plants (Duckett *et. al.*, 1975). It was reported that endophyte – free cultures can be generated by germinating fertile megasporocarps from which the apical cap has been removed (Lin *et. al.*, 1988).

In general, the isolation and growth of cyanobacteria from *Azolla* sp. is achieved by homogenization or microdissection of either surface sterilized *Azolla* fronds or isolated algal packets. Numerous *Azolla* fronds are gently squashed with a rolling pin to isolate large quantities of *Anabaena azollae*. After the cyanobacteria are liberated, they are separated from the plant material by filtration and centrifugation. With this gentle roller technique (Peters and Mayne, 1974), it is possible to obtain nearly pure preparations of cyanobacteria for use in physiological and m olecular studies. Samal and Kannaiyan (1992) used nitrate – free medium supplemented with growth regulators such as 2,4-D, NAA and IBA to isolate *Anabaena azollae.* Two *A. azollae* isolates grew adequately in the N-free Allen and Arnon's medium. Incorporation of biotin and pantothenic acid stimulated the growth and heterocyst frequency of one of the isolates. This isolate also exhibited higher nitrogenase activity and excreted a higher amount of ammonia.

Photosynthetic metabolism

Sucrose has been identified as the major product of photosynthesis in both symbiotic and endophyte – free *Azolla.* However, when CO_2 was added to *Anabaena azollae* that had been isolated from leaf cavities, C-sucrose was not detected as a major product of cyanobacterial photosynthesis. It has been shown that if the intact association was pulsed with CO_2 and the cyanobacteria susequently isolated, then a time-dependent accumulation of C-sucrose occurred in the endophyte (Peters *et. al.*, 1985). Evident for the cross-feeding of carbon compounds has also been obtained from other experiments (Peters, 1975; Peters and Mayne, 1974). It is believed that cyanobacterial CO_2 fixation occurs primarily within vegetative cells located at the apical meristem of *Azolla* (Hill, 1977) and that *Anabaena* changes to a mixotrophic (Peters, 1975) or photoheterotrophic (Peters and Mayne, 1974) type of metabolism in older leaf cavities. Photosynthesis is the ultimate source of all the ATP and reductant required for the conversion of nitrogen to ammonia in *Azolla-Anabaena* symbiosis. Simultaneous measurements of photosynthesis, respiration and acetylene reduction in *Azolla imbricata* demonstrated the direct utilization of photosynthetically trapped solar energy in nitrogen fixation (Shi *et. al.*, 1981). The strong interaction between photosynthesis and N_2 fixation has also been demonstrated by determining the action spectra for nitrogenase catalyzed acetylene reduction (Kumar and Kannaiyan, 1999; Tyagi *et. al.*, 1981).

Photorespiration

The process of photorespiration has been studied in the *A. caroliniana* symbiosis (Ray *et. al.*, 1979) when either the intact association or endophyte free plants were assayed under atmospheric conditions. The CO_2 compensation point was approximately 40 µl/l. This high value is similar to those reported for the plants which utilize the Calvin cycle. An increase in the concentration of O_2 from 2 to 20% resulted in a 40% inhibition of the rate of CO_2 fixation. These results indicated that *Azolla* carries out typical photorespiratory metabolism.

In contrast, the cyanobacterial CO_2 compensation point was less than 5 µl/l. This compensation point was insensitive to the concentration of O_2 and rates of CO_2 fixation and O_2 evolution were the same under aerobic or microaerobic conditions. Therefore, as in other cyanobacteria (Bidwell, 1977; Lloyd *et. al.*,1977), photorespiratory metabolism is not detectable in *Anabaena azollae* (Ray *et. al.*, 1979).

Photochemical activity

Photochemical activity of both photosystems I and II has been demonstrated spectrophotometrically in chloroplasts from symbiotic and endophyte – free *A. caroliniana* and cell –free extracts of the endosymbiont. Estimates of PSI and PS II activities were obtained by estimating the chlorophyll to P700 ratio and delayed light emission, respectively. The results of delayed light emission studies indicated that PS II activity was approximately ten times greater in vegetative cells than in heterocysts. These results are in agreement with the observation that heterocysts lack PS II reaction centers (Haselkorn, 1978).

Pigments

The cynobacterial pigments used for harvesting light energy for photosynthesis include chlorophyll *a*, phycobiliproteins and carotenoids (Peters and Mayne, 1974; Ray *et. al.*, 1978; Tyagi *et. al.*, 1981). The complement of pigment produced by *Azolla* included chlorophyll *a*, *b* and carotenoids (Ray *et. al.*, 1978). The cyanobacterial phycobiliproteins, C-phycocyanin, allophycocyanin, and phycoerythrocyanin have been isolated and characterized from the *A caroliniana* symbiosis (Tyagi *et. al.*, 1980). Anthocyanins including luteolinidin-5 glucoside (Holst, 1977; Pieterse *et. al.*, 1977) are produced by *Azolla* (expect *A. nilotica*) (Lumpkin, 1981) under conditions of high light intensity (Moore, 1969; Martin, 1976; Holst and Yopp, 1979; Watanabe, 1982) or low temperature (Pieterse *et. al.*,

1977). The cyanobacterial phycobiliprotein content is also reportedly variable with culture conditions (Peters *et. al.*, 1979; Ray *et. al.*, 1979).

However, it has been estimated that approximately 12 to 22% of the total protein and 11 to 21% of the total chlorophyll in the symbiosis is contributed by the phycobiont (Ray *et. al.*, 1978). The actual contribution by the endophyte is expected to be less than this percentage, however, because upto 50% of the cyanobacterial cell population may be either heterocysts or akinetes (Peters, 1975) which are incapable of photosynthesis (Ray *et. al.*, 1979).

Phycobiliproteins in heterocysts

In free- living heterocystous cyanobacteria phycobiliproteins are primarily found in vegetative cells where they reportedly serve as the exclusive light harvesting pigments for PSII (Haselkorn, 1978). In contrast, epifluorescence microscopy studies on the endophytes from *A. caroliniana* (Tyagi *et. al.*, 1981) and *A. filiculoides* (Becking and Donze, 1981) have suggested that phycocyanin contents are high within heterocysts of *Anabaena azollae.*

It has been suggested that the occurrence of phycobiliproteins in heterocysts may result in a less favourable environment for nitrogenase activity (Hill, 1975). In contrast it has been argued that the environment retention of phycobiliproteins by heterocyst is a logical adaptation for *Anabaena azollae* which is exposed only to light that has been filtered through a green leaf (Kaplan *et. al.*, 1986).

Immunofluorescence studies

Immunofluorescence studies have shown that cyanobionts of various *Azolla* species are antigenically similar but they are quite distinct from free-living strains including those isolated from *Azolla*. Antibodies raised against *Anabaena azollae*, freshly isolated from *A. caroliniana* leaf cavities, cross- reacted strongly with cyanobacteria freshly isolated from *A. pinnata* (Gates *et. al.*,1980). The antibodies against the endophytes of all 4 species reacted strongly with freshly isolated endophytes of 6 *Azolla* species from 32 samples collected throughout the world (Ladha and Watanabe, 1982). Antibodies raised against freshly isolated phycobionts reacted weakly or not at all with free-living isolates of *Anabaena azollae*. In addition, these antibodies reacted poorly with numerous strains of *Anabaena* (Ladha and Watanabe, 1982), two strains of *Nostoc* and one each of

Cylindrospermum (Gates *et. al., 1980), Gloeotrichia* and *Oscillatoria* (Ladha and Watanabe, 1982). Conversely, antibodies specific for the free-living isolate of Newton and Herman (Newton and Herman, 1979) failed to react with *Anabaena variabils (Anabaena flos-aquae)*, and a free-living strain that was also isolated from *A. caroliniana* (Gates *et. al.*, 1980; Ladha and Watanabe, 1982). The results of these studies were substantiated by cross-absorption tests (Gates *et. al.*, 1980; Ladha and Watanabe, 1982). It has also been reported that a free-living culture of *Anabaena azollae*, that was originally isolated from *A. filiculoides*, is immunologically distinct from its symbiotic counterpart (Arad *et. al.*, 1985).

It has been suggested that isolation and prolonged subculturing of *Anabaena azollae* results in either alteration of the surface antigens and morphology of the free-living isolates (Ladha and Watanabe, 1982; Arad *et. al.*, 1985) or selection for a mutant (Gates *et. al.*, 1980). The results from restriction mapping studies (Nierzwicki-Bauer and Haselkorn, 1986) have indicated that Newton's isolate has different DNA hybridization banding patterns than those of *Anabaena azollae* freshly isolated from leaf cavities.

Lectin studies

Lectins have been suggested to play a role in a recognition process between *Azolla* and *Anabaena*. Lectins of *Azolla* specifically recognize their endosymbiont (Sara *et. al.*, 1987). *Azolla pinnata* lectins agglutinated cyanobiont filaments from *A. pinnata* but not from *A. filiculoides*, and *A. filiculoides* lectin agglutinated cyanobiont filaments from *A. filiculoides* and not *A. pinnata*. The report provided strong suggestive evidence for specificity in the N-fixing symbiosis between aquatic fern and cyanobacterium. Alternatively, it has been suggested that this partnership may simply result from the entrapment of cyanobacterial cells within mucilage, rather than from the specific interactions (Robins *et. al.*, 1986). Nevertheless, conventional haemagglutination techniques have been used to demonstrate the presence of lectins in cyanobacteria freshly isolated from three *Azolla* species. Although it has been suggested that lectins are associated with the cyanobacterial partner (Kobiler *et. al.*, 1981, 1982), the majority of reports have indicated that haemagglutinating activity is specifically assoicated with the fern (Mellor *et. al.*, 1981; Ladha and Watanabe, 1984; McCowen *et. al.*, 1987). Notably, the amount of lectin produced appears to be dependent upon growth conditions. In *A. caroliniana*, the lectin titer decreased when plants were grown on media containing nitrate (Mellor *et. al.*, 1981). McCowen *et. al.* (1987) demonstrated that haemagglutinating activity could be inhibited by preadsorption of the *A. pinnata* lectin to cyanobacteria freshly isolated from *A. pinnata*, but not to cyanobacteria isolated from *A. filiculoides*. Likewise, the haemagglutinating activity of the *A. filiculoides* lectin could only be inhibited by preadsorption to cyanobacteria freshly isolated from *A. filiculoides*.

Genetic diversity of *Azolla*

Unfortunately, most accessions of *Azolla* never become fertile under maintenance conditions of germplasm collections. The identity of any accession may also become suspect after entering a germplasm collection because of cross-contamination with other accessions. Recent applications of molecular technology, including isozyme characterization, Random Amplified Polymorphic DNA (RADP) and Restriction Fragment Length Polymorphism (RFLP) analysis suggest some arrangements in the *Azolla* taxonomy particularly in the section *Azolla*.

Zimmerman *et. al.* (1989) fingerprinted and classified the *Azolla* accessions from the germplasm collections of the International Rice Research Institute and Washington State University by enzyme electrophoresis and leaf trichome morphology. Eskew *et. al.* (1993) have established that DAF produces unique reproducible fingerprints for the intact *Azolla-Anabaena* symbiosis and have defined the contribution of both the micro and macro symbiont to the fingerprints. Coppennole *et. al.* (1993) demonstrated the use of RAPD's to study phylogenetic relationship among accession of the aquatic fern *Azolla* by using random arbitrary primers of 10 nucleotide in length with 70 per cent G + C content. Kim *et. al.* (1997) have generated fingerprints of cyanobionts and its host *Azolla* fingerprints by using specific primers for the chloroplast-encoded intron of the tRNA-leucine (VAA) gene. The random primer 0PK 01 generated maximum of 81 RAPD products among the primers analysed, whereas the primer OPF 12 produced maximum of 26 polymorphic RAPD products which could be recorded as RAPD markers, and these also indicated maximum polymorphism among the *Azolla* cultures tested (Konde, 2000). The similarity index values between *Azolla* cultures indicated that the *A. microphylla* and *A.*

filiculoides cultures are distinctly related, whereas the *Azolla* hybrids recorded intermediate similarity indices.

Genetic diversity of algal symbiont

Using the Southern blot hybridization technique, symbiotic and free-living *Anabaena* DNA were digested with the restriction enzyme *HindIII* and hybridized with the probes. The results of these hybridization studies demonstrated that the free-living *Anabaena* strain isolated from *A. caroliniana* by Newton and Herman is not identical to the major endosymbiont of *A. caroliniana*. These results show that the *quln*A, *psb*A, *rbc*S and *nif*H genes are contained on DNA fragments of different size in free-living and freshly isolated *Anabaena azollae*. These results are consistent with those of the aforementioned study in that a free-living strain of *Anabaena azollae* was shown to be different from its symbiotic counterpart.

Using the Southern blot hybridization technique and restriction fragment length polymorphism (RELP's) it has been shown that the free-living cyanobacterial isolates from *A. mexicana* and *A. pinnata*, are not identical to the major cyanobionts of these associations. These results are similar to those previously obtained using isolates from *A. caroliniana* (Nierzwicki-Bauer and Haselkorn, 1986) and *A. filiculoides* (Franche and Cohen – Bazire, 1985) which differed from their symbiotic counterparts. In addition, Plazinski *et. al.* (1988) have carried out extensive analyses using monoclonal antibodies and RELP's to demonstrate diversification of the cyanobionts of different *Azolla* species.

In the serological surveys, using polyclonal antibodies, the similarities among 32 isolates outweighed the difference (Ladha and Watanabe, 1982). Quantitative difference were demonstrated in antigenic structures between fresh and cultured isolates of *A. azollae* (Gates *et. al.*, 1980). Plazinski *et. al.* (1990) characterized 22 isolates of *A. azollae* derived from 7 *Azolla* species from various geographical and ecological sources by DNA-DNA hybridization. Jackman and Mulligan (1995) have characterized the nitrogen fixation (*nif*) gene cluster from *Anabaena azollae* and demonstrated that the closely related cyanobacteria have highly variable but structured intergenic regions.

Lotti *et. al.* (1996) have analysed the diversity in symbiotic *Nostoc* using DNA methylation and showed that a high specificity between the host plant and the cyanobacterium associations involving *Nostoc*.

Rudi *et. al.* (1997) have devloped a diagnostic system using the DNA sequence polymorphism in the 16S rRNA regions V0 to V8 for individual strains. This is important characterization and identification of cyanobacterial strains in culture collections (*Nostoc/ Anabaena/Microcystis*). Konde (2000) showed that the phylogenetic relationship (dendrogram) clearly distinguished the algal symbionts from the free-living cyanobacteria by showing·three groups. Subhashini *et. al.* (2000 a, b) characterized *Anabaena azollae* isolated from four different *Azolla* cultures based on their RAPD profile to study the variation existing among them. The RAPD profile generated the finger printing pattern characteristic of each strain based on the sequence of the primers used. Common band sharing observed between the strains *A. azollae* – RS-KK-SK-AM and *A. azollae* – RS-KK-SK-RP probably represents maternal inheritance of DNA to the progeny. The polymorphic bands were generated specifically for the isolates *A. azollae* – RS-KK-SK-RP and *A. azollae* – RS-KK-SK-AM with two primers which could be developed as possible markers for these isolates.

A commonly accepted method for identifying unknown bacteria is to use universal 16S rRNA primers to selectively amplify the 16S rRNA molecules from DNA extracted from samples. This yields a large amount of specific product that can be readily subcloned into a vector for sequencing. Once the sequence is obtained, it can be classified using a variety of databases, which provide information about the identity at the group, genus, or species level, of the unknown organisms. This data can also be used to develop 16S rRNA targeted organism specific probes that can be used in confirming that a sequence obtained by PCR is actually represented in the original sample. Using this approach, attempts have been made to identify the endosymbionts present in *Azolla*. DNA was extracted from freshly – isolated cyanobionts using standard methods (Gebhardt and Nierzwicki – Bauer, 1991), or from whole plants (Van Coppenolle *et. al.*, 1993), and then amplified using universal 16S rRNA-PCR primers analysed by agarose gel electrophoresis and positive products subcloned. Using a newly developed cyanobacterial group – specific rRNA target probe, plasmids were screened for the presence of cyanobacterial sequences using a modified method. Clones which contained a cyanobacterial 16S rRNA-like sequence (based on the hybridization results) were further screened by partially sequencing the clone. This allowed for the identification of unique sequence, which were then completely sequenced and phylogenetic analysis performed.

Clones of 16S rRNA products from freshly – isolated cyanobionts from *Azolla caroliniana* were sequenced and manually aligned to the 16S rRNA sequence of *Nostoc (Anabaena)* PCC 7120, using secondary structures considerations (Fig. 13.1). Additional aligned sequences were obtained from the Ribosomal Database Project (RDP) (Maidak *et. al.*, 1996). Phylogenetic analysis was then performed using parsimony. (using PAUP 3.1.1) (Swofford, 1993) and distance (PHYLIP 2.5: (Felsenstein, 1989) methods. In each case, the data was bootstrapped 100 times, and *Escherichia coli* used as an outgroup to root the tree. Additionally, the free-living cyanobacterial isolate, (Newton's isolate) from *A. caroliniana* was sequenced. As shown in Fig. 1, Newton's isolate clearly groups with *Nostoc* (*Anabaena*) PCC 7120, while the major sequences obtained from the freshly – isolated cyanobacteria, form a distinct group, related but not identical to the free-living isolate. These data can be used to develop molecular probes that will be used for *in situ* hybridization studies to determine the abundance, distribution and diversity of the cyanobacteria in the leaf cavities of *Azolla*, and to confirm the presence of free-living isolates in leaf cavities.

Bacterial endosymbiont

In addition to the cyanobionts, eubacteria are present in leaf cavities of *Azolla*. In the *A. caroliniana* association, five ultrastructurally distinct types of bacteria were observed. Bacterial types could be distinguished on the basis of cell shape, cell wall structure, and cytoplasmic organization. The total number of bacteria increased with leaf age and quantitative analyses of the relative contribution of each bacterial type to the total population (within individual leaves) were performed. Based on the ultrastructural diversity of the bacteria, including cell wall structures, and lack of snapping division (for some bacterial types), it seems unlikely that only *Arthrobacter* spp. are present within *Azolla* leaf cavities.

Preliminary studies, using a fluorescent *in situ* hybridization (FISH) procedure in conjunction with 16S rRNA targeted group – specific oligonucleotide probes, have been performed to determine the community structure of the bactobionts. Bactobionts were freshly isolated from *A. caroliniana* using previously described methods (Peters and Mayne, 1974), and then used *in situ* hybridization studies. The results are shown in Table 13.2.

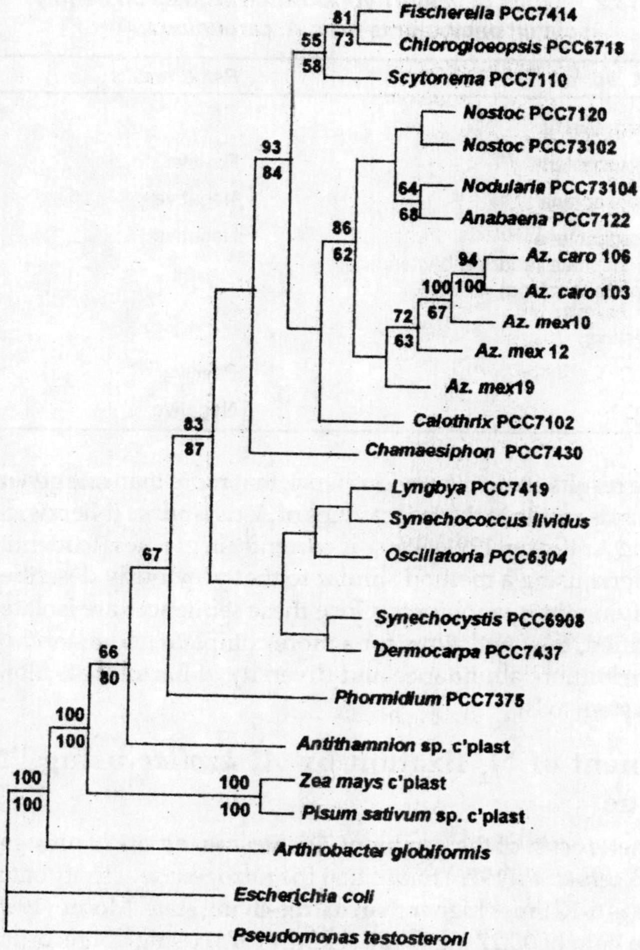

Fig. 13.1: Phenogram based on partial 16S rRNA sequence data obtained from the three *A. mexicana* and two *A. caroliniana* PCR clones, as well as 21 cyanobacterial, chloroplast, and other bacterial 16S rRNA sequence from RDP. Bootstrap values from both neighbour-joining distance (upper) and parsimony (lower) analyses are shown (when greater than 50%). *E. coli* was used as an outgroup to root the tree.
Abbreviations: *Az. Mex. – A. mexicana; Az. Car. – A. caroliniana; C'plast – chloroplast*

Table 13.2. Results of *in situ* hybridization studies on freshly isolated bactobionts from *A. caroliniana*.

16S rRNA proble	*Probe results*
Gram – negative	
Alpha proteobacteria	Positive
Beta proteobacteria	Negative
Delta proteobacteria (including sulphate reducing bacteria; Gamma proteobacteria)	Negative
Gram- positive	
High G + C	Positive
Low G + C	Negative

These results confirm previous work that more than one genera of bactobionts reside in the leaf cavities of *A. caroliniana* (Nierzwicki – Bauer and Aulfinger, 1991). Work is currently in progress to identify these bacteria using a method similar to that previously described for identifying the cyanobionts. Once these sequences are isolated and identified, they will allow for a more complete understanding of the distribution, abundance and diversity of bactobionts along the *Azolla* stem axis.

Assessment of N_2 fixation by *A. azollae* using ^{15}N technique

The heterocyst of the symbiont *Anabaena* is the site of nitrogen fixation. Peters *et. al.* (1976) found that the nitrogenase activity of the intact alga is 6-10 times higher than in free-living state. Moore (1969) reported a gain of 0.222 mg and Becking (1976) using ^{15}N recorded 4.3-5.2 mg N/day/gm dry weight in water culture. The endogenous supply of fixed nitrogen gives *Azolla* a competitive advantage over other floating hydrophytes in many environments. Nitrogen fixation associated with high growth rate can enable *Azolla* to accumulate more than 10kg N/ha/day under optimal growth conditions (Shi and Hall, 1988).

Nitrogen fixation by *Azolla* – *Anabaena* symbiosis has been demonstrated indirectly by use of nitrogen free nutrient solution, acetylene reduction assay by gas chromatography and directly by use of ^{15}N technique (Lumpkin and Plucknett, 1980). The ammonium

N absorbed from the medium is preferably transported to the symbiont (Ito and Watanabe, 1985). The ^{15}N values of the endophyte *Anabaena azollae* isolated from symbiotic complex were higher than those of host fern (Ray *et. al.*, 1978). The availability of NH_4^+ form of N derived from the incorporation of *A. caroliniana* was 19 per cent compared to 61 per cent in ammonium sulphate. Using ^{15}N dilution technique, Okoronkwo *et. al.* (1989) calculated the per cent of N derived from air, which was between 29.5 – 79.9 per cent, whereas Kumarasinghe *et. al.*(1986) and Yoneyama *et. al.* (1987) obtained 76-90 per cent and 59 – 99 per cent respectively. When exposed to ^{15}N, the endophytic *Anabaena* released upto 50 per cent of the fixed N into the incubation medium as ammonium (Peters *et. al.*, 1980). Kumarasinghe *et. al.* (1986) reported that ^{15}N labelled method is preferable than ^{15}N dilution technique for assessing the nitrogen availability from *Azolla*.

Thangaraju (1990) used ^{15}N dilution technique in *Azolla* inoculation to rice crop and reported that the incorporation of *Azolla* equivalent to 30 kg N/ha recorded more ^{15}N recovery per cent both in grain (17.78%) and straw yield (10.81%) in rice var. IR 20. By using ^{15}N isotope dilution (Kumarasinghe *et. al.*, 1986) and ^{15}N natural abundance technique (Yoneyama *et. al.*, 1987), it has been shown that 80-95 per cent of the N accumulated by *Azolla* is derived from N_2 fixation. Eskew *et. al.* (1993) compared the response and ^{15}N recovery by rice for 30 kg N/ha applied as urea into field with and without 2 tonnes of fresh *Azolla* inoculated at transplating. The ^{15}N uptake data showed that urea N was more efficiently used in the mixed *Azolla*/rice crop compared to the sole rice crop. De *et. al.* (1995) measured the nitrogen fixing activity of BGA in field using the isotopic ^{15}N trace assay and ARA (acetylene reduction technique) and reported that about 1.09 per cent of total nitrogen content of rice plant was obtained through N-fixation by inoculated algae. The utilization ratio of the N-fixed by algae per crop was estimated to be 11-93 per cent. Most of the nitrogen fixed by cyanobacteria remained in the soil as hydrolysable form (Ghosh and Saha, 1977). The ^{15}N activity in the algal symbionts *Anabaena azollae* isolated from *A. filiculoides, A. microphylla* and *Azolla* hybrids AH-RS-SK. and TNAU-1 after ^{15}N urea enrichment was studied and the presence of ^{15}N activity in the *A. azollae* isolates was observed , which indicated their true symbiontic nature with the host *Azolla* (Sasikumar, 1998; Balasubramani, 1999; Konde and Kannaiyan, 2000).

Ammonia excretion

Ammonia being the only inorganic form of nitrogen that is directly incorporated into organic linkage and thus an obligate intermediate in the utilization of other inorganic nitrogen sources. The excretion of ammonia is directly related to the activity of the enzyme, nitrogenase. Talley and Rains (1980) reported that during growth of *Azolla*, there was a very slow release of ammonical nitrogen directly into the water. Branched hair in leaf cavities are throught to be involved in the uptake of ammonia which moves from matured cavities of the growth apical regions of *Azolla* in the form of glutamate, glutamine and ammonia (Peters *et. al.*, 1985). Ito and Watanabe (1985) reported that fresh *Azolla* releases ammonia more rapidly than dried *Azolla*. The algal symbionts from *A. mexicana, A. caroliniana, A filiculoides, A pinnata* and a free-living cyanobacterium *Anabaena variabilis* were tested for their growth rate and ammonia excretion (Kannaiyan and Goyal, 1988). They found that all the algal symbionts have excreted ammmonia in the culture medium and the excretion was higher on 14th day and lower on 28th day. Kumar and Kannaiyan (2000) observed that the nitrogenase activity, ammonia assimilating enzymes major and micro-nutrient content were high in the *Azolla* cultures exposed to chemical mutagens such as EMS and colchicine.

Anabaena azollae releases around 40% of fixed N_2 in the form of ammonia which is translocated from the mature cavities to the stem apices of *Azolla* to support the undifferentiated filaments devoid of nitrogenase activity (Kaplan *et. al.*, 1986). Kannaiyan (1991) demonstrated that ammonia formation is generally observed when assimilation is blocked by pretreatment of *A. azollae* cells with MSX to covalently inhibit glutamine synthetase. Immobilization of the algal symbionts in PUF has activated the ammonia excretion than the freely growing algal symbionts (Merina Prem Kumari, 1995). Bavistin treatment stimulated the growth and nitrogenase activity of the algal symbiont *A. azollae* (Uma and Kannaiyan, 1996).

Immobilization of the algal symbiont in solid matrices

Immobilization is the process of attaching cells or their constituent biocatalysts to a solid matrix such that the cells do not move independently when placed in a fluid environment. Immobilization can be carried out either by physical means, such as adsorption or entrapment in a gel or foam matrix or onto hollow fibre, or by chemical methods such as covalent binding. The advantages of immobilization are stabilization of catalytic activity, often enhanced reaction rates due to concentration of catalyst surface, ease of separation and reuse of catalysts from soluble products, and processes control in a bioreactor. The main disadvantages are loss of activity during immobilization, limitation of reaction rates due to diffusional resistance and higher costs. Gels (acrylamide, agar, alginate, carrageenan, chitosan), glutaraldehyde – albumin spongers, reticulate foams (polyvinyl and polyurethane), rayon, polyproplene and polysulfone hollow fibres and ceramics (silica) are the materials which have been used as transparent or reflecting supports for immobilization.

Entrapment in porous gels or foam matrices is the favoured technique for the immobilization of plant and algal cells, as it does not modify or cause any damage to the cell. Since light is a direct substrate for photosynthesis, the immobilization matrix must be reasonably transparent as well as porous for penetration of the cells. Hall *et. al.*(1998) used alginate gels, polyurethane and polyvinyl foams and membrane hollow fibres for immobilization of cyanobacterial cultures. In all these matrices immobilization results in enhancing the stability of enzymes and proteins in the cells compared to their stability in free cell suspension.

The criteria adopted to test for the success of immobilization of microalgae are retention of photosynthetic electron transport (measured as O_2 exchange and PSII fluorescence), retention of enzymatic activities such as of nitrogenase and hydrogenase (cyanobacteria) and cell viability determined by growth in the matrix. It has been observed that in the symbiotic cyanobacterium, *Anabaena azollae*, cells immobilized in polymer foam showed an increase in the frequency of heterocysts as is observed with *A. azollae* in symbiotic association with the *Azolla* fern (Kannaiyan *et. al.*, 1994). Immobilized cyanobacterial cultures were able to excrete more ammonium into the medium and *A. azollae* immobilized in polymer foam revealed that the surface morphology of the immobilized cells closely resemble that of the symbiotic cyanobacterium growing inside the N_2 fixing water fern *Azolla* (Kannaiyan,1994; Hall *et. al.*, 1995). The immobilized cyanobacteria were intact in the solid matrix of PU (polyurethane) foam, which provided the cyanobiont a niche similar to the cavity of *Azolla* fronds in the natural association. Hall *et. al.* (1995) observed that the production of NH_4^+ from immobilized *A. azollae* mimicks the natural *Anabaena – Azolla* symbiosis.

Recently Kanniayan (1998) has shown that the cyanobacterial cultures immobilized in PU foam increased the quality and viability of the cyanobacteria and could survive for a period of 2-3 years. The algal symbiont *A. azollae* is known to colonize and immobilize well in polyvinyl foam (Kannaiyan *et. al.*, 1992). They have also demonstrated the ammonia excretion by *A. azollae* (AS-DS) and *Nostoc muscorum* (DOH) immobilized in polyvinyl foam at different foam density. Kannaiyan *et. al.*(1997) showed that the immobilized cyanobacteria treated with systemic fungicide bavistin, stimulated the nitrogenase activity and inhibited the GS-activity. Similarly, free-living as well as symbiotic cyanobacterial cultures under immobilized state in polyurethane foam excrete significantly higher amounts of ammonia into the flood water (Uma and Kannaiyan, 1994) and this was reflected on the higher ammonia level in flood water during the growth of the seedlings (Mahesh and Kannaiyan,1994; Aruna and Kannaiyan,1995). The cyanobacterial culture is well protected inside the pores of PU foam and also increases the nitrogen use efficiency (NUE) of rice owing to its continuous photoproduction of ammonia in rice field.

A carrier based immobilized cyanobacterial inoculant was developed to maintain better quality. The polyrethane foam, sugarcane waste and paper waste and immobilized cyanobacteria were the components of the inoculant. Rice husk and soil mixture act as a carrier for these immobilized cyanobacteria. PU foam immobilized cyanobacterial cultures recorded higher growth, nitrogenase activity, ammonia excretion and heterocyst frequency than free-living cultures. Sugarcane waste (SCW) and paper waste (PW) were used for immobilization of cyanobacteria in combination with rice husk and rice soil for improving the quality of cyanobacterial biofertilizers (Kanniayan and Kumar, 2000).

Conclusion

The information on the *Azolla – Anabaena* symbiosis is still inadequate to have a complete understanding about the association. More information on the diversity of the symbionts should further our understanding of the intricate interactions of the association. The potential benefits of the system can be further improved by manipulating the nitrogen fixing ability of the algal symbionts in this agronomically useful association. Hence, the following fields need detailed study: -

1. Recognition processes (or lack thereof) between *Anabaena* and *Azolla* ,
2. The characterization of the leaf cavity millieu,
3. Control of heterocyst and akinete differentiation,
4. The physiological capabilities of *Anabaena* present in cavities of different age,
5. Increasing the usefulness of the *Azolla – Anabaena* association as a biofertilizer,
6. Factors that induce sporulation and germination,
7. Production of *Azolla* hybrids with desirable agronomic characters,
8. Selection of genetically superior *Azolla – Anabaena* strains.

Acknowledgement

The authors are thankful to the Indian Council of Agricultural Research, New Delhi for financial support extended to the Azolla laboratory, TANU, Coimbatore to continue the basic and applied research on *Azolla anabaena* symbiosis under the NATP-team of Excellence on Biofertilizers for rice-based cropping system.

References

Arad, H., Keyasari, A., Tel-Or, E. and Kobiler, D. 1985. A comparison between cell antigens in different isolates of *Anabaena azollae*. *Symbiosis* **1**: 195-99.

Aruna, S. J. and Kannaiyan, S. 1995. Effect of inoculation of immobilized cyanobacteria on continouous supply of ammonia to the rice seedlings. In: *National seminar on Azolla and algal biofertilizers for rice*. Edited by Kannaiyan, S., Tamil Nadu Agric. Univ. Coimbatore, India. pp. 10.

Ashton, P. J. and Walmsley, R. D. 1976. The aquatic fern *Azolla* and its *Anabaena* symbiont. *Endeavour* **35**: 39-43.

Balasubramani, G. 1999. *Effect of immobilization of six isolates of Anabaena azollae in polyurethane foam (PUF) on the growth, nitrogen fixing activity, ammonia excretion and phylogenetic analysis by PCR*. Ph. D. Thesis, Centre for Plant Molecular Biology, Tamil Nadu Agri. Univ., Coimbatore, India.

Becking, J. H. 1976. Nitrogen fixation in some natural escosystem in Indonesia. In: *Symbiotic Nitrogen fixation in Plants*. Edited by Nutman, P. S., Cambridge Univ. Press, London. pp 539-557.

Becking, J. H. 1978. Ecology and physiological adaptations of *Anabaena* in the *Azolla- Anabaena azollae* symbiosis. *Ecol. Bull.* **26**: 266-281.

Becking, J. H. 1979. Environmental requirements of *Azolla* for use in tropical rice production. In: *Nitrogen and Rice.* Int.Rice Res. Inst., Los Banos, Philippines, pp. 345-373.

Becking, J. H. 1987. Endophyte transmission and activity in the *Anabaena – Azolla* association. *Plant Soil* **100**: 183-212.

Becking, J. H. and Donze, M. 1981. Pigment distribution and nitrogen fixation in *Anabaena azollae. Plant Soil* **61**: 203.

Bidwell, R. G. S. 1977. Photosynthesis and light and dark respiration in freshwater algae. *Can. J. Bot.* **55**: 89.

Buckingham, K. W., Ela, W. S., Morris, J. G. and Goldmanr, C.R. 1978. Nutritive value of the nitrogen fixing aquatic fern, *Azolla filiculoides. J. Agri. Food Chem.* **26**: 1230-1234.

Braun-Howland, E. B. and Nierzwicki-Bauer, S. A. 1990. Occurrence of the 32-kDa QB – binding protein of photosystem II in vegetative cells, heterocysts and akinetes of *Azolla caroliniana* cyanobionts. *Planta* **180**: 361-371.

Braun – Howland, E. B., Lindblad, P., Nierzwicki – Bauer, S.A. and Bergman, B. 1988. Dintrogenase reductse (Fe –protein) of nitrogenase in the cyanobacterial symbionts of three *Azolla* species: localization and sequence of appearance during heterocyst differention. *Planta* **176**: 319-322.

Calvert, H. E., Pence, M. K. and Peters, G. A. 1985. Ultrastructural ontogeny of leaf cavity trichomes in *Azolla* implies a functional role in metabolite exchange. *Protoplasma* **129:** 10-27.

Calvert, H. E., Perkins, S. K. and Peters, G. A. 1983. Sporocarp structure in the heterosporous water fern *Azolla mexicana* Presl. *Scan. Electron Microsc.* **3**: 1499-1510.

Calvert, H. E. and Peters, G. A. 1981. The *Azolla – Anabaena* relationship. Morphological analysis of leaf cavity hair population. *New Phytol.* **89**: 327-335.

Campbell, D. H. 1983. On the development of *Azolla filiculoides* Lam. *Ann. Bot.* **7**: 155-187.

Coppenolle, B.V., Watanabe, I., Van Hove, C., Second, G., Huang, N. and McCouch, S.R. 1993. Genetic diversity and phylogeny analysis of *Azolla* based on DNA amplification by arbitrary primers. *Genome* **36**: 686-693.

Dao, T.T. and Tran, Q.T. 1979. Use of *Azolla* in rice production in Vietnam. In: *Nitrogen and Rice*. Int. Rice. Res. Inst., Los Banos, Philippines, pp. 395-405.

De, D. J., Tian, S. S. and Baolin, S. 1995. Studies on the nitrogen fixation of algae in Beijing rice field. III. The use of ^{15}N trace assay estimated BGA nitrogen fixing contribution. *Acta Agri. Univ. Pekinensis* **21**: 7-11.

Duckett, J. G., Toth, R. and Soni, S. L. 1975. An ultrastructual study of the *Azolla- Anabaena azollae* relationship. *New Phytol*. **75**: 111-118.

Dunham, D. G. and Fowler, K. 1987. Megaspore germination, embryo development and maintenance of the symbiotic association in *Azolla filiculoides* Lam. *Bot. J. Linn. Soc*. **95**: 43-53.

Eskew, D. L., Caetano-Anolles, G., Bassam, B. J. and Gresshoff, P. M. 1993. DNA amplification fingerprinting of the *Azolla-Anabaena* symbiosis. *Plant Mol. Biol*. **21**: 363-373.

Felsenstein, J. 1989. PHYLIP- Phylogenetic inference package. *Caldistics* **5**: 164-166.

Fowler, K. and Stennet-Willson, J. 1978. Sporoderm architecture in modern *Azolla*. *Fern Gaz*. **11**: 405-412.

Franche, C. and Cohen – Bazire, G. 1985. The structural *nif* genes of four symbiotic *Anabaena azollae* show a highly conserved physical arrangement. *Plant Sci*. **39**: 125-131.

Franche, C. and Cohen-Bazire, G. 1987. Evolutionary divergence in the *nif* K,D,H gene region among nine symbiotic *Anabaena azollae* and between *Anabaena azollae* and some free-living heterocystous cyanobacteria. *Symbiosis* **3**: 159-178.

Gates, J. E., Fisher, R. W., Goggin, T. W. and Azrolan, N. I. 1980. Antigeneic difference between *Anabaena azollae* fresh from the *Azolla* fern leaf cavity and free-living cyanobacteria. *Arch. Microbiol*. **128**: 126.

Gebhardt, J. S. and Nierzwicki – Bauer, S. A. 1991. Identification of a common cyanobacterial symbiont associated with *Azolla* spp. through molecular and morphological characterization of free-living symbiotic cyanobacteria. *Environ. Microbiol*. **57**: 2141-2146.

Ghosh, T. K. and Saha, K. C. 1977. Effects of inoculation of cyanobacteria on nitrogen status and nitrogen of rice in an entisol amended with chemical and organic sources of N. *Biol. Fertil. Soil* **24**: 123-128.

Gopalaswamy, G. and Kannaiyan, S. 1998. Utilization of *Azolla* hybrids for bioreclamation of sodic soil and rice (*Oryza sativa*) growth. *Indian J. Agro.* **43**: 511-517.

Groswell, E. T. and Godwin, D. C. 1984. The efficiency of nitrogen fertilizers applied to cereals in different climates. In: *Advances in Plant Nutrition.* Edited by Tinker, P. B. and Lauchli, A., Prager Scientific, New York. pp. 1-55.

Hall, D. O., Lichtl, R., Markov, S. A. and Rao, K. K. 1998. The ammonia and hydrogen production by N_2^- fixing cyanobacteria immobilized on polymer foams and hollow fibres In: *Cyanobacterial Biofertilizer for Rice Crops.* Edited by Kannaiyan, S. Tamil Nadu Agri. Univ., Coimbatore, India. pp. 122-129.

Hall, D. O., Markov, S. A., Watanabe, Y. and Rao, K. K. 1995. The potential applications of cyanobacterial photosynthesis for clean technologies. *Photosynth. Res.* **46**: 159-167.

Haselkorn, R. 1978. Heterocysts. *Annu. Rev. Plant Physiol.* **29**: 319-344.

Herd, Y. R., Cutter, E. G. and Watanabe, I. 1986. An ultrastructural study of post-meiotic development in the megasporocarp of *Azolla microphylla. Can. J. Bot.* **64**: 822-833.

Hill, D. J. 1975. The pattern of development of *Anabaena* in the *Azolla-Anabaena* symbiosis. *Planta* **122**: 179-184.

Hill, D. J. 1977. The role of *Anabaena* in the *Azolla-Anabaena* symbiosis. *New Phytol.* **78**: 611-616.

Holst, R.W. 1977. Anthocyanins of *Azolla. Amer. Fern J.* **67**: 99-100.

Holst, R. W. and Yopp, J. H. 1979. Studies of the *Azolla-Anabaena* symbiosis using *Azolla mexicana.* I Growth in nature and laboratory. *Amer. Fern J.* **69**: 17-24.

Ito, O. and Watanabe, I. 1985. Availability to rice plants of nitrogen fixed by *Azolla. Soil Sci. Plant Nutr.* **31**: 91-104.

Jackman, D. M. and Mulligan, M. E. 1995. Characterization of a nitrogen-fixation (*nif*) gene cluster from *Anabaena azollae* 1a shows that closely related cyanobacteria have highly variable but structured intergenic regions. *Microbiology* **141**: 2235-2244.

Kannaiyan, S. 1985. Potentiality of *Azolla* biofertilizer for rice crop. *Proc. Soil Microbiol. Symp*. HAU, Hisar, Haryana. pp. 253-259.

Kannaiyan, S. 1986. Studies on the growth, nitrogen fixation and sporulation in certain cultures of *Azolla*. In: *FAO/UNDP/ICAR Research Programme – Report on Biological Nitrogen fixation Azolla-Anabaena system*, C. F. Kettering Research Lab, Yellowspring, Ohio, USA. pp. 120.

Kannaiyan, S. 1988. Studies on salinity, temperature and pesticide interaction on the growth and nitrogen fixation in *Azolla*. *Proc. Int. Conf. Biol. Nitrogen Fixation*, Central Rice Res. Inst., Cuttack, Orissa, India.

Kannaiyan, S. 1989. Effect of benlate on growth and nitrogen fixation in three strains of *Azolla pinnata*. *J. Soil. Biol. Ecol.* **9**: 97-103.

Kannaiyan, S. 1990. Effect of sodium chloride on the growth, nitrogenase and photosynthesis of *Azolla*. *Indian J. Plant Physiol.* **33**: 125-129.

Kannaiyan, S. 1991. *Studies on the immobilization of nitrogen fixing symbiotic cyanobiont Anabaena azollae and free living cyanobacteria in polyvinyl and polyurethane foams for their growth behaviour and ammonia production*. Tech. Report, Division of Biosphere Sciences, King's college, Univ. of London, U. K.

Kannaiyan, S. 1992. *Azolla Biofertilizer Technology for Rice*. Tech. Bull., Tamil Nadu Agricultural University, Coimbatore, India. pp. 56.

Kannaiyan, S. 1994. Immobilization of *Anabaena azollae* in soild matrix on ammonia excretion and its effect on rice crop. *National Seminar Azolla and Algal Biofertilizers for Rice*, Tamil Nadu Agric. Univ., Coimbatore, Tamil Nadu, India. pp. 25-26.

Kannaiyan, S. 1998. Mass production of polyurethane foam immobilized cyanobacterial biofertilizer inoculants for rice crop. *Int. Conf. Agric. Biotechnol.*,Univ. Saskatchewan, Canada. pp. 4.

Kannaiyan, S., Aruna, S. J., Merina Prem Kumari, S. and Hall, D. O. 1997. Immobilized cyanobacteria as a biofertilizer for rice crop. *J. Appl. Phycol.* **9**: 167-174.

Kannaiyan, S. and Goyal, S. S. 1988. Effect of light intensity and salt stress on the growth and ammonia excretion by algal symbionts of *Azolla*. 29th *Ann. Conf. Association of Microbiologists of India*, HAU, Hisar, Haryana, India.

Kannaiyan, S. and Kumar, K. 2000. PUF immobilized cyanobacterial biofertilizer for rice. *First Int. Global Biotech. Cong.*, Nagpur University, Nagpur, India.

Kannaiyan, S., Kumar, K. and Pandiyarajan, P. 1992. The use of saline tolerant cyanobacteria for salt affected lands. *Proc. Int. Symp. on Strategies for utilizing salt affected lands*, Bangkok, Thailand. pp. 394-404.

Kannaiyan, S., Rao, K. K. and Hall, D. O. 1994. Immobilization of *Anabaena azollae* from *Azolla filiculoides* in polyvinyl foam for ammonia production in a photobioreactor system. *World J. Microbiol. Biotechnol.* **10**: 55-58.

Kannaiyan, S. and Somporn, C. 1988. Studies on the growth, nitrogen fixation, chlorophyll content and photosynthesis in *Azolla*. *Indian J. Plant Physiol.* **31**: 134-139.

Kaplan, D., Calvert, H. E. and Peters, G. A. 1986. Nitrogenase acitivity and phycobiliproteins of the endophyte as a function of leaf age and cell type. *Plant Physiol.* **80**: 884-890.

Kaplan, D. and Peters, G. A. 1981. The *Azolla – Anabaena* relationship X. $^{15}N_2$ fixation and transport in main stem axes. *New Phytol.* **89**: 337-346.

Kim, J. J., Krawczyk, N., Lorenty, K. and Zimmerman, W. P. 1997. Fingerprinting cyanobionts and hosts of the *Azolla* symbiosis by DNA amplification. *World J. Microbiol. Biotechnol.* **13**: 97-101.

Kobiler, D., Cohen-Sharon, A. and Tel-Or, E. 1981. Recognition between the N_2-fixing *Anabaena* and the water fern *Azolla*. *FEBS Lett.* **133**: 157-160.

Kobiler, D., Cohen-Sharon, A. and Tel-Or, E. 1982. Lectins are involved in the recognition between *Anabaena* and *Azolla*. *Isr. J. Bot.* **31**: 324.

Konar, R. N. and Kapoor, R. K. 1974. Embryology of *Azolla pinnata*. *Phytomorphology* **24**: 228-261.

Konde, V. B. and Kannaiyan, S. 2000. *Azolla-Anabaena* symbiosis: Establishment of true symbiotic nature as detected by ^{15}N studies. *J. Maharastra Agri. Univ.* **25** (In press).

Kulasooriya, S. A. 1985. Establishement and management of *Azolla* in rice fields. In: *The role of isotopes in studies on nitrogen fixation and nitrogen cycling by BGA and Azolla – Anabaena azollae association*. IAEA-TEC. DOC. 325., IAEA. Vienna, Austria. pp. 21-33.

Kumar, K. and Kannaiyan, S. 1999. Characterization of mutants of *Azolla microphylla* and *Azolla filiculoides* obtained through chemical mutagenesis. *Indian J. Plant Physiol.* **4**: 242-245.

Kumar, K. and Kannaiyan, S. 2000. Selection of mutants of *Azolla* for higher nitrogen fixing potential, ammonia assimilation and nutrient content. *J. Microb. World* **2**: 35-41.

Kumarasinghe, K. S., Zapata, F., Kovacs, G., Eskew, D. L. and Danso, S. K. A. 1986. Evaluation of the availability of *Azolla*-N and urea-N to rice crop using ^{15}N. *Plant Soil* **90**: 293-296.

Ladha, J. K. and Watanabe, I. 1982. Antigenic similarity among *Anabaena azollae* separated from different species of *Azolla*. *Biochem. Biophys. Res. Commun.* **109**: 675.

Ladha, J. K. and Watanabe, I. 1984. Antigenic analysis of *Anabaena azollae* and the role of lectin in the *Azolla-Anabaena* symbiosis. *New Phytol.* **98**: 295-300.

Lampe, K. 1988. *Rice think tanks review IRRI strategy*. Int. Rice Res. Inst., Los Banos, Philippines. pp.1.

Lin, C. and Watanabe, I. 1988. Study on the association between *Anabaena azollae* and *Azolla microphylla* during the germination of megasporocarps. *Symbiosis* **5**: 199-208.

Lin, C., Watanabe, I., Liu, C. C., Zheng, D.V. and Tang, L. F. 1988. Re-establishment of symbiosis to *Anabaena* – free *Azolla*. In: *Nitrogen fixation: Hundred Years After.* Edited by Bothe,H., de Brujin, F. J. and Newton ,W. E. Gustav Fischer, Stuttgart, pp. 223.

Lindblad, P., Bergman, B. and Nierzwicki – Bauer, S. A. 1991. Immunocytochemical localization of nitrogenase in bacteria symbiotically associated with *Azolla* spp. *Environ. Microbiol.* **57**: 3637 – 3640.

Lloyd, N. D. H., Canvin, D. J. and Culver, D. A. 1977. Photosynthesis and photorespiration in algae. *Plant Physiol.* **59**: 936.

Lotti, F., Glovannetti, L., Margheri, M. C., Ventura, S. and Materassi, R. 1996. Diversity of DNA methylation pattern and total DNA restriction pattern in symbiotic *Nostoc*. *World J. Microbiol. Biotechnol.* **12**: 38-42.

Lumpkin, T.A. 1981. An introduction to *Azolla nilotica*. *Acta Bot. Sin.* **23**: 492.

Lumpkin, T. A. and Plucknett, D. L. 1980. *Azolla* – Botany, physiology and use as a green manure. *Econ. Bot.* **34**: 111-153.

Lumpkin, T. A. and Plucknett, D. L. 1982. *Azolla as a green manure: Use and management in crop production*. Westview Tropical Agriculture Series No. 5, West view Press, Colorado, USA. pp. 230.

Mahesh, G. and Kannaiyan, S. 1994. Effect of inoculation of immobilized cyanobacteria on rice seedlings growth. *National seminar on Azolla and algal biofertilizer for rice*. Tamil Nadu Agric. Univ. Coimbatore, India. pp. 28.

Maidak, B. L., Olsen, C. J., Larsen, N., Overbeek, R., BcCaughey, M. J. and Woess, C. R. 1996. The ribosomal database project. *Nucl. Acids Res.* **24**: 82-86.

Martin, A. R. H. 1976. Some structures in *Azolla* megaspores and an anomalous fertilization. *Rev. Paleobot. Polynol.* **21**: 141-145.

McCowen, S. M., MacArthur, L. and Gates, J. E. 1987. *Azolla* fern lectins that specifically recognize endosymbiotic cyanobacteria. *Curr. Microbiol.* **14**: 329-335.

Meeks, J. C., Joseph, C. M. and Haselkorn, R. 1988. Organization of the *nif* genes in cyanobacteria in symbiotic association with *Azolla* and *Anthoceros*. *Arch. Microbiol.* **150**: 61-71.

Mellor, R. B., Gadd, G. M., Rowell, P. and Stewart, W.D.P. 1981. A. phytochaemagglutinin from the *Azolla* – *Anabaena* symbiosis. *Bichem. Biophys. Res. Commun.* **99**: 1348-1354.

Merina Prem Kumari, S. 1995. *Studies on the production of growth promoting substances, nitrogen fixing activity and ammonia excretion by the cyanobiont, Anabaena azollae Strasburger, isolated from Azolla microphylla Kaulfuss and Azolla filiculoides Lamarck*. Ph. D. Thesis, Centre for Plant Molecular Biology, Tamil Nadu Agril. Univ., Coimbatore, India.

Mian, M. H. 1981. *Biofertilizer and rice production – a ^{15}N tracer study*. Ph. D. Thesis, Univ. Dundee, Scotland, UK.

Moore, A. W. 1969. *Azolla* biology and agronomic significance. *Bot. Rev.* **35**: 17-35.

Newton, J. W. and Cavins, J. F. 1976. Altered nitrogenous pools induced by the *Azolla – Anabaena azollae* symbiosis. *Plant Physiol.* **58**: 798-799.

Newton, J. W. and Herman, A. I. 1979. Isolation of cyanobacteria from the aquatic fern, *Azolla. Arch. Microbiol.* **120**: 161-165.

Nierzwicki – Bauer, S. A. and Aulfinger, H. 1991. Occurrence and ultrastructural characterization of bacteria in association with and isolated from *Azolla caroliniana. Appl. Environ. Microbiol.* **57**: 3629 –3636.

Nierzwicki-Bauer, S. A. and Haselkorn, R. 1986. Differences in mRNA levels in *Anabaena* living freely or in symbiotic association with *Azolla, EMBO J.* **5**: 29-35.

Okoronkwo, N., Van Hove, C. and Eskew, D. L. 1989. Evaluation of nitrogen fixation by different strains of the *Azolla – Anabaena* symbiosis in the presence of the high level of ammonium. *Biol. Fertil. Soils* **7**: 275-278.

Perkins, S.K. and Peters, G.A. 1992. The *Azolla- Anabaena* symbiosis: Endophyte continuity in the *Azolla* life-cycle is facilitated by epidermal trichomes. I. Partitioning of the endophytic *Anabaena* into developing sporocarps. *New Phytol.* **123**: 53-64.

Peters, G. A. 1975. The *Azolla - Anabaena azollae* relationship. III. Studies on metabolic capabilities and further characterization of the symbiont. *Arch. Microbiol.* **103**: 113-122.

Peters, G. A. 1977. The *Azolla – Anabaena azollae* symbiosis. In: *Genetic Engineering and Nitrogen Fixation*. Edited by Hollaender, A., New York, USA. pp. 231-258.

Peters, G.A. and Calvert, H.E. 1983. The *Azolla – Anabaena* symbiosis. In: *Algal Symbiosis.* Edited by Goff, C.J. Cambridge University Press, New York. pp. 109-145.

Peters, G. A. and Mayne, B.C. 1974. The *Azolla – Anabaena* relationship. I Initial characterization of the association. *Plant Physiol.* **53**: 813-819.

Peters, G.A. and Perkins, S.K.1993. The *Azolla – Anabaena* symbiosis endophyte continuity in the *Azolla* life cycle is facilitated by epidermal trichomes. *New Phytol.* **123**: 65-75.

Peters, G.A., Kaplan, D., Meeks, J.C., Buzby, K.M., Marsh, B.H. and Corbin, J.L. 1985. Aspects of nitrogen and carbon interchange in the *Azolla – Anabaena* symbiosis. In: *Nitrogen fixation and CO_2 Metabolism*. Edited by Ludden, P.W. and Burris, J.E., Elsevier Science, New York. pp. 213-232.

Peters, G.A., Evans, W.R. and Toia, R.E. 1976. *Azolla – Anabaena azollae* relationship. IV. Photosynthetically driven, nitrogenase catalysed H_2 production. *Plant Physiol.* **58**: 119-126.

Peters, G.A., Mayne, B.C., Ray, T.B. and Toia, R.E., Jr. 1979. Physiology and biochemistry of the *Azolla-Anabaena* symbiosis. In: *Nitrogen and Rice.* Int.Rice Res. Inst., Los Banos. Philippines. pp. 325.

Peters G.A., Ray T.B., Mayne, B.C.and Toia, R.E. 1980. *Azolla-Anabaena* association morphological and physiological studies. In: *Nitrogen Fixation*. Edited by Newton, W.E. and Orme Johnson,W.H., Univ. Park Press, Baltimore, USA. pp. 293-309.

Peters, G.A., Toia, R.E. Jr., Raved, D. and Levine, N.J. 1978. The *Azolla-Anabaena azollae* relationship VI. Morhpological aspects of the association. *New Phytol.* **80**: 583-593.

Pieterse, A.H., De Lange, L. and Van Vliet, J.P. 1977. A comparative study of *Azolla* in the Netherlands, *Acta Bot. Neerl.* **26**: 433.

Plazinski, J., Franche, C., Liu, C.C., Lin, T., Shaw, W., Ginning, B.E.S. and Rolfe, B.G. 1988. Taxonomic status of *Anabaena Azollae* an overview. *Plant Soil.* **108**: 185-190.

Plazinski, J., Zheng, Q., Taylor, R., Croft, C., Rolfe, B.G. and Gunning, B.C.S. 1990. DNA probes show genetic variation in cyanobacterial symbionts of the *Azolla* fern and a closer relationship to free-living *Nostoc* strains than to free-living *Anabaena* strains. *Appl. Environ. Microbiol.* **56**: 1263-1270.

Rao, H.S. 1935. The structure and life history of *Azolla pinnata* R. Brown with remarks on fossil history of the hydropteridat. *Proc. Indian Acad. Sci.* **2B**: 175-200.

Ray, T.B., Mayne, B.C., Toia, R.E. and Peters, G.A. 1979. *Azolla-Anabaena* relationship. VIII. Photosynthetic characterization of the association and individual partners. *Plant Physiol.* **64**: 791-795.

Ray, T.B., Peters, G.A., Toia, R.E., Jr. and Mayne, B.C. 1978. *Azolla-Anabaena* relationship. VII. Distribution of ammonia assimilating enzymes, protein and chlorophyll between host and symbiont. *Plant Physiol.* **62**: 463-467.

Reynaud, P.A. and Franche, C. 1986. *Azolla pinnata var. africana: from molecular biology to use as a green manure.* Technical bulletin, ORSTOM, Senegal, W., Africa. pp. 15.

Robins, R.J., Hall, D.O., Shi, D.J., Tunrer, R.J. and Rhodes, M.J.C. 1986. Mucilage acts to adhere cyanobacteria and cultured plant cells to biological and inert surface. *FEMS Microbiol. Lett.* **34**: 15

Rudi, K., Skulberg, O.M., Larsen, F. and Jakobsen, K.S. 1997. Strains characterization and classification of oxyphotobacteria in clone cultures on the basis of 16S rRNA sequences from the variable regions V6, V7 and V8. *Appl. Environ. Microbiol.* **63**: 2593-2599.

Samal, K.C. and Kannaiyan, S. 1992. Isolation of the algal symbiont *Anabaena azollae* and the role of vitamins in growth, heterocyst development and nitrogen fixing activity. *Folia Microbiol.* **37**: 421-426.

Sara, M.M., Arthur, L.M. and Gates, J.E. 1987. *Azolla* fern lectins that specifically recognize endosymbiotic cyanobacteria. *Curr. Microbiol.* **14**: 329-333.

Sasikumar, S. 1998. *Studies on the immobilization of nitrogen fixing cyanobacteria on ammonia production in flood water and its effect on rice yield.* M.Sc. (Agr.) Thesis, Deptt. of Agrl. Microbiology, Tamil Nadu Agri. Univ., Coimbatore, Tamil Nadu, India. pp.180.

Shi, D.J. and Hall, D.O. 1988.The *Azolla-Anabaena* association: Historical perspective, symbiosis and energy metabolism. *Bot. Rev.* **54**: 353-386.

Shi, D.J., Li, J.G., Zheng, Z. P., Wang, F.Z., Zhu, L.P. and Peters, G.A. 1981. Nitrogen fixation and photosynthesis in *Azolla imbricata* and *Azolla filiculoides*. *Acta Bot. Sin.* **23**: 306-315.

Siddiq, E.A. 1999. Rice – Not a distance dream. In: *The Hindu Survey of Indian Agriculture.* Edited by Ravi, N. The Hindu, Chennai. pp. 95-118.

Singh, A.L. 1994. Use of *Azolla* cyanobacterial symbiosis in rice cultures. In: *Biology and application of N_2 fixing organisms: Problems and Prospects.* Edited by Prasad, A.B. and Vaishampayan, A., Scientific Publ., Jodhpur, India. pp. 95-118.

Singh, P.K. 1977. Multiplication and utilization of fern *Azolla* containing nitrogen fixing algal symbiont a green manure in rice cultivation. *Il Rizo* **26**: 125-137.

Singh, P.K. 1979. Symbiotic algal N_2 fixation and crop productivity. In: *Ann. Rev. Plant Sci. Vol. I*. Edited by Malik, C.P., Kalyani Pub., New Delhi. pp. 37-69.

Strasburger, E. 1973. *Uber Azolla Jena*: Hermann Dabis Verlag.

Stergianou, K.K. and Fowler, K. 1990. Chromosome numbers and taxonomic implications in the fern genus *Azolla (Azollaceae)*. *Plant Syst. Evol.* **173**: 223-229.

Subhashini,R., Kumar K. and Kannaiyan, S. 2000a. Variation revealed by RPD-PCR profile of microsymbiotic *Anabaena azollae* strains from four N_2 fixing *Azolla* cultures. *Acta Biotechnol.* **20**: 169-174.

Subhashini,R., Kumar K. and Kannaiyan, S. 2000b. Biodiversity of *Anabaena azollae* isolates from different *Azolla* cultures. *Indian J. Exp. Biol.* **38**: 1168-1171.

Svenson, H.K. 1944. The New world species of *Azolla*. *Amer. Fern. J.* **34**: 69-84.

Sweet, A. and Hills, L.V. 1971. A study of *Azolla pinnata* R. Brown. *Amer. Fern. J.* **61**: 1-13.

Swofford, D.L. 1993. *PAUP. Phylogenetic analysis parsimony, version 3.1 computer program.* Illinios Natural History Survey campaign, Illinosis, USA.

Talley, S.N. and Rains, D.W. 1980. *Azolla filiculoides* Lam. as a fallow season green manure for rice in temperate climate. *Agron.* J. **72**: 11-18.

Talley, S.N., Talley, B.J. and Rains, D.W. 1977. Nitrogen fixation by *Azolla* in rice fields. In: *Genetic engineering for nitrogen fixaiton.* Edited by Hollaender, A., Plenum Publ. Co., New York, USA. pp. 250-282.

Tel-Or, E., Sandovsky, T., Kobiller, D., Arad, C. and Weinberg, R. 1983. The unique symbiotic property of *Anabaena* in the water fern *Azolla*. In: *Photosynthetic Prokaryotes: Cell Differentiation and Function.* Edited by Papageorgiou, G.C. and Packer, L., Elsevier Sci., New York. pp. 303.

Tel-Or, E. and Stewart, W.D.P. 1977. Photosynthetic components and activities of nitrogen-fixing isolated heterocysts of *Anabaena cylindrica*. *Proc. R. Soc. Lond*. **198:** 61-86.

Thangaraju, M. 1990. *Studies on the nitrogen fixing water fern Azolla and its use as biofertilizer for rice*. Ph.D Thesis, Deptt. of Agrl, Microbiology, Tamil Nadu Agrl. Univ., Coimbatore, Tamil Nadu, India, pp. 254.

Tran, T. and Dao, T. 1973. *Azolla*, a green compost (Vietnamese studies). *J. Agric. Prod*. **38**: 119-217.

Tung, H.F. and Watanabe, I. 1983. Different response of *Azolla-Anabaena* associations to high temperature and minus phosphorus treatments. *New Phytol*. **93**: 423-431.

Tyagi, V.V.S., Mayne, B.C. and Peters, G.A. 1980. Action spectra of acetylene reduction in the *Azolla-Anabaena* association and in the isolated endophyte. *Plant Physiol*. **65**: 109.

Tyagi, V.V.S., Ray, T.B., Mayne, B.C. and Peters, G.A. 1981. The *Azolla-Anabaena* relationship. XI. Phycobilins in the action spectrum for nitrogenase – catalyzed acetlene reduction. *Plant Physiol*. **68**: 1479-1484.

Uheda, E. 1980. Isolation of empty packets from *Anabaena* – free *Azolla*. *Plant Cell Physiol*. **27**: 1187.

Uma, D. and Kannaiyan, S. 1994. Effect of the insecticide carbofuran on ammonia production by cyanobacteria under immobilized state in PU foam. *National Seminar on Azolla and algal biofertilizers for rice*. Tamil Nadu Agric. Univ., Coimbatore, India.

Uma, D. and Kannaiyan, S. 1996. Effect of systemtic fungicide bavistin on the nitrogen status of cyanobacteria under immobilized state in polyurethane foam. *S. Afr. J. Bot*. **62**: 127-132.

Van Commenolle, B., Watanabe, I., Van Hove, C. and McCouch, S.R. 1993. Genetic diversity and physiology analysis of *Azolla*, based on DNA amplification for phylogenetic study. *J. Bacteriol*. **173**: 697-703.

Walmsley, R.D., Breen, C.M. and Kyle, E. 1973. Aspects of the fern-algal relationship in *Azolla filiculoides*. *Limnol. Soc. S. Afr*. **20**: 13.

Watanabe, I. 1978. *Azolla* and its use in lowland rice culture. *Soil Micro*. **20**: 110.

Watanabe, I., 1982. *Azolla-Anabaena* symbiosis – its physiology and use in tropical agriculture. In: *Microbiology of Tropical Soils and Plant Productivity*. Edited by Dommergues, Y.R. and Diems, H.G., Martinus Nijhoff, The Hague. pp. 169-185.

Watanabe, I. 1990. Biological sources of nitrogen and the improvement of nitrogen fixing organisms. *Int. Symp. on Rice Res.– New frontiers*. Directorate of Rice Research, ICAR, Rajendra Nagar, Hyderabad, India. pp. 1-18.

Watanabe, I.. Espinas, C.R., Berja, N.S. and Alimagno, B.V. 1977. Utilization of *Azolla-Anabaena* complex as a nitrogen fertilizer for rice. *Int. Rice Res. Inst. Pub. Series* **11**: 115.

Yoneyama, T., Ladha, J.K. and Watanabe, I. 1987. Nodule bacteroids and *Anabaena:* Natural ^{15}N enrichment in the legume-*Rhizobium* and *Azolla-Anabaena* symbiotic systems. *J. Plant Physiol.* **127**: 251-259.

Zimmerman, W.J., Lumpkin, T.A. and Watanabe, I. 1989. Classification of *Azolla* spp. section *Azolla*. *Euphytica* **43**: 223-232.

Chapter 14

AKINETES: STRUCTURE, DIFFERENTIATION AND GERMINATION

Manjit Kaur and Amrik S. Ahluwalia

Abstract

Akinetes are the specialized structures in blue–green algae and these germinate to give rise to new filaments of parental genetic make up. Their role as perennating structures is still disputed. Many physiological and environmental factors have been known to regulate akinete differentiation but more studies are required for deciphering their role. Pattern of akinetes in the filaments is also a complex process and till date no single parameter can explain the pattern formation in different algal species.

Key words: Akinetes, algae, *Anabaena*, cyanobacteria, differentiation, heterocysts.

Introduction

Blue-green algae constitute one of the largest, most diverse and widely distributed group of photosynthetic prokaryotes. These organisms show examples of differentiation of one cell type to another and offer the possibility of extending molecular biology into those fields which could not be tackled using bacteria as an experimental material (Fogg *et. al.*, 1973).

Cellular differentiation, in general, may be defined as the process that involves the emergence or appearance (development) of distinctiveness in structure and function among the cellular population (progeny) of identical genetic make up. Differentiation, in all biological systems, entails selective activation and expression of genetic potential in a temporal sequence, which permits an orderly morphogenesis or the modification of cellular and biochemical capabilities. In the strict sense, enzyme induction represents a simple case of cell differentiation.

Blue–green algae produce usually a variety of unicellular structures for asexual reproduction such as spores, nannocytes endospores, exospores etc. Many blue–green algae belonging to order nostocales produce akinetes (Nichols and Adams, 1982). Akinetes are morphologically and functionally unique cells and are also known as spores. Their ability to perennate for many years and subsequently germinate, has led to the belief that akinetes are resting cells. However, little evidence is available to show that akinetes are important in the life cycle of nostocacean algae as many species overwinter as vegetative filaments (Kappers, 1976; Barbiero and Welch, 1992). During akinete differentiation, a temporal sequence of events leads to the development of a cellular form distinct from the vegetative cell in morphological, physical and biochemical properties and physiological functions.

An akinete is essentially a vegetative (somatic) cell that has thickened its wall and thus has got the ability to withstand desiccation and other conditions hostile to vegetative development. Microfossils indicate that akinetes and heterocysts were present at the very early stages of the evolution of cyanobacteria (Nagy, 1978). Akinetes enclose an oxygen-labile enzyme capable of assimilating a form of phosphorus other than orthophosphate (Van Dok and Hart, 1996). They contain a large amount of cyanophycean granules. Their shape, size and the position in the filament in relation to heterocyst constitute important taxonomic characters at the species level (Desikachary, 1959; Watanabe, 1992). The pattern of akinete formation is also important in distinguishing various blue-green algal species and genera (Rippka *et. al.*, 1979; Ahluwalia and Kumar, 1982a). The formation and germination of the akinetes are affected by different nutritional, physiological, environmental and physical factors. There is still little understanding about the various factors that regulate akinete differentiation and germination.

Morphology and ultrastructure

The akinetes are readily distinguishable under the light microscope due to their large size, characteristic shape, modified pigmentation and presence of many prominent cytoplasmic granules. The akinete envelope completely encloses the cell and thus separates it from the adjacent cells (Fritsch, 1945). In general, concomitant with an increase in cell size, is found disappearance of gas vacuoles, compression of intrathylakoidal spaces and increased number of cyanophycin and polyglucan granules. Miller and Lang (1968) and Clark and Jensen (1969) have reported the developmental changes occurring during akinete formation in *Cylindrospermum* sp. The envelope sometimes contains more than one layer formed presumably by condensation of the gelatinous sheath of the trichome (Fritsch, 1945).

Grilli Caiola and De Vecchi (1980) recognized various stages during sporulation in *Nostoc* sp. isolated from cycad coralloid roots. They found that the culture medium becomes turbid due to consistent production of mucilaginous substance. Afterwards, the colour of the culture changes from green to yellow. Simultaneously cell enlarges and produces a thick envelope around the cell wall. There is deposition of numerous large sized granules in the cytoplasm, and finally the envelope thickens and the cell becomes quiescent. Two types of akinetes were observed ultrastructurally. The first type is characterized by a homogeneous internal envelope, electron-dense cytoplasm with thylakoids distributed throughout the whole cell, numerous polyglucoside and small cyanophycin granules and lipid bodies. Polyphosphate bodies were also located. The second type of akinete has two more distinct envelopes; the inner is more electron dense and thicker, while the outer one is reduced in width. The cytoplasm shows a large nucleoplasmic area, ribosomes, few polyhedral granules, concentric or branched thylakoids dispersed at the periphery and numerous cyanophycin granules. First type was found in *Nostoc* sp. isolated from *Dioon* and the second type, isolated from *Macrozamia communis* (Grilli Caiola and De Vecchi, 1980).

Chemical composition

Most of the work on biochemical composition of akinetes have come from the laboratories of Wolk and Simon (Cardemil and Wolk, 1976; Hirosawa and Wolk, 1979; Simon, 1980; Zhou and Wolk, 2002). The sheath and envelope of akinetes in *Cylindrospermum* sp. stain readily with ruthenium red and hence considered to be composed of proteinaceous substance (Miller and Lang, 1968). Dunn and Wolk (1970) analysed the wall fraction of akinetes of *Anabaena cylindrica* and found 41% carbohydrate, 24% amino compounds, 11% lipids, 2% ash and 2% moisture on dry weight basis. Envelopes of akinetes and heterocysts in *Anabaena variabilis* have been shown to consist of repeating units, containing 1 mannosyl, and 3 glucosyl residues, all linked by β 1-3 glucosidic bonds, with glucose, xylose, galactose and mannose present in side branches (Cardemil and Wolk, 1981). The akinete envelope contains several times more amino compounds than the heterocyst envelope.

The prominent granules found extensively in akinetes during their formation are known as cyanophycin granules. They have been isolated from *Anabaena cylindrica* and chemically identified as polypeptides of two amino acids namely aspartic acid and arginine in a 1: 1 ratio (Simon, 1973). Sutherland *et. al.* (1979) reported that during akinete formation, there was an 8-fold increase in cyanophycin and 16–fold increase in glycogen contents. Nitrogen content of akinetes showed 17% increase while the total dry weight, carbon and chlorophyll *a* contents approximately doubled than those of vegetative cells in *Nostoc* PCC 7524. During akinete differentiation, increased synthesis of cyanophycin granules is probably a device that permits intracellular stockpiling of nitrogen as Simon (1973) has commented that such a polymer would make an excellent storage protein. The high intensity of alcian blue staining in young akinetes suggest the presence of high quantities of pectic acid and polysaccharides in these cells (Leak, 1967).

Polyphosphate granules are found to be of rare occurrence in the mature akinetes of *Anabaena cylindrica* (Talpasayi, 1963), *Cylindrospermum* sp. (Miller and Lang, 1968; Clark and Jensen, 1969), *Anabaena fertilissima* (Reddy, 1983) and *Aphanizomenon flos–aquae* (Wildman *et. al.*, 1975) although these were commonly present in vegetative cells during early stages of akinete differentiation (Sutherland *et. al.*, 1979; Ahluwalia, 1980).

During sporulation, there is no visible change in polyhedral bodies, polyglucoside and lipid body contents of the cells (Sutherland *et. al.*, 1979), however, Wildman *et. al.* (1975) reported decrease in number of lipid bodies during differentiation of akinetes. Lipid and fatty acid contents in akinetes were shown to be lower than those of vegetative cells in *Anabaena cylindrica* (Yamamoto, 1972; Ahluwalia and Paul, 1987). Large quantities of glycogen accumulate in cells undergoing sporulation and the amount increases with the onset of maturation of spores (Sarma and Kanta, 1980). The photosynthetic lamellae become contorted and get distributed throughout the protoplast. Polyglucan granules are present but their numbers gradually decrease in the resting period (Miller and Lang, 1968). There is alteration in the pigment contents of akinetes at various developmental phases. An increase in contents of xanthophylls was found while β-carotene decreased in akinetes of *Anabaena cylindrica* as compared to vegetative cells (Fay, 1969). Although there is considerable reduction in the pigments of akinetes in *Nostoc ellipsosporum*, *Anabaena cylindrica* and some other algae yet akinetes of *Nostoc* PCC 7524 show approximately double the amount of chlorophyll *a* and unchanged phycocyanin (Sutherland *et. al.*, 1979; Ahluwalia and Kumar, 1980).

Certain glycolipids and polysaccharides have been found in the cell envelopes of akinetes and heterocysts of *Anabaena cylindrica* (Cardemil and Wolk, 1976) and *Cyanospira rippkae* (Soriente *et. al.*, 1993), which are not present in the vegetative cells. In akinetes of *Anabaena variabilis* (Bjorn *et. al.*, 1983) and *Fischerella muscicola* (Singh and Kashyap, 1988), the oxygen evolving PSII remains suppressed. The akinetes of *A. cylindrica* have less superoxide dimutase activity than the vegetative cells (Grilli Caiola *et. al.*, 1991).

DNA content in an akinete is double than that of a vegetative cell during the late growing phase, while it is approximately equal to exponential phase of vegetative cells (Simon, 1977). Akinetes have been shown to be metabolically active and synthesize their own proteins and fats (Thiel and Wolk, 1983). Van Dok and Hart (1996) hypothesized that akinetes enclose an oxygen labile enzyme system (analogous to the nitrogenase complex in the heterocyst) that is capable of utilizing a form of phosphorus other than orthophosphate.

Genomic composition

Herdman and Carr (1974) estimated that a vegetative cell of *Anabaena cylindrica* has 25 genome copies during late logarithmic growth phase whereas an akinete contains an average of 57 genome copies. Ueda and Sawada (1972) using the flourochrome coriphosphin, cytochemically measured the relative amounts of deoxyribonucleic acid (DNA) in vegetative cells and akinetes of many filamentous heterocystous cyanobacteria and found that akinetes appear to contain 5 to 30 times the amount of vegetative cell DNA. However, Simon (1977) using *Anabaena cylindrica* found that DNA content per akinete is approximately equal to that from an exponentially growing vegetative cell, whereas the ratios of protein per akinete and RNA per akinete are respectively three and four times to that of exponentially growing vegetative cell. The ratio of RNA: DNA (54: 1) in the akinetes is found to be much greater than that of vegetative cells (13: 1) of *Anabaena cylindrica* (Simon, 1980). In *Nostoc* PCC 7524, the exponentially growing vegetative cells and akinetes contained approximately the same amount of DNA and RNA (Sutherland *et. al.*, 1979).

Metabolism of akinetes

Low rate of photosynthesis or its complete absence in akinetes is evident due to the changes in the pigments during differentiation. Fay (1969) found that isolated akinetes of *Anabaena cylindrica* fixed carbon dioxide approximately 20% of the rate obtained with whole filaments, and showed a higher respiratory rate than that of intact filaments. Nitrogenase activity reduced with the progress of akinete differentiation and was found to be absent in mature akinetes (Ahluwalia and Kumar, 1982a). Sutherland *et. al.* (1979) have also observed a decrease in photosynthetic CO_2 fixation during sporulation in *Nostoc* PCC 7524, and the rate of acetylene reduction became negligible with the maturation of the akinetes. Nichols and Carr (1978) demonstrated significant [3H] leucine uptake by isolated akinetes in *Anabaena cylindrica* indicating that the rate of protein synthesis in akinetes is comparable to those of vegetative cells (Sarma and Kanta, 1982).

Microcycle akinetes

The akinetes of certain cyanobacteria germinate readily in the medium of their formation (Fay, 1969; Singh *et. al.*, 1972) or upon transfer to fresh medium (Miller and Lang, 1968; Stulp and Stam, 1982). The combination of N and C sources induced high frequency of akinete formation in *Anabaena torulosa* in a 10 days old culture whereas nitrogen free medium supported a low frequency of akinetes by 20th day. The akinetes in these sources undergo germination too readily and the hormogones accumulated granular contents resulting in the formation of akinete like cells. This has been designated as microcycle akinete formation (Sarma, 1993).

Akinete differentiation

During differentiation, the vegetative cell increases several times in size and accumulates food reserves accompanied by the secretion of a thick multilayered spore wall. The presence of heterocysts in the filament facilitates/regulates the formation of akinetes (Wolk, 1965,1966), but their presence is not a pre-requisite for sporulation. Akinetes have even been observed in the absence of heterocysts (Ahluwalia and Kumar, 1982b).

Genes control akinete differentiation but these need to be triggered by some environmental stimulus (Singh,1967). All cyanobacteria do not have the same environmental trigger. A wide range of physico-chemical factors have been reported to stimulate akinete differentiation in blue–green algae (Tyagi, 1974; Rother and Fay, 1977; Sutherland *et. al.*, 1979; Van Dok and Hart, 1996; Li *et. al.*, 1997; Adams and Duggan, 1999; Adams, 2000). Differentiation of akinetes in the symbiotic cyanobacterium *Anabaena azollae* has also been characterized (Hardin and Fisher, 1995).

Akinete differentiation needs to commence while the trichome is still metabolically active so that reserve of carbon and nitrogen can be synthesized or mobilized (Sutherland *et. al.*, 1979; Whitton, 1992). Phosphate deficiency, low temperature and reduction in the availability of light energy can prove to be limiting (Van Dok and Hart 1996; Li *et. al.*, 1997). Akinete formation is characterized by an increase in cell size, gradual disappearance of gas vacuoles and polyphosphate bodies, increase of cytoplasmic density, cyanophycin and glycogen granules and decrease in lipid bodies. All these changes are reversed during akinete germination (Wildman *et. al.*, 1975). Akinetes generally form in stationary phase of culture in response to some unfavourable conditions. However, from field and laboratory studies on *Gloeotrichia echinulata*, Roelofs and Oglesby (1970) concluded that it is not a result of unfavourable conditions, as the akinetes are also seen during the logarithmic phase of growth. The storage of nitrogen as cyanophycin granules may be the initial inductive step for sporulation. Inhibition of sporulation at higher concentrations and enhancement at lower concentrations of nitrate-nitrogen, indicates that exhaustion or availability of carbon and nitrogen may act synergistically as well as antagonistically depending upon other factors possibly resulting from nitrogen stress (Singh and Srivastava, 1968; Nichols and Carr, 1978; Van Dok and Hart, 1996).

The cell division gets inhibited as the differentiation initiates, and an akinete is formed from the dividing cell. Replication of genetic material, which might have initiated in dividing cells, remains together in a single cell (akinete) due to the failure of cell division and hence akinetes may sometimes be found as polyploid structures (Fogg *et. al.*, 1973).

Factors controlling akinete differentiation

Phosphate concentration

This is considered to be one of the important parameters in akinete formation. In *Anabaena cylindrica*, a much lower concentration of phosphate (0.057 mM K_2HPO_4) was sufficient to inhibit sporulation (Wolk, 1965). Phosphate deficiency stimulates sporulation in *Cylindrospermum licheniforme* (Fisher and Wolk, 1976). But in certain other blue-green algae like *Aphanizomenon flos-aquae* (Rother and Fay, 1979), *Nostoc* PCC 7524 (Sutherland *et. al.*, 1979) and *Nostoc ellipsosporum* (Ahluwalia, 1980), sporulation was not observed in the cultures, inoculated directly into phosphate deficient medium. Many other blue–green algae (*Anabaena circinalis, A. flos-aquae, A. spiriodes, A. solitaria*) of eutrophic Kataglacial lakes were found sporulating even with higher levels of intracellular phosphorus (Rother and Fay, 1979). Polyphosphate granules are absent in the developing akinetes. Accumulation of cyanophycin and carbon reserve in akinetes of *A. circinalis* (Van Dok and Hart, 1996), *A. cylindrica* (Talpasayi, 1963), *A. fertilissima* (Reddy and Talpasayi, 1981) and *Aphanizomenon flos-aquae* (Cmiech *et. al.*, 1988) have been reported.

Optimum concentration of phosphate in the medium aided in quicker sporulation in *Anabaena doliolum* and *Fischerella muscicola*, and the phosphate concentration beyond a particular level inhibited the process (Kaushik *et. al.*, 1971). However, one species i.e., *Gloeotrichia echinulata* accumulated cyanophycin and glycogen but was observed to contain only one molecule of polyphosphate (Cmiech *et. al.*, 1984).

Phosphorus deficiency did not stimulate the formation of akinetes in nitrate free medium in *A. torulosa*. The organism was unable to fix nitrogen and differentiate akinetes in phosphorus deficient nitrate free medium. The regulation of akinete differentiation in phosphorus deficiency proceeds through the regulation of nitrogen assimilating systems (Sarma and Khattar, 1992, Sarma, 1993).

Temperature

In *Anabaena cylindrica,* temperature above 20°C was found to augment sporulation (Wolk, 1965). However, with higher concentration of calcium ion (2.5 mM) in the medium, even temperatures upto 30°C also showed significant sporulation. It was found that species of *Nostoc, Anabaena* and *Cylindrospermum* could sporulate in basal medium between 25-32°C. Li *et. al.* (1997) reported that none of the species of *Anabaena* grew above 30°C and no akinete developed even at 20° C in *A. mucosa, TBC 426, A. crassa* TAC 436, *A. spiriodes* TAC 443 though these species grew well at 10-20°C.

Nutrient ions

Various nutrient ions have been thought to affect growth and differentiation in blue-green algae. Rother and Fay (1979) suggested that the induction of sporulation may be due to extreme conditions at the water surface during summer to which the algal bloom is exposed rather than to nutrient deficiency.

Calcium plays an important role in sporulation. Normal growth and sporulation of *Anabaena cylindrica* occur in medium containing 0.106 mM calcium ion (Wolk, 1965). It was also found that sodium acetate significantly enhanced/increased the process of sporulation(Wolk, 1965). Several amino acids showed a stimulatory effect. In case of *A. torulosa,* maximum frequency of akinetes was found with L-glutamine (Sarma and Kanta, 1980). Akinete formation was stimulated by Fe-deficiency in *Anabaena cylindrica* whereas in *Gloeotrichia echinulata* deficiencies of Fe, Mg, Ca and SO_4 result in the reduction of akinete units (akinete frequency). Compounds of other components of the culture medium, such as Mn, Zn, Co, Cu and Mo, failed to induce sporulation. Few akinetes formed in *Nostoc* sp. even in unionised distilled water but these lacked the thick walls and extensive granulation, typical of normal akinetes (Ahluwalia, 1980).

Culture density

Sporulation in *Anabaena cylindrica* occurred only when the algal inoculum was added in larger amounts; a reduction in the inoculum resulted in curtailed sporulation (Wolk, 1965). It has also been found to affect spore initiation and maturation in many species of *Anabaena* and *Nostoc.* However, there was no effect of culture concentration on akinete formation in algae grown on agar plates (Ahluwalia, 1980).

pH (Hydrogen ion concentration)

Wolk (1965) used DL-alanyl glycine and DL-alanyl –DL- alanine as buffers and found that there was no sporulation in unbuffered medium due to change in pH of the medium. Similarly, Rother and Fay (1979) reported stimulated sporulation with N-2-hydroxy-ethylpiperazine–N'-2-ethanolsulphonic acid (HEPES) buffer, while no akinete was found in unbuffered cultures. However, few species of *Nostoc* and *Anabaena* have been found sporulating normally in unbuffered medium as well (pH range between 7.00-9.5), otherwise at extreme (very high and low) pH values, even growth of the algae stopped and therefore, there is no question of akinete formation (Ahluwalia, 1980).

Cultural filtrate

Sporulation can be avoided for many days by regularly transferring the cultures to fresh medium much before the initiation of sporulation (Ahluwalia, 1980). In *Cylindrospermum licheniforme*, Fisher and Wolk (1976) reported the production of a substance by the alga which greatly stimulated sporulation in the same alga. Hirosawa and Wolk (1979) purified the substance from the supernatant of sporulating cultures and proposed its structure, In other blue-green algae such as *Aphanizomenon flos-aquae* (Rother and Fay, 1979) and *Nostoc* PCC 7524 (Sutherland *et. al.*, 1979), culture filtrate did not show such stimulating effect on the process of akinete differentiation. Culture filtrate from sporulating cultures hastened the formation of akinetes in *Nodularia spumigena* (Pandey and Talpasayi, 1980).

Carbon and nitrogen levels

It has been suggested that the ability of a cell to sporulate is controlled by a C: N ratio and that heterocyst regulates this ratio in differentiating cells *via* its metabolism (Tyagi, 1974) .In *Aphanizomenon flos-aquae*, there was an appreciable nitrogenase activity preceding and during differentiation of akinetes; and the cellular C: N ratio of sporulating populations was lower than in the non sporulating algae. In cyanobacteria, the breakdown of carbon source is known to occur through oxidative pentose phosphate (OPP) pathway with glucose–6-phosphate dehydrogenase (G6PDH) as the first enzyme involved in this route (Stewart, 1980; Smith, 1982). The levels of G6PDH in relation to nitrogen source and phase of akinete differentiation revealed that nitrogen fixing culture exhibited a very high activity than in the presence of combined nitrogen source (nitrate) in photoautotrophic, phototrophic, photoheterotrophic and heterotrophic conditions and maximum levels were attained at the maturation phase of akinete formation (Sarma and Khattar, 1992). It is suggested that the onset of sporulation in surface blooms is initiated by an imbalance of the normal vegetative growth of the algal cells and of their carbon and nitrogen metabolism (Rother and Fay, 1979). Decreased sporulation was found with an increased concentration of glucose in the medium (Kaushik *et. al.*, 1971). However, Tyagi (1974) found that promotion of sporulation depended upon carbon–nitrogen balance of the cells and any change in nitrogen being more important.

Hydrogen stimulation

Hydrogen gas is found stimulating akinete differentiation adjacent to heterocysts in *Cylindrospermum* sp. It is believed that the effect of hydrogen on sporulation is apparently mediated by heterocysts because the 'uptake hydrogenease' appears to be restricted to heterocysts only (Hirosawa and Wolk, 1979).

Light (Photoregulation)

Light is of much importance in affecting differentiation of blue-green algae (Tyagi and Ahluwalia, 1978). The optimal light intensity for sporulation was found to be about 80 foot candles, however, it was broadly supportive in the range of 40-160 foot candles. At very low or much higher values i.e. 20 and 240 foot candles, sporulation was significantly suppressed (Wolk, 1965), The coincidence of sporulation with dense surface accumulations of blue-green algae in nature indicated that light may be playing a critical role. Contrary to their expectation from observations of natural populations of algae, high light intensity did not induce sporulation in indoor experiments and laboratory cultures of *Aphanizomenon flos–aquae* (Rother and Fay, 1979). In cultures incubated under monochromatic light of various wavelengths, a cessation of growth and of morphological differentiation was observed like that of low light intensity conditions (Rother and Fay, 1979). However, formation of akinetes was accelerated by regimes of white, red or yellow light (Kaushik and Kumar, 1970). Akinetes do not form in darkness but differentiate in red, orange and white light regimes (Ahluwalia, 1980). It appears that photosynthetic role of light may be important during akinete differentiation. Wyman and Fay (1986) observed that akinete differentiation occurs at a lower culture density in green light. Exposure of cells to UV-B significantly delayed the differentiation of vegetative cells into heterocysts and akinetes in *A. aequalis* (Mary and Harris, 1994).

Role of heterocysts in sporulation

Carter (1856) was the first to suggest the role of heterocyst in the control of sporulation and this hypothesis has been considered subsequently by Geitler (1925), Fritsch (1951), Wolk (1966), Tyagi (1974) and Ahluwalia (1980). This is based on the fact that akinetes are found usually in heterocystous algae and normally develop adjacent to heterocysts on either one or both sides in the filament, although in a few genera they develop away from the heterocyst and this spatial relationship may not always exist (Tyagi, 1978).

The role of heterocyst in sporulation was strengthened by the work of Wolk (1966) and Tyagi (1974). It was found that sporulation in *Anabaena cylindrica,* adjacent to heterocyst, is prevented by the detachment of the two cell types (Wolk, 1966). Akinete differentiation initiates in the cells lying away from the heterocyst in *Anabaena doliolum (*Tyagi, 1974) thus indicating a role for heterocysts in stimulation or inhibition of akinete formation in the filaments of various algae. However, many species of *Raphidiopsis* undergo sporulation in the absence of heterocysts (Eberly, 1966; Hill, 1970).

Some mutants (nonheterocystous) of blue-green algae have been reported to sporulate while a number of non-sporulating mutants with heterocysts have also been isolated (Singh, 1967; Ladha and Kumar, 1978). Sutherland *et. al.* (1979) reported that heterocysts were not necessary for akinete differentiation in *Nostoc* PCC 7524. Akinetes can form in filaments of *Anabaena* and *Nostoc* even without heterocysts whereas *Cylindrospermum* formed akinetes only in the presence of heterocysts in the filaments. Their presence, however, in *Anabaena, Nostoc* and *Cylindrospermum* resulted in a more regular and predictable initial spatial pattern of akinete differentiation.

Pattern formation

Akinetes and heterocysts form at specific positions in the filaments of different blue-green algal species. Akinetes always formed adjacent to heterocysts in *Anabaena cylindrica* (Wolk, 1966;

Simon 1977). However, a centrifugal pattern of akinete formation in *Anabaena doliolum*, *Nostoc* PCC 7524 and *Nostoc* sp. respectively was reported (Tyagi,1974; Sutherland *et. al.*,1979; Ahluwalia,1980). In *Anabaena circinalis*, cells between two intercalary heterocysts differentiate into akinetes at random (Tyagi, 1978). In *Cylindrospermum licheniforme* and *Gloeotrichia* sp., a conspicuous akinete (rarely more) differentiates in the immediate vicinity of the terminal heterocyst. Different studies have shown that the pattern of sporulation is very specific for a species. The sporulation pattern in *Anabaena cylindrica* and *Cylindrospermum licheniforme* supports the hypothesis of akinete induction by heterocysts (Wolk, 1966 ; Fisher and Wolk,1976; Hirosawa and Wolk, 1979) whereas in *Anabaena doliolum* and *Nostoc* sp., inhibition of akinete development near the heterocyst is evident (Tyagi, 1974). Nichols *et. al.* (1980) studied the pattern of akinete formation using canavanine and other amino acid analogues at different intervals during sporulation. The differentiation of akinetes in *Scytonema fritschii* occurred adjacent to the newly developed heterocyst in the late exponential phase (Sarma and Ghai, 1998). The regulation of cell fate and pattern formation is a fundamental problem in developmental biology. Cell–cell communication often plays a key role in controlling such development.

In view of these reports, it is quite evident that there is no single factor that can be held responsible for initiation of sporulation or pattern formation but many factors either separately or in combination, control the steps of sporulation in blue-green algae.

Akinetes as perennating structures

Akinetes are relatively resistant cells and their ability to survive desiccation, cold and other harsh conditions has been much better than the vegetative cells (Fritsch, 1945; Yamamoto, 1975; Sutherland *et. al.*, 1979). However, Jones (1979) questioned these structures as perennating bodies on the basis of his studies with *Aphanizomenon flos-aquae.*

Ahluwalia (1980) using three different algal genera found that there was a gradual increase in lag period and fall in germination percentage, with the increase in age of the akinetes. Their study, further suggests that these are definitely more resistant bodies than the vegetative cells towards temperature, light intensity and ultraviolet radiation.

Akinete germination

Akinetes germinate immediately after their transfer to favourable conditions. Various workers have studied the germination of blue green and green algal akinetes under defined growth conditions (Miller and Lang, 1968; Kaushik and Kumar, 1970; Wildman *et. al.*, 1975; Agrawal, 1984; Huber, 1985; Neely-Fisher *et. al.*, 1989; Van Dok and Hart, 1997). Synchronous germination of akinetes and development of heterocysts in such germlings (2-3 celled stage) have been reported in *Anabaena* PCC 7937 and *Nostoc* PCC 6720 (Skill and Smith, 1987). Dry akinetes of *Anabaena cylindrica* have been found to retain the capacity to germinate even after their storage for 5 years in the dark, while the vegetative cells, under similar conditions, fail to rejuvinate even after 15 days of storage (Yamamoto, 1975). Akinetes have been regarded as perennating structures; however, Jones (1979) suggested that production of akinetes was not necessary for perennation of species in *Aphanizomenon flos-aquae*. During germination, the akinetes enlarge and their contents become homogeneous and blue-green in colour. One, two or rarely more divisions occurred prior to liberation of the germling from the akinete envelope. Very little information is available to date regarding the factors, which induce or regulate germination of akinetes in various blue-green algae. Most of the factors affecting formation of akinetes also influenced the germination of akinetes.

Nitrogen

Various nitrogen sources normally do not affect the germination, however, sometimes a slight enhancement occurs with the supplementation of NH_4^+ and NO_3^-. Urea inhibited akinete germination in species of *Anabaena*, *Nostoc* and *Cylindrospermum* (Ahluwalia, 1980). Addition of nitrate increases akinete germination in *Anabaena circinalis* and *A. vaginicola* (Rai and Pandey, 1981; Fay, 1988). Nitrate is important for akinetes of some species to germinate while in others it is not required (Rai and Pandey, 1981; Reddy, 1983).

Yamamoto (1976) found no effect of nitrate on the germination of akinetes (of *Anabaena cylindrica*) incubated in air, however, the germination decreased in the presence of nitrate under anaerobic conditions. Cyanobacterial akinetes generally contain large number of cyanophycin granules as the primary storage material. In case of *A. circinalis* and *Nodularia* sp., there was no requirement for nitrogen during germination as they had sufficient store of nitrogen in the form of cyanophycin granules (Huber, 1985; Van Dok and Hart, 1997). In some algae, addition of nitrate increased the per cent germination of akinetes (Reddy, 1983). Thus, nitrate seems to be important for some species while it is not required in others.

Phosphorus

Phosphorus is needed for healthy growth of the germlings. About 40% germination has been recorded in akinetes of *Anabaena vaginicola* without phosphate, compared to 98% with addition of phosphate (Kaushik *et. al.*, 1971; Rai and Pandey, 1981). The akinetes of *Nodularia spumigena* and *Ańabaena circinalis* required small concentrations of phosphorus for germination (Huber, 1985; Van Dok and Hart, 1997). Some akinetes showed germination and further growth on bactoagar without any exogenous addition of nutrients indicating thereby that either the phosphorus is available in the agar or carried over in the inoculum.

Light

This has been regarded as an essential requirement for germination of akinetes in *Anabaena cylindrica* (Fay, 1969; Yamamoto, 1976), *Anabaena doliolum*, *Fischerella muscicola* (Kaushik and Kumar, 1970), *Anabaena* sp., *Anabaenopsis arnoldii* (Reddy *et. al.*, 1975), *Anabaena variabilis* (Braune, 1979), and in *Nostoc ellipsosporum* (Ahluwalia and Kumar, 1980). According to Kaushik and Kumar (1970) akinetes germinated only in the light, irrespective of the presence or absence of organic nutrients in the medium. They found red light to be ineffective in germination of akinetes of *Anabaena doliolum* and *Fischerella muscicola* and this light effect was not to be associated with photosynthesis. But Reddy *et. al.* (1975) found a red: far red photoreversal for akinete germination in *Anabaena fertilissima* suggesting the presence of a phytochrome action in blue-green algae. Yamamoto (1976) also noticed stimulation in germination by red light but blue or green light did not support the germination. Germination was inhibited by exogenous addition of 3-(3, 4-dicholorophenyl) 1, 1- dimethyl urea (DCMU), an inhibitor of photosynthesis. Thus to date there are conflicting reports regarding the role of light in akinete germination. The most active spectral range for the germination process in akinetes coincided well with the maximum of the light absorption by C-phycocyanin in *Anabaena variabilis*. Further, since there was slight impairment in germination, when electron transport was blocked by DCMU, the formation of photosynthetic products as pre-requisite of germination is rather unlikely (Braune, 1979,1980; Braune *et. al.*,1986).

Ahluwalia and Kumar (1980) have found that red and orange light supported an increased akinete germination while blue-green and far-red light did not support the process. They did not notice any photoreversal in germination with red: far red and red: green light regimes. They concluded from various experiments that the effect of light on akinete germination is cumulative and favour photosynthetic role of light, yet at the same time, photomorphogenetic role or some other role of light (through induction of certain enzymes or hormones) cannot be ruled out (Ahluwalia, 1980). Singh and Sunita (1974) studied different biochemical changes during germination of akinetes in *Anabaena doliolum* in light and dark-conditions. They found that the biosynthetic events following RNA synthesis during akinete germination show a requirement for light. The emergence of germling, which was a criterion for scoring germination, is a very late event. It would be difficult to detect, if some type of phytochrome (phycochrome) was involved at a very early stage, because once the germination is triggered, the emergence of germlings may be driven by photosynthesis, since green light, which is non-photosynthetic, do not support germination (Singh and Sunita, 1974).

Temperature

Temperature broadly suitable for akinete germination is the same as required for optimal vegetative growth (25°C). Lower and higher temperatures inhibited the germination and both cold and heat shocks delayed the process (Yamamoto, 1976; Rai and Pandey, 1981; Agrawal, 1984; Li *et. al.*, 1997). The lag period for akinete germination was shortened as the temperature was elevated from 20°C to 27°C. There was no germination above 35°C and hence optimum temperature for germination was in good agreement with that of vegetative growth. The viability of akinetes of *Anabaena cylindrica* was lost completely at temperature higher than 55°C (Yamamoto, 1976; Ahluwalia, 1980).

pH

High pH may alter the permeability of the cell membrane influencing the uptake of ions or suitable metabolites. Brock (1973) observed death and decay of blue-green algal cells in acidic conditions caused by the breakdown of chlorophyll to phaeophytin. The pH optima for akinete germination were consistent corresponding to their optima for growth (Van Dok and Hart, 1997). In *Anabaena cylindrica*, germination of the akinetes tended to decrease in strong acidic and basic conditions. At an alkaline pH (10), most of the germinating akinetes failed to grow beyond a 4-celled stage (Yamamoto, 1976). Inability of akinetes to germinate or death of the germlings in highly acidic and alkaline pH may be attributed to the toxic effect of such conditions on the photosynthetic pigments, non availability of carbondioxide and liberation of organic acids in the culture medium (Rai and Pandey, 1981).

Effect of desiccation

The akinetes of *Anabaena vaginicola* showed a lag period and decrease in germination with the increase in desiccation time (Rai and Pandey, 1981). Akinetes of two alkaliphilic cyanobacteria *Cyanospira rippkae* and *C. capsulata* proved to be highly resistant to desiccation being able to germinate in the presence of either nitrogen or nitrate as nitrogen source with a frequency of more than 90 per cent after seven years of storage in dry state (Claudio *et. al.*, 1994). Akinetes of *Aphanizomenon flos-aquae* have been found in 18 years old sediments while *Anabaena* sp. may survive for more than 60 years (Livingstone and Jaworski, 1980). Dry akinetes of *Anabaena cylindrica* retained the capacity to germinate even after five years of storage in the dark (Yamamoto, 1975).

Antibiotics

Germination of akinetes is dependent upon protein synthesis, which can be inhibited by chloramphenicol, streptomycin and penicillin (Singh and Sunita, 1974). Kumar and Kaushik (1971) found penicillin to be much more effective than the other two antibiotics in inhibiting germination.

Metabolic changes associated with germination

Cyanobacteria producing their metabolically quiescent structures, called akinetes, must have developed control to switch off various metabolic events during the onset of akinete development or germination. The nature of these switches is unknown. The less of nitrogenase and glutamine synthetase (GS) activity is known to precede akinete formation, thus suggesting a role for the GS mediated pathway of nitrogen assimilation (Rao *et. al.*, 1987; Sarma, 1993). During germination, there is development of photosynthetic pigments, the carbon decline, protein increase and nitrogen content remain unchanged (Thiel and Wolk, 1983; Rai *et. al.*, 1988).

In the opinion of Fogg (1969) "the impression at present is that the formation of akinetes is relatively unimportant means of survival of adverse conditions by most algae." This is also supported by the observation on sub-aerial blue-green algae which do not form akinetes but survive the environmental stress due to highly reducing state of the cytoplasm coupled with the presence of thick coloured sheath e.g. *Scytonema* sp. In cases where akinetes were reported to have survived extreme conditions, it could be due to the high reducing state of the protoplast. The thick coloured akinete is also considered as a means of withholding cell division, rather than of resistance to environmental extremes (Nichols and Carr, 1978).

Acknowledgement

Authors thank The Chairman, Department of Botany, Panjab University, Chandigarh for research facilities during the course of this work.

References

Adams, D. G. 2000. Cyanobacterial phylogeny and development: questions and challenges. In: *Prokaryotic Development*, Edited by Brun, Y. V. and Shimkess, L. J., Amer. Soc. Microbiol. pp.109-148.

Adams, D. G. and Duggan, P. S. 1999. Heterocyst and akinete differentiation in cyanobacteria. *New Phytol.* **144**: 3-33.

Agrawal, S. C. 1984. Effects of different factors on the akinete germination of the green alga *Stigeoclonium pascheri* (Vischer) Cox and Bold. *Microbios Lett.* **27**: 141-144.

Ahluwalia, A.S. 1980. *Studies on cellular differentiation in some nitrogen fixing blue-green algae.* Ph.D. thesis, Banaras Hindu Univ., Varanasi. pp. 178.

Ahluwalia, A.S. and Kumar, H.D. 1980. Akinete germination in blue green alga *Nostoc ellipsosporum* Desm Rabenh Ex Born. *et* Flah. *Indian J. Exp. Biol.* **18**: 663-664.

Ahluwalia, A.S. and Kumar, H.D. 1982a. Cellular differentiation and nitrogenase activity in the cyanobacterium *Anabaena. Biol. Plant.* **24**: 136-141.

Ahluwalia, A.S. and Kumar, H.D. 1982b. Pattern of akinete differentiation in the blue–green alga *Nostoc* sp. *Beitr. Biol. Pflanzen.* **57**: 459-467.

Ahluwalia, A.S. and Paul, R. 1987. Fatty acid composition of a nitrogen-fixing blue-green alga, *Nostoc ellipsosporum. Ann. Biol.* **3**: 56-58.

Barbiero, R.P. and Welch, E.B. 1992. Contribution of benthic blue-green algal recruitment to lake population and phosphorus translocation. *Freshwater Biol.* **27**: 249-260.

Bjorn, G.S., Braune, W. and Bjorn. L.O. 1983. Photochromic pigments in akinetes and pigment characteristics of akinetes in comparison with vegetative cells of *Anabaena variabilis. Physiol. Plant.***59**: 493-500.

Braune, W. 1979. C- phycocyanin, the main photoreceptor in the light dependent germination process of *Anabaena* akinetes. *Arch. Microbiol.* **122**: 289-295.

Braune, W. 1980. Structural aspects of akinete germination in the cyanobacterium *Anabaena variabilis. Arch. Microbiol.* **126**: 257-261.

Braune, W., Wiedemann, B. and Bisen,P.S.1986. *In vivo* microspectrophotometry of *Anabaena* akinetes in different stages of their formation and during germination. *Cytobios* **47**: 53-63.

Brock, T. D. 1973. Lower pH limit for the existence of blue green algae: Evolutionary and ecological implications. *Science* **179**: 480-483.

Cardemil, L. and Wolk, C.P. 1976. The polysaccharides from heterocyst and spore envelope of blue green algae. Methylation analysis and structure of the back bone. *J. Biol. Chem.* **251**: 2967-2975.

Cardemil, L. and Wolk, C.P. 1981. Polysaccharides from the envelopes of heterocysts and spores of the blue-green algae

Anabaena variabilis and *Cylindrospermum licheniforme*. *J. Phycol.* **17:** 234-240.

Carter, H.J. 1856. Notes on freshwater infuroria of the island of Bombay No. 1 organization. *Ann. Mag. Nat. Hist. Ser.* II. **18:** 115–132.

Clark, R.L. and Jensen, T.E. 1969. Ultrastructure of akinete development in a blue–green alga, *Cylindrospermum* sp. *Cytologia* **34:** 439–448.

Claudio, S., Ena, A., Materassi, R. and Viacenzini, M. 1994. Germination of desiccated aged akinetes of alkaliphilic cyanobacteria. *Arch. Microbiol.* **162**: 20-25.

Cmiech, H.A., Leedale, G.F. and Reynolds, C.S. 1984. Morphological and ultrastructural variability of planktonic cyanophyceae in relation to seasonal periodicity. I. *Gloeotrichia echinulata*. *Br. Phycol. J.* **19** : 259–275.

Cmiech, H.A., Leedale, G.F. and Reynolds, C.S. 1988. Morphological and ultrastructural variability of planktonic cyanophyceae in relation to seasonal periodicity. IV *Aphanizomenon flos–aquae*. *Br. Phycol. J.* **23** : 239–250.

Desikachary, T.V. 1959. *Cyanophyta*. Indian Council of Agricultural Research, New Delhi, India, pp. 686.

Dunn, J.H. and Wolk, C.P. 1970. Composition of the cellular envelopes of *Anabaena cylindrica*. *J. Bacteriol.* **103**: 153–157.

Eberly, W.R. 1966. Notes on some new and rare myxophyceae in laboratory culture. *Trans. Amer. Microsc. Soc.* **85**: 130–138.

Fay, P. 1969. Metabolic activities of isolated spores of *Anabaena cylindrica*. *J. Exp. Bot.* **20** : 100-109.

Fay, P. 1988. Viability of akinetes of the planktonic cyanobacterium *Anabaena circinalis*. *Proc. R. Soc. London* B **234:** 283-301.

Fisher, R.W. and Wolk, C.P. 1976. Substances stimulating the differentiation of spores of the blue green alga *Cylindrospermum licheniforme*. *Nature* **259**: 394-395.

Fogg, G.E. 1969. The physiology of an algal nuisance. *Proc. Soc. London* **173**: 175-189.

Fogg, G.E., Stewart, W.D.P., Fay, P. and Walsby, A.E. 1973. *The Blue Green Algae*. Academic Press, London. pp. 459.

Fritsch, F.E. 1945. *The Structure and Reproduction of the Algae, Vol. II.* Cambridge University Press. pp. 796.

Fritsch, F.E. 1951. The heterocysts: A botanical engima. *Proc. Linn. Soc. London* **162:** 194–211.

Geitler, L. 1925 Synoptische Darstellung der Cyanophycean in Morphologischer and Systematischer Hinsicht. *Beih. Bot. Zentralbl.* **41:** 163–294.

Grilli Caiola, M., Canini, A. and Rotilio, G. 1991. Superoxide dismutase in vegetative cells, heterocysts and akinetes of *Anabaena cylindrica. FEMS Microbiol. Lett.* **80** : 161-166.

Grilli Caiola, M. and Vecchi, L. De. 1980. Akinete ultrastructure of *Nostoc* species isolated from cycad coralloid roots. *Can. J. Bot.* **58:** 2513-2519.

Hardin, S. C. and Fisher, R. W. 1995. Characterization of akinete differentiation in the cyanobacterium *Anabaena azollae. Curr. Microbiol.* **31:** 265-269.

Herdman, M. and Carr, N.G. 1974. Estimation of the genome size of blue-green algae from DNA renaturation rates. *Arch. Microbiol.* **99**: 251–254.

Hill, H. 1970. A new form of *Raphidiopsis mediterranea* Skuja found in Minnesota lakes. *Phycologia* **9:** 73–77.

Hirosawa, T. and Wolk, C.P. 1979. Factors controlling the promotion of akinetes adjacent to heterocyst in the cyanobacterium, *Cylindrospermum licheniforme* Kutz. *J. Gen. Microbiol.* **114**: 423-432.

Huber, A.L. 1985. Factors affecting the germination of akinetes of *Nodularia spumigena. Appl. Environ. Microbiol.* **49:** 73-78.

Jones, R.I. 1979. Notes on the growth and sporulation of a natural population of *Aphanizomenon flos-aquae. Hydrobiologia* **62:** 55–58.

Kappers, F.I. 1976. Presence of blue-green algae in sediments in lake Brielle. In: *Interactions between Sediments and Freshwater*. Edited by Golterman, H.L., W. Junk, The Hague. pp 383-386.

Kaushik, M. and Kumar, H.D. 1970. The effect of light on growth and development of two nitrogen fixing blue-green alga. *Arch. Microbiol.* **74**: 52-57.

Kaushik, M., Kumar, H.D. and Singh, H.N. 1971. Studies on growth and development of two nitrogen fixing blue-green algae. I carbon and phosphorus. *Z. Pflanzenphysiol.* **65:** 432-442.

Kumar, H.D. 1970. Ultraviolet lethality and spore survival in blue-green algae. *Phykos* **9**: 104-115.

Kumar, H.D. and Kaushik, M. 1971. Studies on growth and development of two nitrogen fixing blue green algae.11. Effects of antibiotics. *Z. Pflanzenphysiol.* **65:** 443-452.

Ladha, J.K. and Kumar, H.D. 1978. Some characteristics of two morphological mutants of *Nostoc linckia* induced by nitrosoguanidine. *Z. Allg. Microbiol.* **17 :** 513–520.

Leak, L.V. 1967. Fine structure of the mucilaginous sheath of *Anabaena* species. *J. Ultrastru. Res.* **21:** 61-74.

Li, R., Watanabe, M. and Watanabe,M.M. 1997. Akinete formation in planktonic *Anabaena* species by treatment with low temperature. *J. Phycol.* **33 :** 576-584.

Livingstone, D. and Jaworski, G. H. M. 1980. The viability of akinetes of blue green algae recovered from sediments of Rostherne Mere. *Br. Phycol. J.* **15**: 357-364.

Mary, K.B. and Harris, O.D. 1994. Delay of cell differentiation in *Anabaena aequalis* caused by UV–B radiation and the role of photoreactivation and excision repair. *Photochem. Photobiol.* **59**: 204–208.

Miller, M.M. and Lang, N.J. 1968. The fine structure of akinete formation and germination in *Cylindrospermum. Arch. Microbiol.* **60** : 303 -313.

Nagy, L.A. 1978. New filamentous and cystous microfossils,~ 2300 MY old, from the Transvaal sequence. *J.Paleontol.* **52**: 141-154.

Neely-Fisher, D., White, W. and Fisher, W. 1989. Fructose- induced dark germination of *Anabaena* akinetes. *Curr. Microbiol.* **19**: 139-142.

Nichols, J.M. and Adams, D.G. 1982. Akinetes. In: *The Biology of Cyanobacteria.* Edited by Carr, N.G. and Whitton, B.A. Blackwell Scientific Publ. Oxford. pp. 387–412.

Nichols, J.M., Adams, D.G. and Carr, N.G. 1980. Effect of canavanine and other amino acid analogues on akinete formation in the cyanobacterium *Anabaena cylindrica. Arch. Microbiol.* **127:** 67–75.

Nichols, J.M. and Carr, N.G. 1978. Akinetes of cyanobacteria. In: *Spores VII.* Edited by Chambers, G. and Vary, J.M., Amer. Soc. Microbiol., Washington DC. pp. 335–343.

Pandey, R.K. and Talpasayi ,E.R.S. 1980. Control of sporulation in a blue-green alga, *Nodularia spumigena* Mertens. *Indian J. Bot.* **3:** 128–133.

Rai, A.K. and Pandey, G.P. 1981. Influence of environmental stress on the germination of *Anabaena vaginicola* akinetes. *Ann. Bot.* **48** : 361–370.

Rai. A.N., Rao ,V.V. and Singh,H.N. 1988. Metabolic changes associated with akinete germination in the cyanobacterium *Anabaena doliolum. New Phytol.* **109:** 133–138.

Rao, V.V., Ghose, R. and Singh,H.N. 1987. Diazotrophic regulation of akinete development in the cyanobacterium *Anabaena doliolum. New Phytol.* **106** : 161–168.

Reddy, P.M. 1983. Changes in the phycobiliproteins during spore (akinete) differentiation in a cyanobacterium, *Anabaena fertilissima. Biochim. Biophys. Acta* **761:** 191–195.

Reddy, P. M., Rao, P. S. N. and Talpasayi, E.R. S. 1975. Effect of red and far red illuminations on the germination of two blue green algae. *Curr. Sci.* **44:** 678-679.

Reddy, P. M. and Talpasayi, E.R. S. 1981. Some observations related to red-far red antagonism in germination of spores of the cyanobacterium *Anabaena fertilissima. Biochem. Physiol. Pflanz.* **176:** 105-107.

Rippka, R.,Deruelles, J., Waterbury, J.B., Herdman, M. and Stanier ,R.Y. 1979. Generic assignments, strain histories and properties of pure cultures of cyanobacteria. *J. Gen. Microbiol.* **111:** 1–61.

Roelofs, T.D. and Oglesby, R.T. 1970. Ecological observations on the planktonic cyanophyte *Gloeotrichia echinulata. Limnol. Oceanogr.* **15** : 224–229.

Rother, J.A. and Fay, P. 1977. Sporulation and development of planktonic blue green algae in two Salopian meres. *Proc. R. Soc. London B* **196:** 317-332.

Rother, J.A. and Fay, P. 1979. Blue green algal growth and sporulation in response to simulated surface bloom conditions. *Br. Phycol. J.* **14**: 59–68.

Sarma, T.A. 1993. Akinete differentiation in the cyanobacterium *Anabaena torulosa*. In: *Nat. Sem. Cyanobacterial Res.– Indian Scene.* NFMC,BARD, Tiruchirapalli. pp. 50–62.

Sarma, T. A. and Ghai, R. 1998. Pattern of akinete differentiation in cyanobacterium *Scytonema fritschii. Folia Microbiol.* **43:** 649–656.

Sarma, T.A. and Kanta, S. 1980. Biochemical studies on sporulation in blue –green algae. III Effect of amino acids on glycogen accumulation. *Z. Allg. Microbiol.* **20** : 653–655.

Sarma, T.A. and Kanta, S. 1982. Variation in enzyme levels during sporulation in blue-green algae. *Biochem. Physiol. Pflanz.* **177:** 278-282.

Sarma, T.A. and Khattar, J.I.S. 1992. Phosphorus deficiency, nitrogen assimilation and akinete differentiation in the cyanobacterium *Anabaena torulosa. Folia Microbiol.* **37:** 223–226.

Simon, R.D. 1973. Measurement of the cyanophycin granule, polypeptide content in the blue–green alga *Anabaena cylindrica. J. Bacteriol.* **114:** 1213–1216.

Simon, R.D. 1977. Sporulation in the filamentous cyanobacterium *Anabaena cylindrica*: The course of spore formation. *Arch. Microbiol.* **111:** 283–288.

Simon, R.D. 1980. DNA content of heterocysts and spores of the filamentous cyanobacterium *Anabaena variabilis. FEMS Micribiol. Lett.* **8:** 241-246.

Singh, H. N. 1967. Genetic control of sporulation in the blue green alga *Anabaena doliolum* Bharadwaja. *Planta* **75:** 33-38.

Singh, H. N. and Srivastava, B. S. 1968. Studies on morphogenesis in a blue green alga . 1. Effect of inorganic nitrogen sources on developmental morphology of *Anabaena doliolum. Can. J. Microbiol.* **14**: 1341-1346.

Singh, H. N. and Sunita, K. 1974. A biochemical study of spore germination in the blue green alga *Anabaena doliolum. J. Exp. Bot.* **25:** 837-845.

Singh, R.N., Tiwari, D.N. and Singh, V.P. 1972. Genetic basis of cellular differentiation in blue-green algae. In: *Taxonomy and Biology of Blue-green algae.* Edited by Desikachary, T.V., Univ. Madras, Madras, pp. 27-37.

Singh, S. and Kashyap, A.K. 1988. Metabolic characteristics of akinetes of the cyanobacterium *Fischerella muscicola*. *New Phytol*. **110**: 97-107.

Skill, S. C. and Smith, R. J. 1987. Synchronous akinete germination and heterocyst differentiation in *Anabaena* PCC 7937 and *Nostoc* PCC 6720. *J. Gen. Microbiol*. **133:** 299-303.

Smith, A.J. 1982. Modes of cyanobacterial carbon metabolism. In: *The Biology of Cyanobacteria*. Edited by Carr, N.G. and Whitton, B.A., Blackwell Scientific Publ., Oxford. pp. 47–85.

Soriente, A., Gambacorta, A., Trincone, A., Sili, A., Vincenzini, M. and Sodano,G. 1993. Heterocyst glycolipids of the cyanobacterium *Cyanospira rippkae*. *Phytochemistry* **33** : 393–396.

Stewart, W.D.P. 1980. Some aspects of structure and function in nitrogen fixing cyanobacteria. *Ann. Rev. Microbiol*. **34:** 497–536.

Stulp, B.K. and Stam, W.T. 1982. General morphology and akinete germination of a number of *Anabaena* strains in culture. *Arch. Hydrobiol*. **63:** 35–40.

Sutherland, J.M., Herdman, M. and Stewart, W.D.P. 1979. Akinetes of the cyanobacterium *Nostoc* PCC 7524: Macromolecular composition, structure and control of differentiation. *J. Gen. Microbiol*. **115:** 273–287.

Talpasayi, E.R.S. 1963. Polyphosphate containing particles of blue-green algae. *Cytologia* **28:** 76–80.

Thiel, T. and Wolk, C.P. 1983. Metabolic activities of isolated akinetes of the cyanobacterium *Nostoc spongiaeforme*. *J. Bacteriol*. **156** : 369–374.

Tyagi, V.V.S. 1974. Some observations on the pattern of sporulation in the blue-green alga *Anabaena doliolum*. *Ann. Bot*. **38:** 1107–1111.

Tyagi, V.V.S. 1978. The pattern of sporulation in *Anabaena circinalis* and comments on the role of heterocysts in sporulation in blue-green algae. *Z. Allg. Mikrobiol*. **18**: 281-286.

Tyagi, V.V.S. and Ahluwalia, A.S. 1978. Heterocyst formation in the blue-green alga *Anabaena doliolum*- a study of some aspects of photoregulation. *Ann. Bot*.**42**: 1333-1341.

Ueda, K. and Sawada, M. 1972. DNA content of akinete cells in *Cylindrospermum. Cytologia* **37** : 519–523.

Van Dok, W. and Hart, B.T. 1996. Akinete differentiation in *Anabaena circinalis* (cyanophyta). *J. Phycol.* **32:** 557–565.

Van Dok, W. and Hart, B.T. 1997. Akinete germination in *Anabaena circinalis* (cyanophyta). *J. Phycol.* **33:** 12–17.

Watanabe, M. 1992. Studies on planktonic blue–green algae 4. Some *Anabaena* species with straight trichomes in Japan. *Bull. Nat. Sci. Ser. B.* **18:** 123–137.

Whitton, B.A. 1992. Diversity, ecology and taxonomy of the cyanobacteria In: *Photosynthetic Prokaryotes*. Edited by Mann, N.H. and Carr, N.G., Plenum Press, New York.

Wildman, R.B., Loescher, J.H. and Winger, C.L. 1975. Development and germination of akinetes of *Aphanizomenon flos-aquae*. *J. Phycol.* **11:** 96–104.

Wolk, C.P. 1965. Control of sporulation in a blue green alga. *Dev. Biol.* **12:** 15–35.

Wolk, C.P. 1966. Evidence of a role of heterocysts in the sporulation of a blue-green alga. *Amer. J. Bot.* **53** : 260–262.

Wyman, M. and Fay, P. 1986. Interaction between light quality and nitrogen availability in the differentiation of akinetes in the planktonic cyanobacterium *Gloeotrichia echinulata. Br. Phycol. J.* **21:** 147–153.

Yamamoto, Y. 1972. The fatty acid composition of akinetes, heterocysts and vegetative cells in *Anabaena cylindrica. Plant Cell Physiol.* **13:** 913–915.

Yamamoto, Y. 1975. Effect of desiccation on the germination of akinetes of *Anabaena cylindrica. Plant Cell Physiol.* **16:** 749–752.

Yamamoto, Y. 1976. Effect of some physical and chemical factors on the germination of akinetes of *Anabaena cylindrica. J. Gen. Appl. Microbiol.* **22:** 311–323.

Zhou, R. and Wolk, C.P. 2002. Identification of an akinete marker gene in *Anabaena variabilis. J. Bacteriol.* **186**: 2529-2532.

Section III

MOLECULAR BIOLOGY AND BIOTECHNOLOGY

Chapter 15

MOLECULAR PROFILING AND GENETIC TRANSFORMATION OF CYANOBACTERIA - CURRENT STATUS AND PROSPECTS

Radha Prasanna and P.K. Singh

Abstract

Cyanobacteria comprise a large group of structurally complex and ecologically important gram negative prokaryotes whose taxonomy and characterization have generated much interest and controversy. The development of a taxonomy for cyanobacteria was primarily based on the use of clonal, axenic cultures and some of the cytological, morphological, genetical, chemical and physiological characteristics.

In recent years, molecular tools such as 16S rRNA cataloguing and DNA based markers have shown to provide efficient and rapid methodologies for identification and characterization of cyanobacteria. The integration of information based on purely morphological, physiological features with molecular data can provide a reliable methodology for classification of these ubiquotous organisms. Gene transfer mechanisms such as transformation and conjugation can effectively aid in improving our understanding of these organisms, their genotypic and evolutionary relationships and provide a basis for a more realistic evolutionary scheme and classification of these oxygenic photosynthetic bacteria. This will not only help in silencing doubts regarding the authenticity of several isolates, all over the world but also accelerate the rate of exploitation of these organisms in agriculture, industry and environmental management.

Key words: Cyanobacteria, electroporation, genetic transformation, molecular profiling, RFLP.

Introduction

Cyanobacteria are an unique assemblage of organisms, which occupy and predominate a vast array of habitats, as a result of several general characteristics - some belonging to bacteria and others unique to higher plants (Wilmotte, 1991). These organisms have been extensively exploited in agriculture, industry and environmental management, besides being used as model systems for studying diverse physiological and biochemical processes.

The presence of pigments and photosynthesis similar to algae coupled with their prokaryotic nature, similar to bacteria, have made their classification ambiguous. Although, initially classified by botanists, on the basis of their morphological characteristics, they are now classified in the Bergey's Manual of Systematic Bacteriology (volume 3) under oxygenic photosynthetic bacteria and guidelines have been provided for characterising them both as algae and bacteria. It is important, at this juncture, therefore, to identify and utilise genetic and molecular tools for characterization of cyanobacteria, which will be more reliable and silence doubts/ uncertainities of both botanists and bacteriologists (Singh and Pandey, 1985). Cyanobacterial transformation, though still in its infancy, is an excellent tool to manipulate these organisms for their better utilisation in agriculture and value addition. In the following compilation, an attempt has been made to provide an overview of the techniques utilised for characterization and transformation of cyanobacteria. Emphasis on the limitations associated with the methods presently being used are also provided with an aim to obtain useful pointers for better and more reliable definitions of cyanobacterial strains and effective exploitation in the years to come.

Cyanobacterial repositories in the world

The task of bringing all or most cyanobacteria into culture and development of a taxonomic system for all cultured material, is a long term ongoing process, requiring permanent culture collections of cyanobacteria especially in view of the growing importance of these organisms as food/feed in developing countries. There are a number of culture centres which hold collections of cyanobacterial accessions. But at present, there is a definite need for atleast one internationally acknowledged well funded culture collection having a trained work force and adequate financial resources. Among the recognised culture collections housing cyanobacterial strains, ATCC

(American Type Culture Collections), Maryland, CCAP (Culture Centre of Algae and Protozoa), Cambridge, UTEX (Culture Collection of Algae at the University of Texas) are well known. The principal collection of cyanobacteria at the Pasteur Institute (PCC), Paris maintains over 350 strains, which were included in the taxonomic analysis of Rippka *et. al.* (1979). In addition to these, there are a number of smaller collections all over the world. In India, there are two repositories for cyanobacteria - NCCUBGA (National Centre for Conservation and Utilization of Blue Green Algae) at New Delhi and NFMC (National Facility for Marine Cyanobacteria) at Tiruchirapalli, Tamil Nadu. The NCCUBGA, established by Department of Biotechnology (DBT), India, in 1986, at Indian Agricultural Research Institute and presently funded by both DBT and IARI holds more than 500 unialgal isolates of cyanobacteria from all diverse agroecological regions of India. The repository not only conserves germplasm of cyanobacterial strains but also honours requests from research workers all over the world. Routine characterization of strains for their growth attributes, pigment composition and nitrogen fixing ability are being carried out for their effective utilization as biofertiliser strains and sources of value added products.

Routine methods utilised for characterization of cyanobacteria

Cyanobacteria are thought to have evolved in the Pre-Cambrian, well before the Paleozoic boundary, which is supported by microfossils of late Proterozoic, that are nearly identical morphologically, to some present day living cyanobacteria (Knoll, 1985). But their classification has been clouded with uncertainty due to attempts by both botanists and bacteriologists to define them. The building of current morphology based taxonomy of blue green algae or cyanobacteria, began in earnest with the investigations of Bornet and Flahaurt (1886-1888) following the International Code of Botanical Nomenclature. The gathering of genera and species continued, culminating with Geitler's comprehensive treatise (1932) which recognized approximately 1300 species, 145 genera, 20 families and 3 orders. Later Desikachary (1959), Komarek (1970), following "Geitlerian" system, classified a number of fresh water algal species, although a major revision and consolidation was done by Drouet (1968). These classification systems differentiated species and genera on the basis of a single difference such as trichome width, presence and nature of sheath, number of trichomes per sheath, colour and other features which were later shown to exhibit a great flexibility within cloned populations of same species, under varying environmental conditions. In the early 1970s, Stanier and associates (1971) developed a system of classification, based on the use of axenic clonal cultures and morphological, cytological, genetical, chemical and physiological characteristics. Morphological attributes such as type of cellular division, plane of subsequent divisions, formation and structural features of 'baeocytes'/ hormogonia, presence and type of glycocalyx, shape and size of cells and presence/absence of constrictions, nature of trichomes, presence and location of heterocysts/akinetes and gas vacuoles, which were observable under light/ electron microscopy were taken into consideration. Chemical, genetical and physiological characteristics including motility and phototaxis, base composition (%G + C) of DNA, nutritional requirements, substrate specificity, oxygen, trace metals, and vitamin requirements, pigment composition, N_2 fixation ability and sensitivity to cyanophages, were all utilised for distinguishing genera (Komarek and Anagnostides, 1986). Although the photoautotrophic nature of these organisms diminishes the utility of physiological characteristics, this system in combination with genetical/ biochemical attributes, provides a realistic description of a strain. Rippka and co-workers (1979) published the first treatise; based on bacteriological taxonomy which divided cyanobacteria into five sections. The unicellular/ clonal aggregates reproducing by binary fission/ budding or by multiple fission/ binary fission were placed in Section I and II respectively while Sections III and IV comprised filamentous forms having single type of cells or differentiated cells besides simple vegetative cells. A modified version of this system is given in the appropriate sections of Bergey's Manual of Systematic Bacteriology (Staley *et. al.*, 1989). Characteristics and criteria for new species descriptions of cyanobacteria according to both the Bacteriological and Botanical code have been provided in the form of guidelines, inclusive of cell morphology, ultrastructure, genetic characteristics, physiology/ biochemistry, culture conditions and habitat/ecology.

More recently 16S rRNA cataloguing has provided interesting clues regarding the evolution and phylogeny of cyanobacteria. The first global evolutionary scheme of cyanobacteria comprising 29 partial 16S rRNA sequences (about 700 positions) was published, which indicated that sections III, IV and V corresponded to coherent phylogenetic clusters but sections I and II comprising unicellular and filamentous cyanobacteria were scattered in different lineages and were sometimes mixed together (Giovannoni *et. al.*, 1988). The rise of oxygen concentration in the pre Cambrian atmosphere is thought to have allowed colonisation of new biotypes and led to extensive divergence of cyanobacteria, as observed through 16S rRNA cataloguing.

At the present juncture, it is essential to first analyse the advantages and problems associated with the different methodologies currently utilised for molecular profiling of cyanobacteria.

Modern tools for molecular profiling

DNA/ DNA hybridization

This method was widely utilised in the late 1970s and 1980s as it allows determination of sequence similarity between genomes of different organisms without prior knowledge of nucleotide sequence (Herdmann *et. al.*, 1979). It is based on the feature that denatured, single stranded DNA will renature under suitable conditions, and the extent of renaturation/ hybridization is proportional to the complementarity of the DNA strands. Wayne (1987) proposed the use of DNA-DNA hybridization as a criterion for the definition of eubacterial species, with 70% hybridization as a minimum value for defining the members of the same species and 20% for congeneric strains. Stam (1980) pioneered the use of this method for cyanobacteria using a filter hybridization technique. The genotypic relationships among twenty nine filamentous strains belonging to the LPP group *sensu*. Rippka *et. al.* (1979) were determined and observed for the homogeneity with fresh water strain, *Plectonema boryanum* (Stam and Venema, 1975; Stam, 1980). Later Lachance (1981) and Stulp and Stam (1984) investigated the relationships among heterocystous forms such as *Nodularia*, *Cylindrospermum* and *Anabaena* strains and observed good homology with morphological data.

Refinement of this technique by measuring optical renaturation rates has shown that five toxic strains of *Microcystis* sp., isolated in Europe and North America belong to the same species and show no similarity with *Synechocystis* sp., to which genus *Microcystis* was originally assigned (Rippka *et. al.*, 1979; Wilmotte, 1994).

Ribosomal RNA cataloguing

Woese (1987) pointed out that 16S rRNA is a good "molecular chronometer" and can be effectively utilised for measuring the overall rate of evolutionary change in a line of descent. Its universal presence, essential role in the cell and lack of evidence of its lateral gene transfer make it a valuable tool for analysis of relationships - from phylum to species (Golden *et. al.,* 1989). It has a highly conserved secondary structure while a mosaic of evolutionarly conserved and variable regions are observed in the primary structure. It is approximately 1500 nucleotides long.

The first global evolutionary scheme of cyanobacteria comprising 29 partial 16S rRNA sequences was published by Giovannoni *et. al.* in 1988. This scheme suggested that simple, unicellular and filamentous forms have multiple evolutionary origins. Reverse transcriptase sequencing of ribosomal RNA for phylogenetic analysis has also been utilised in cyanobacteria (Lane *et. al.,* 1985).

But recent studies reveal that 16S rRNA gene sequences are probably too conserved to merit attention for studies on intraspecies variability (Ward *et. al.,* 1992). The sequence of internal transcribed spacer (ITS), situated between 16S rRNA and 23S rRNA genes has shown to be a useful marker (Wilmotte, 1994). Species-specific probes inferred from the ITS sequence have been found to be useful in eubacterial identification. The length variability combined with the high degree of conservation, make it a suitable marker for detection of more than one cyanobacterium in a culture. A number of conserved structural and regulatory domains have been observed within divergent 16S rRNA-23S rRNA spacer sequences which can be potential targets of PCR primers and oligonucleotide probes for detection and identification of cyanobacteria at different taxonomic levels (Iteman *et. al.,* 2000).

Nubel and co-workers (1997) have developed a seat of PCR primers for amplification of 16S rRNA genes from cyanobacteria which was found to be valuable in probing cyanobacteria in mixed populations, occurring in natural environments.

A rapid technique, consisting of digestion of amplified portions of rRNA cistrons, with restriction enzymes can provide a useful modification of RFLP and has been utilised in eubacterial taxonomy (Vaneechoutte *et. al.,* 1992). Ward and co-workers (1992) have suggested that 23S rRNA being more variable than 16S rRNA, may be a better candidate for resolving close relationships. The entire 23S rRNA sequence of *Synechococcus* sp. PCC 6301 is available, upto 2869 positions (Kumano *et. al.*, 1983). Worden and co-workers (2000) have developed a simple method for whole cell hybridisation using fluorescently labelled rRNA targetted peptide nucleic acid probes which can be very useful for phylogenetic probing of whole cells in the natural environment.

DNA based marker technologies

For effective exploitation of cyanobacteria, an authentically identified and characterized collection of strains using reliable parameters is very essential. This can be facilitated through the use of molecular markers which can reliably characterize strains, as compared to routine morphological/ physiological parameters, which are prone to get altered due to even slight changes in the environment.

DNA based molecular markers have become important tools for identifying strains and for studying phylogenetic relationships. DNA based marker technologies can be conveniently divided into three categories :

— DNA hybridization based assays e.g. RFLP

— DNA-PCR methods utilising arbitrary primers e.g. RAPD, DAF, AFLP

— DNA-PCR methods employing primers synthesised specific to complementary sequences in the target DNA e.g. SCARs, STS, SSRs, STRRs, L-TRRs, Microsatellites.

The major advantages of these systems are that they are not affected by environmental changes, found abundantly in the genome and exhibit a high degree of discrimination, even between closely related strains/ isolates. Also, they require very minimal amounts of DNA and the first two methods do not require major prior investments in terms of sequence analysis.

Restriction fragment length polymorphism (RFLP)

RFLP provide a powerful technique for taxonomy, as determined by genetic relatedness, in cyanobacteria. Non-identity of two populations of cyanobacteria, especially for symbiotic forms and their "so called" counterparts in free living conditions, can be established by DNA restriction examining length polymorphisms. A restriction site with a six base pair recognition sequence occurs by chance at a particular location with a probability of one in 4^6 or 4096. The probability that two such sites occur a certain distance apart by chance is one in 1.7×10^7, which exceeds the number of base pairs in *Nostoc / Anabaena* chromosome (approximately 8×10^6). Therefore, if two cyanobacterial isolates have several restriction fragments in common, in the absence of horizontal gene transfer, they are likely to be closely related. But if two isolates contain the same genetic information on different restriction fragments, they cannot be isolates of the same organism. RFLP analysis of *Anabaena* isolates from *Azolla* have revealed that the cultured isolates from *Azolla filiculoides* and *A. caroliniana* are not the major symbionts (Franche and Cohen-Bazire, 1987). Comparison of restriction fragment length polymorphism of *Nostoc* strains in and from coralloid roots of natural populations of cycad species using *Anabaena* PCC 7120 *nif* K and *gln* A probes (Lindblad *et. al.*, 1989) indicated that a wide range of *Nostoc* strains are associated with coralloid roots of *Cycas*. More than one *Nostoc* species is also found to be involved in the symbiosis with *Gunnera* sp.

RFLP technique has also been used for aiding in the taxonomy of small planktonic phycoerythrin containing *Synechococcus* sp. (Douglas and Carr, 1988; Wood and Townsend, 1990) and in characterizing heterocystous and non-heterocystous strains, with respect to *nif* and related genes (Prasanna and Kaushik, 1995). The small size of non-heterocystous forms and minor differences in morphological/physiological parameters make their identification and documentation difficult but RFLP based methods have helped to a large extent in resolving such problems.

With the advent of PCR (Polymerase Chain Reaction) technology and its subsequent improvement, using *Taq* DNA polymerase, exponential amplification of target sequences, and generation of specific sequences for use as probes has become easier and quicker. Although *Taq* polymerase can be sometimes problematic, due to relatively high rate of misincorporation of nucleotides, identification of newer polymerases such as *Pfu* I will help in improved proof reading and reliable amplification. Most of the recent techniques developed for DNA fingerprinting, rely on PCR and modifications in protocols/ regeants for improving its sensitivity and application for specific purposes.

Random amplified polymorphic DNA (RAPD)

This is a technique which is based on PCR, but its most important facet is that it does not require any prior knowledge of target DNA sequence. It involves amplification of genomic DNA using a single arbitrary primer (10 bp) and generation of amplification products, provided binding sites for primers on opposite strands of template exist within 5000 base pairs of each other. Electrophoretic analysis of the products reveals the DNA polymorphism and these have also been used as genetic markers for construction of genetic maps.

RAPD has not been so widely utilised in cyanobacterial characterization, due to its sensitivity to variation in DNA concentrations and contaminants. However, discrimination of five species of *Microcystis*, classified on the basis of morphology, toxicity and allozyme diversity, was carried out using RAPD analysis. A good accordance between allozyme divergence and RAPD patterns and genetic homogeneity / heterogeneity among species of *Microcystis* could be established (Nishihara *et. al.*, 1997).

A number of modifications of RAPD assay have been made for improving its performance. These include techniques such as DAF (DNA Amplification Fingerprinting), DGGE (Denaturing Gradient Gel Electrophoresis) and TSGE (Temperature Sweep Gel Electrophoresis). DAF has been utilised for reliable identification of *Anabaena azollae.* Primer sequences of 5-10 bp were utilised to produce complex but unique and reproducible fingerprints for the intact *Azolla-Anabaena* symbiosis and contribution of micro and macro symbionts to the fingerprints (Eskew *et. al.*, 1993). The speciality of DAF technique lies in its ability to produce reliable fingerprints for both prokaryotic and eukaryotic system and therefore can be valuable in improving our understanding of symbiotic systems, especially those involving cyanobacteria.

Amplified fragment length polymorphism(AFLP)

This technique is based on selective PCR amplification of restriction fragments from a total digest of genomic DNA and is a novel and powerful DNA fingerprinting technique for DNAs of any origin or complexity (Vos *et. al.*, 1995). It has been widely utilised for typing of bacterial strains and resolving taxonomical problems (Sevelkoul *et. al.*, 1999). Recent modifications include use of fluorescent probes and three restriction enzymes instead of routinely employed two enzymes (Lindstedt *et. al.*, 2000). AFLP analysis has not been as yet utilised in cyanobacteria, but it is a powerful high resolution tool which can resolve a number of doubts in the classification and in evolutionary studies on these organisms.

AFLP has been shown to demonstrate a superior discriminative power towards the differentiation of highly related bacterial strains that belong to the same species or even biova. (i.e. to characterise strains at the infrasubspecific level).

Use of repeat sequences in molecular profiling

Repetitive DNA sequences have shown to constitute an important part of the prokaryotic genome. These are generally observed in the form of short sequence repeats (SSR) and such loci are commonly present in prokaryotic and eukaryotic genomes. They are encountered in many different branches of the prokaryote kingdom and found in genes encoding products as diverse as microbial cell surface components to specific bacterial virulence factors, such as adhesions (Van Beikum *et. al.*, 1994). They function at various levels of gene expression, regulation and offer the organism genetic and consequently phenotypic flexibility. The occurrence, evolution and function of SSRs and their molecular analysis is known to have significant consequences in the context of responsiveness to environmental factors, pathogenicity and evolutionary processes.

Characteristic prokaryotic repeats such as enterobacterial repetitive intergenic consensus (ERIC) sequences and repetitive extragenic palindrome (REP) sequences have been found in many bacterial species (Hulton *et. al.*, 1991; Versalovic *et. al.*, 1991). Although dispersed repeats were initially deemed to be species-specific, analogous sequences have also been detected in other microbial species including cyanobacteria (Asayama *et. al.*, 1996). The discovery of DNA rearrangements occurring during heterocyst differentiation, and presence of some mobile DNA elements, brought to light, the plasticity and dynamic nature of the cyanobacterial genome (Golden, 1988; Bancroft and Wolk, 1989). Further investigations revealed the presence of a distinct family of repetitive sequences - STRRs (Short Tandemly Repeated Repetitive Sequences) which were first reported in heterocystous cyanobacteria (Mazel *et. al.*, 1990). Three distinct families of repeat sequences were observed in *Calothrix* sp. PCC 7601, which were present at a level of about 100 copies per *Calothrix* genome and consisted of tandemly amplified heptanucleotides. Later a 37-bp long tandemly repeated repetitive (LTRR) sequence was identified in *Anabaena* sp. 7120 (Masepohl *et. al.*, 1996).

Asayama and co-workers (1996) found two novel sets of repetitive sequences A (REP A) and B (REP B) on the *Microcystis aeruginosa* K-81 genome, which was composed of distinct motifs of tandem repeated sequences located in the up and downstream regions of *orf* I structural gene respectively. These sequences, when used as probes in Southern hybridization with genomic DNA of *Synechocystis* sp. PCC 6803, *Synechococcus* sp. PCC 7942 and *Anabaena* sp. PCC 7120 showed no signal, indicating the utility of such sequences as species-specific probes.

In the recent years, the STRR sequences have been demonstrated to be a valuable tool for identification and characterization of cyanobacteria. A species-specific probe using this sequence has been developed for identifying toxin producing cyanobacteria from a Finnish Lake. Restriction fragment length polymorphism analysis, probing with a short tandemly repeated DNA sequence (found at many locations in the chromosome of *Anabaena* 7120), distinguished hepatotoxic *Anabaena* isolate from neurotoxin producing strains and a *Nostoc* sp. (Rouhiainen *et. al.*, 1995).

Rasmussen and Svenning (1998) utilised different oligonucleotides from repetitive sequences as primers in the PCR and developed a valuable and rapid fingerprinting technique not only for identification purpose but also for studying diversity of cyanobacteria. They found it to be a powerful technique for studying the ecological distribution of cyanobacteria in a natural environment, and also for identification of symbiotic isolates.

An octameric highly iterated palindrome (HIP1) was found to occur on an average, every 664 bp within the genome of freshwater strains of *Synechococcus* sp. and with a slightly lower frequency within the genomes of other cyanobacteria and species, suggesting a widespread role for HIP1 in genome plasticity and adaptation in these organisms (Gupta *et. al.*, 1993).

Genetic transformation of cyanobacteria

The usefulness of cyanobacteria as model systems for the study of higher plant processes such as photosynthesis and inorganic nitrogen assimilation can be greatly extended by the availability of systems for genetic manipulation and exchange such as transformation, conjugation and transduction. This also requires 'genes' which can help in monitoring expression of transferred segments of DNA. A number of such 'platforms' have been constructed which can not only serve as sites for insertion of gene of interest but also report the transfer/ integration of the foreign genes into the host system.

Reporter systems in cyanobacteria

The analysis of gene expression requires genes whose products can be easily detected/ monitored and quantitated. For this purpose, 'reporter genes' fused to various regulatory sequences have proved very useful. This involves, the use of the gene, trimmed of its promoter and 5' end of this truncated gene being fused to the promoter/ transcriptional/ translational signals of interest. It is very essential to use such reporter genes whose product(s) are readily quantitable by simple enzymatic assays and its synthesis should allow scoring/ selection of cells expressing the gene. Another additional desirable feature is the "*in situ* localization of product" in particular cells, especially of filamentous forms (Friedberg, 1988). Although quantification of product is convenient, it should be kept in mind that gene fusions may not produce the same transcript as desired and its stability may differ from that of the native transcript (Bustos and Golden, 1992).

Recent developments in gene cloning strategies in cyanobacteria permit the use of several reporters including *cat* (chloramphenicol acetyl transferase) *lac*Z (galactosidase), GUS (glucuronidase) and *lux* AB (luciferase). The *cat* gene was first used as a reporter for deletion analysis of *psb* B promoter in *Anabaena* sp. strain PCC 7120 (Lang and Haselkorn, 1991). Expression of *lac* Z from a replicating plasmid was first demonstrated in *Synechococcus* sp. strain PCC 7002 (Buzby *et. al.*, 1985) and later as a transcriptional reporter to measure expression of *nif* H in *Anabaena* sp. strain PCC 7120 (Elhai and Wolk, 1990). For overcoming problems of gene replacement, integration platforms or 'neutral sites' (which are fragments of cyanobacterial chromosomal DNA cloned in a non-replicating vector) in vectors such as NSI and NSII have been constructed which serve as regions of insertion *in vitro*, of any gene (foreign/ cyanobacterial) destined for the cyanobacterial chromosome. These sites also have antibiotic resistance genes cloned in them for identifying recombinants. During recombination, the presence of homologous sites in the host chromosome, allows integration of gene of interest into the chromosome, *Lac* Z has been utilised as a reporter gene in NSI series of vectors (Andersson *et. al.*, 2000).

Luciferase is an ideal reporter for studying several light responsive processes, including circadian rhythms in cyanobacteria. Its short half life, real time reporting and non-destructive assay technique make it a valuable tool in improved understanding of the cyanobacterial metabolism. The most elegant use of this reporter has been the work of Wolk and co-workers who used it for *in situ* reporting of gene expression in vegetative cells and heterocysts (Elhai and Wolk, 1990). The luciferase reporter system has undergone several refinements since its first utilisation. Originally the endogenous luciferase of *Gonyaulax* was used, later firefly luciferase and more recently the bacterial luciferases have been utilized. The first reporter constructs using bacterial *lux* AB from *Vibrio harveyi* proved slightly problematic due to requirement for exogenous supply of decanal as a substrate and its high concentrations proving lethal in many cases. Also, gradual reduction in the concentration of decanal led to reduced amplitude of bioluminescence rhythms. Therefore, the search for autoluminescent assays lead to the use of *Xenorhabdus luminescens lux* C-DE genes encoding enzymes of the fatty acid reductase complex, which assembles aldehyde for the luciferase reaction. Therefore, high level of expression of *lux* C-DE genes in a strain rhythmically expressing *lux* AB produces autonomous bioluminiscence. This reporter system was utilised in the construction of a more refined series of NSII vectors which carry *lux* AB *lux* C-DE along with suitable transcriptional promoters.

In vitro experiments suggested that an *E. coli* consensus promoter would be an excellent cyanobacterial promoter (Schneider *et. al.*, 1991). Later a number of cyanobacterial genes were identified as a source of strong promoters such as *cpc*B2A2 (light regulated) *pho* A (phosphatase), and *rbc*LS (from *Anabaena* sp. PCC 7120). Other promoters such as *tac*, *psb* A (from *Amaranthus hybridus*) have also shown promise with different reporter systems (Thiel, 1994).

Transformation

Among the different methods of transfer and exchange of genetic material, DNA mediated transformation in cyanobacteria has been well investigated and widely utilised in cyanobacteira. Studies on transformation of cyanobacteria can be categorised as firstly those wherein the incoming DNA undergoes recombination with the genome of the recipient cell leading to alterations in the genetic composition of these genomes and the second group wherein the incoming DNA is sorted out in such a fashion as to give rise to an independent replicon in the recipient cell. The first category of genomic transformations are useful in genomic mapping studies and for introducing mutations in the genome of recipient cell. On the other hand, the introduction of plasmid cloning vehicles is more often utilised for introduction of foreign genes, mainly for commercial purpose. Genetic transformation in cyanobacteria may be carried out using purified nucleic acid fragments or plasmid DNA and remains the primary means for gene transfer in unicellular cyanobacteria. After the first convincing report of transformation of *Anacystis nidulans* 602 (presently known as *Synechococcus* PCC 7943) by Shestakov and Khuyen in 1970, a number of unicelluar / colonial nonheterocystous cyanobacteria have shown to be naturally transformable (Herdman and Carr, 1971; Stevens and Porter, 1980; Grigorieva and Shestakov, 1982). Most of these strains shown to be naturally transformable are classified as *Synechococcus* or *Synechocystis* and demonstrate % G + C content in the range of 47-55 % (Herdman *et. al.*, 1979; Wilmotte and Stam, 1984).

Interspecific transformation has been demonstrated among species of *Synechococcus* (Grigorieva and Shestakov, 1982) and *Synechocystis* (Grigorieva, 1985) besides intergeneric transformation between *Synechococcus* and *Synechocystis* (Devilly and Houghton, 1977; Stevens and Porter, 1986). Such studies have been widely utilized for taxonomic purposes. Transformation of *Gloeocapsa alpicola* (having 35% G + C) has shown to require pretreatment with $CaCl_2$ (Devilly and Houghton, 1977). Studies carried out regarding various factors affecting transformation rate and efficiency have indicated that DNA saturating concentrations range from 1-2 µg/ml to 20-40 µg/ml (Shestakov and Khuyen, 1970; Wilmotte and Stam, 1984) and $1.4 \times 10^3 \times 10^5$ transformants per microgram can be observed (Grigorieva and Shestakov, 1982; Golden and Sherman, 1984). It has been shown that transformation is dependent on DNA concentration showing single hit kinetics (Porter, 1986).

Cyanobacteria appear to be competent during all phases of growth (Stevens and Porter, 1980), however, cells are transformed usually during mid-late exponential growth phase (Porter, 1988). Competence cycle determinations with genomic transformation have yielded basically two types of results. In *Anacystis nidulans* UTEX 625, maximal transformation efficiency was observed at the late stage of logarithmic growth, with virtually no transformation after cells enter stationary phase, which is very similar to that observed in *B. subtilis* (Orkwiszwski and Kanaf, 1974). On the other hand, many cyanobacterial strains *viz., Anacystis nidulans* 602, *Agmenellum quadruplicatum, Synechocystis* PCC 6803, do not exhibit any defined competence cycle for genomic transformation (Shestakov and Khuyen, 1970; Stevens and Porter, 1980; Grigorieva and Shestakov, 1982).

According to stringent definition of DNA mediated transformation, this process should be totally sensitive to the addition of extracellular DNase, but several of the transformation systems do not meet this criterion. Although both chromosomal and plasmid DNA have been utilised for transforming cyanobacteria in the last few years, much of the work in the area of cyanobacterial transformation has focused on plasmid transformation.

Plasmid transformation work, till the early 1980s was restricted to strains such as *A. nidulans* R2 or *Agmenellum quadruplicatum* PR-6, which contain 2 and 6 plasmids respectively ranging in size from 3.0 Mdal upwards (van den Hondel *et. al.*, 1980; Buzby *et. al.*, 1983). As most of the *E. coli* based cloning vectors were not observed to be capable of replication in cyanobacteria, a segment of DNA from cyanobacterial plasmids, including genes required for replication were cloned into mobilisable *E. coli* plasmids. Further, it was found necessary to introduce selectable genetic markers into native cyanobacterial plasmids (which have been mostly observed to be cryptic), in order to provide a basis for selection of cyanobacterial cells that have undergone plasmid transformation. The first cyanobacterial plasmid with a selectable marker was generated by transforming *Anacystis nidulans* R2 with *E. coli* plasmid pR 146, containing a transposon Tn901 coding for ampicillin resistance which transposed to the resident cyanobacterial plasmid pUH 24 (van den Hondel *et. al.*, 1980). This stimulated construction of several biphasic *E. coli* - cyanobacterial plasmids which permitted *in vitro* manipulations and production of large quantities of plasmid DNA (Kuhlemeir *et. al.*, 1981; Buzby *et. al.*, 1983). The original biphasic plasmids have undergone considerable refinement with introduction of additional selectable markers, and restriction enzyme/ polylinker sites for improving their utility and efficiency for cloning (Gendel *et. al.*, 1983; Kuhlemeir *et. al.*, 1987; Shestakov and Reaston, 1987). Most of the selectable markers utilised in cyanobacterial transformation have been antibiotic resistance genes from enterobacteria-ampicillin (Ap^r), chloramphenicol (Cm^r), streptomycin (Sm^r), and kanamycin/ neomycin (Nm^r).

The parameters involved in transformation using plasmids are similar to those mentioned earlier for free nucleic acid fragments. The problems associated with plasmid mediated transformation with respect to host restriction enzyme activity have not proved serious in unicellular forms (Porter, 1986). Recombination during plasmid mediated transformation has been reported when *Anacystis nidulans* R2 was the host (Kuhlemeir *et. al.*, 1981; Gendel *et. al.*, 1983). But if strains are cured of endogenous plasmids and then utilised for transformation, the efficiency of transformation was observed to be much lower than that of wild strain (Chauvat *et. al.*, 1983). The creation of stable merodiploids (strains containing two copies of a segment of DNA) using replicating plasmid vector is rare but "gene conversion" is common. This phenonmenon is a type of non-reciprocal recombination in which there is conversion of a mutant copy of a gene in a replicating plasmid to a wild type copy (Porter *et. al.*, 1986). But for targetted mutagenesis, non-replicating vectors are more useful, with frequency of replacement of chromosomal gene being 100-fold higher, than frequency of addition of plasmid to the genome (Williams and Szalay, 1983).

Electroporation

In the last decade, electroporation has been successfully employed for transformation of cyanobacteria (Thiel and Poo, 1989). Electroporation relies on using a high voltage pulse to temporarily permeabilise cell membranes to exogenous DNA and optimal conditions vary with each particular cyanobacterium. Transformation of *Anabaena* sp. M131 using a replicating plasmid, by electroporation requires a field strength of 8 kV cm^{-1} and a time constant of 5 m sec. (Thiel and Poo, 1989). In comparison with conjugation, electroporation is advantageous as no contamination by *E. coli* cells is possible and vectors lacking *bom* site can also serve as donors. But as is true for conjugation, restriction of DNA is a significant problem. Electroporation has been utilised to create targetted mutations of *frx*C (protochlorophyllide reductase) and *nif* H (dinitrogenase reductase) in *Plectonema boryanum* strain IAM-M101 (Fujita *et. al.*, 1992). But the optimal conditions for electroporation were significantly higher - a field strength of 12.5 kV cm^{-1} and a time constant of 10 m sec. was required along with a high DNA concentration of 25-150 µg ml^{-1}. Chang and co-workers (1992) transformed *Fremeylla diplosiphon* by electroporation, for construction of genomic library, during the course of their studies on chromatic adaptation. Studies on DNA transfer by electroporation to *Calothrix* sp. strain PCC 7601 indicated that optimum conditions were intermediate between those reported for *Anabaena* sp. and for *Plectonema boryanum*. Mutagenesis of this strain resulted in a higher frequency of spontaneous pigmentation mutants (Tandeau de Marsac *et. al.*, 1982; Bruns *et. al.*, 1989), which has also been reported to contain a large number of IS elements. Because electroporation involves large amount of DNA, it may either significantly increase the induction of transposition of IS or SOS response in this strain, resulting in a high frequency of mutants.

Conjugation

Conjugation is one of the most commonly utilised methods of gene transfer in filamentous cyanobacteria and involves cell-cell contact mediated transfer of genes using broad host range conjugative plasmids. Stable conjugal transfer involves three main steps - effective cell-cell contact, transferred DNA escaping degradation or the action of restriction enzymes and autonomous replication of transferred DNA or its integration into the replicon of the recipient (Wolk, 1991). Conjugation is more advantageous if a large segment of DNA is to be transferred or specific merodiploids are sought, without interfering with gene function.

Conjugation is the method of choice for filamentous cyanobacteria, but it is also useful for unicellular cyanobacteria (Wolk *et. al.*, 1984). The process of conjugation is mediated by two groups of plasmid coded genes - *tra* (transfer genes) and *mob* (mobilizing genes). The transfer gene cluster consists of more than 12 genes and is responsible for synthesis of pili and other surface characteristics, while *mob* functions are a combination of *nic/ bom* region or *ori* T (results in nick formation) and a second region *mob* which provides DNA nicking protein that specifically recognizes *bom* site (*ori* T). In general, triparental matings, involving three plasmids are utilised in conjugative transfer of DNA to cyanobacteria - the cargo plasmid, i.e. the plasmid to be transferred, containing *bom/ori* T which can be nicked by the enzyme produced *in trans* by a second helper plasmid in the donor cell and conjugative plasmid. Compatibility among these plasmids is very essential for co-existence in a cell. The transferred DNA may replicate and exist autonomously in the recipient cell or recombine with homologous DNA in either the chromosome or resident plasmid of the recipient cell.

Most of the common cloning vectors are based on pBR 322 replicon but there is no evidence that most *E. coli* replicons function in cyanobacteria. Therefore, a number of shuttle vectors have been constructed for unicellular / filamentous cyanobacteria (Wolk *et. al.*, 1984; Golden *et. al.*, 1987; Thiel and Wolk, 1987; Elhai and Wolk, 1988a).

Vectors may be of 2 types - replicating shuttle vectors and integrative vectors while helper plasmids, are not mobilisable but form a component of the conjugative system. Replicating shuttle vectors contain 2 replicons - one that allows replication in *E. coli* and the other that allows for replication in the host cyanobacterial strains. A large number of vectors have been constructed for unicellular cyanobacteria by cloning into a mobilisable *E. coli* plasmid and a segment of cyanobacterial plasmid that includes the genes required for replication (Elhai and Wolk, 1988b; Houmard and Tandeau de Marsac, 1988). Vectors such as pRL 488, pRL 11 (containing replication origin from pDU 1 a plasmid from *Nostoc* sp. PCC 7524) allow autonomous replication in several strains of cyanobacteria (Wolk *et. al.*, 1984; Flores and Wolk, 1985).

Integrative vectors, are those lacking a cyanobacterial replication origin but allowing selection for recombination between a cloned gene of interest and a region of homology in the chromosome. A number of pRL series of vectors (pRL 1080, pRL 498) with positive selection markers such as antibiotic resistance and some with *sac* B and Km / Sp / Cm have been developed (Black *et. al.*, 1993; Wolk *et. al.*, 1993). Such vectors are typically used either for gene inactivation by targetted mutagenesis or for integration of a second copy of a gene into the chromosome, employing negative and positive selection for identifying whether integration of the plasmid into chromosome via double reciprocal recombination has occurred. This is often referred to as "hit and run allele replacement" technique (Golden *et. al.*, 1987).

A number of vectors have been developed based on a *neutral site* (locus where insertions cause no apparent phenotype). These comprise - Neutral Site I vectors and Neutral Site II vectors. The former series of vectors contain a segment of DNA from the *Synechococcus* sp. strain PCC 7942 chromosome that mediates homologous recombination with the chromosome, leading to apparently a double recombination event that transfers cloned DNA to the chromosome, at this site. They contain cloning sites and selectable markers inserted onto the *Xho* I site of the neutral site. Some common vectors of NSI series include pAM 1303 (5.8 kb), pAM 990 which contain spectinomycin/ streptomycin resistance and *lac* Z as the selectable markers respectively. The more advanced NSII series of vectors, have cloning sites and selectable markers inserted into either the *Bst* E11 or *Bgl*11 site of neutral site besides showing common features such as pBR 322 replicon, with NSI vectors. pAM 1573, pAM 1580 are NSII series of vectors with chloramphenicol resistances and *lux* AB reporter system as selectable markers. Vectors such as pAM 1619 carry P*psb* A1: *lux* CDE to make *Synechococcus lux* AB reporter strains glow, without decanal addition and constitute the improved series of vectors. The NSI and NSII series of vectors have been not only utilised for conjugation but also widely employed in transformation of unicellular forms especially in relation to studies on photosynthesis and circadian rhythms in these organisms.

A commonly encountered problem during transfer of foreign genes is the restriction of the incoming DNA by enzymes in the host cell, which reduces the number of exconjugants (Wolk *et. al.*, 1984). Helper plasmids, which generally provide transacting *mob* gene product have been modified by the addition of gene(s) encoding *Ava*I methylase and *Eco* 4711 methylase (which modifies GGNCC sites and protects against restriction by *Ava*11 (Elhai and Wolk, 1988a, b). In general, vectors destined for cyanobacterial hosts are methylated in the *E. coli* donor carrying the helper plasmid and thus protected from restriction by isoschizomers of *Ava* I, *Ava* II or *Ava* III, which are commonly found in cyanobacteria (Houmard and Tandeau de Marsac, 1988). Such vectors include pRL 625, pRL 1045 and pRL 1124, and were developed by Wolk and co-workers (Wolk *et. al.*, 1991, 1993).

Conjugation, using the developed array of vectors is being widely utilised in *Anabaena* PCC 7120, for gene replacement or transposon mutagenesis (Wolk *et. al.*, 1983; Wolk *et. al.*, 1988). Vectors based on pDU 4 have been transferred in symbiotic strains of *Nostoc* (Wolk *et. al.*, 1991) and shuttle vector type cargo plasmids for *Anabaena* and *Nostoc* have been constructed with tetracycline and ampicillin resistance genes as selectable markers, which did not prove very useful (Wolk *et. al.*, 1984). Further refinement of vectors by introducing chloramphenicol resistance and no *Ava* II sites resulted in the shuttle vector pVW 1. pRL series of vectors such as pRL 5, 6 and 8 exhibit a high conjugal efficiency (Wolk *et. al.*, 1984). Conjugation systems have been utilised for construction of genomic library, studies on chromatic adaptation, photosynthesis and nitrogen fixation in filamentous forms wherein gene transfer through other techniques did not prove useful (Thiel, 1994).

There is a need to develop broader host range plasmids, and better expression systems for genes involved in products of bioindustrial value besides optimizing more efficient protocols for gene transfer through conjugation.

Approaches for the future

The collection and analysis of molecular information has made a beginning only now, and conclusions are still limited by the relatively small number of data available. It is very essential that more number of laboratories all over the world, make concerted efforts towards utilizing molecular techniques such as ribosomal RNA sequences and PCR based markers for determining genotypic relationships and characterization of important attributes, in order to integrate it with other morphological/ biochemical/ physiological markers. Only this can lead to a well defined classification of these ubiquitous organisms, and researchers can be sure of the authenticity of the strains they utilise.

Similarly, there is a definite need to develop genetic systems for cyanobacteria, other than *Anabaena* sp. strain PCC 7120, *Synechocystis* sp. strain PCC 6803 which have proved valuable 'workhorses' for improving our understanding of cyanobacteria in general. This would involve development of new reporter systems, more efficient shuttle vectors which can evade the action of restriction enzymes and identifying optimal conditions for gene transfer and integration in useful strains. Transduction, as a means of genetic transfer needs to be explored, as it can be a powerful tool especially with the growing knowledge on cyanophages. Recently the entire sequence of the genome of *Synechocystis* PCC 6803 has been analyzed (Kaneko *et. al.*,

1996) and the annotations of all 3168 genes have been deposited in the cyanoBase database (http://www.kaz.usa.or.jp: 3080/cyano/.). This is the first phototroph to be sequenced and this has become a major resource for not only cyanobacteriologists, but also for bacteriologists, botanists and evolutionary biologists.

New thinking and innovative means of research may definitely pave the way for more effective exploitations of these cyanobacteria/blue green algae and bring about a "Blue-green" revolution in not only agriculture, but in diverse areas of science and industry.

References

Andersson, C.R., Tsinoremas, N.F., Shelton, J., Lebedeva, N.V., Yarrow, J., Min, H. and Golden, S.S. 2000. Application of bioluminiscene to the study of circadian rhythms in cyanobacteria. *Meth. Enzymol.* **305**: 527-542.

Asayama, M., Kubasawa, I., Takahashi, T., Aida, T. and Shirai, M. 1996. Highly repetitive sequences and characteristics of genomic DNA in unicellular cyanobacterial strains. *FEMS Microbiol. Lett.* **137**: 175-181.

Baneroft, I. and Wolk, C.P. 1989. Characterization of an insertion sequence (IS 891) of novel structure from the cyanobacterium *Anabaena* sp. strain M-131. *J. Bacteriol.* **171**: 5949-5954.

Black, T., Cai, Y. and Wolk, C.P. 1993. Spatial expression and autoregulation of *het* R, a gene involved in the control of heterocyst development of *Anabaena*. *Mol. Microbiol.* **9**: 77-84.

Bornet, E. and Flahaurt, C. 1886-1888. Revision de Nostocacees heterocystees. *Ann. Sci. Nat. Bot.* **7** Ser 3: 3230381; **4**: 343-373, **5**: 52-119, **7**: 178-262.

Bruns, B.U., Briggs, W.R. and Grossman, A.R. 1989. Molecular characterization of phycobilisome regulatory mutants of *Fremyella diplosiphon*. *J. Bacteriol.* **171**: 901-908.

Bustos, S.A. and Golden, S.S. 1992. Light regulated expression of the *psb*D gene family in *Synechococcus* sp. strain PCC 7942: Evidence for the role of duplicated *psb*D genes in cyanobacteria. *Mol. Gen. Genet.* **232**: 221-230.

Buzby, J.S., Porter, R.D. and Stevens, S.E. 1983. Plasmid transformation in *Agmenellum quadruplicatum* PR-6: construction of biphasic plasmids and characterization of their transformation properties. *J. Bacteriol.* **154**: 1446-1450.

Buzby, J.S., Porter, R.D. and Stevens, S.E. 1985. Expression of the *Escherichia coli lacZ* gene on a plasmid vector in a cyanobacterium. *Science* **230**: 806-807.

Chauvat, F., Astier, Vedel, F. and Joset-Espendeillier, F. 1983. Transformation in the cyanobacterium *Synechococcus* R-2; Improvement of efficiency, role of the pUH24 plasmid. *Mol. Gen. Genet.* **191**: 39-45.

Chiang, G.G., Schaefer, M.R. and Grossman, A.R. 1992. Transformation of the filamentous cyanobacterium *Fremyella diplosiphon* by conjugation or electroporation. *Plant Physiol. Biochem.* **30**: 315-325.

Desikachary, T.V. 1959. *Cyanophyta*. Indian Council of Agricultural Research, New Delhi, pp. 686.

Devilly, C.I. and Houghton, J.A. 1977. A study of genetic transformation in *Gloeocapsa alpicola*. *J. Gen. Microbiol.* **98**: 277-280.

Douglas, S.E. and Carr, N. 1988. Examination of genetic relatedness of marine *Synechococcus* spp. by using restriction fragment length polymorphism. *Appl. Environ. Microbiol.* **54**: 3071-3078.

Drouett, F. 1968. *Revision of the classification of the Oscillatoriaceae:* Monograph Acad. Natur. Sci., Philadelphia, Fulton Press, Lancaster, Pennsylvania.

Elhai, J. and Wolk, C.P. 1988a. Conjugal transfer of DNA in cyanobacteria. *Meth. Enzymol.* **167** 747-754.

Elhai, J. and Wolk, C.P. 1988b. A versatile class of positive selection vectors based on the nonviability of palindrome containing plasmids that allows cloning into long polylinkers. *Gene* **68**: 119-138.

Elhai, J. and Wolk, C.P. 1990. Developmental regulation and spatial pattern of expression of the structural genes for nitrogenase in the cyanobacterium *Anabaena*. *EMBO J.* **9**: 3379-3388.

Eskew, D.L., Caelamo-Anolles, G., Bassam, B.J. and Gresshoff, P.M. 1993. DNA amplification fingerprinting of the symbiosis. *Plant Mol. Biol.* **21**: 363-373.

Flores, E. and Wolk, C.P. 1985. Identification of facultatively heterotrophic N_2-fixing cyanobacterium able to receive plasmid vectors from *Escherichia coli* by conjugation. *J. Bacteriol.* **162**: 1339-1341.

Franche, C. and Cohen-Bazire, G. 1987. Evolutionary divergence of *nif* HDK region among nine symbiotic *Anabaena azollae* and between *Anabaena azollae* and some free living heterocystous cyanobacteria. *Symbiosis* **3**: 159-178.

Friedberg, D. 1988. Use of reporter genes in cyanobacteria. *Meth. Enzymol.* **167**: 736-747.

Fijita, Y., Takahashi, Y., Chuganji, M. and Matsuhara, H. 1992. The *nif* H-like (*fix*C) gene is involved in the biosynthesis of chlorophyll in the filamentous cyanobacterium *Plectonema boryanum*. *Plant Cell Physiol*. **33**: 81-92.

Geitler, J. 1932. Cyanophyceae, In: *Kryptogamenflora von Deutschland Ostyereich. and der Schweiz.*Edited by Rabenhorst **14**: 1-1196.

Gendel, S., Straus, N., Pulleyblank, D. and Williams, J. 1983. Shuttle cloning vectors for the cyanobacterium *Anacystis nidulans*. *J. Bacteriol.* **156**: 148-154.

Giovannoni, S.J., Turner, S., Olsen, G.J., Barns, S., Lane, O.J. and Pace, N.R. 1988. Evolutionary relationships among cyanobacteria and green chloroplasts. *J. Bacteriol.* **170**: 3584-3592.

Golden, J.W. 1988. Genome rearrangements during *Anabaena* heterocyst differentiation. *Can. J. Bot*. **66**: 2098-2102.

Golden, S.S., Brusslan, J. and Haselkorn, R. 1987. Genetic engineering of the cyanobacterial chromosome. *Meth. Enzymol.* **153**: 215-231.

Golden, S.S., Naity, M.S. and Chu, D.S. 1989. Genetic relationship of two highly studied *Synechococcus* strains designated *Anacystis nidulans*. *J. Bacteriol.* **171**: 24-29.

Golden, S.S. and Sherman, L.A. 1984. Optimal conditions for genetic transformation of the cyanobacterium *Anacystis nidulans* R2. *J. Bacteriol.* **158**: 36-42.

Grigorieva, G.A. 1985. Heterospecific transformation in the genus *Synechococcus*. *Biol. Nauki*. **9**: 91-94.

Grigorieva, G.A. and Shestakov, S.V. 1982. Transformation of the cyanobacterium *Synechocystis* PCC 6803. *FEMS Microbiol. Lett*. **13**: 367-370.

Gupta, A., Merby, A.P., Turner, J.S., Whelton, B.A. and Robinson, N.J. 1993. Deletion within the metallothionein locus of cadmium tolerant *Synechococcus* PCC 6301 involving a highly iterated palindrome (HIP1). *Mol. Microbiol.* **7**: 189-195.

Herdman, M. and Carr, N.G. 1971. Recombination in *Anacystis nidulans* mediated by an extracellular DNA/RNA complex. *J. Gen. Microbiol.* **68**: 14.

Herdman, M., Janvier, M., Waterbury, J.B., Rippka, R., Stanier, R.Y. and Mandel, M. 1979. Deoxyribonucleotide acid base composition of cyanobacteria. *J. Gen. Microbiol.* **111**: 63-71.

Houmard, J. and Tandeau de Marsac, N. 1988. Cyanobacterial genetic tools: Current status. *Meth. Enzymol.* **167**: 808-847.

Hulton, C.S.J., Higgins, C.F., and Sharp, P.M. 1991. ERIC sequences: a novel family of repetitive elements in the genome of *Escherichia coli, Salmonella typhimurium* and other enterobacteria. *Mol. Biol.* **5**: 825-834.

Iteman, I., Rippka, R., Tandeau de Marsac, N. and Herdman, M. 2000. Comparison of conserved structural and regulatory domains within divergent 16S rRNA - 23S rRNA space sequences of cyanobacteria. *Microbiology* **146**: 1275-1286.

Kaneko, T., Sabo, S., Kotane, H., Tanaka, A., Asamizu, E., Nakamura, Y., Miyajima, N., Hirosawa, M., Sugwa, M., Sasamole, S., Kinura, T., Hosouchi, T., Matsuno, A., Maraki, A., Nakasaki, N., Namo, K., Okumua, S., Shimpo, S., Takeuchi, C., Wada, T., Watanabe, A., Yamada, M. and Tabata, S. 1996. Sequence analysis of the genome of the unicellular cyanobacterium. *Synechocystis* sp. strain PCC 6801. II. Sequence determination of the entire genome and assignment of potential protein coding regions. *DNA Res.* **3**: 109-136.

Knoll, A.H. 1985. The distribution and evolution of microbial life in the late Proterozoic era. *Ann. Rev. Microbiol.* **39**: 391-417.

Komarek, J. 1970. Genetic identity of the "*Anacystis nidulans*" strain Kratz-Allen/Bloom 625 with *Synechocystis* Nag 1849. *Arch. Protistenk.* **112**: 343-364.

Komarek, J. and Anagnostidis, K. 1986. Modern approach to the classification system of cyanophytes. 2. Chroococcales. *Arch. Hydrobiol.* Suppl. 73, *Algological Studies* **43**: 157-226.

Kuhlemeier, C.J., Borrias, W.E., Van den Hondel, C.A.M.J.J. and Van Arkel, O.A. 1981. Vectors for cloning in cyanobacteria: construction and characterization of two recombinant plasmids capable of transforming *Escherichia coli* K-12 and *Anacystis nidulans* R2. *Mol. Gen. Genet*. **184**: 249-254.

Kumano, M., Tomoika, N. and Suguira, M. 1983. The complex nucleotide sequence of a 23S rRNA gene from a blue green alga, *Anacystis nidulans*. *Gene* **24**: 219-225.

Lachance, M.A. 1981. Genetic relatedness of heterocystous cyanobacteria by deoxyribonucleic acid-deoxyribonucleic acid reassociation. *Int. J. Syst. Bacteriol*. **31** : : 139-147.

Lane, D.J., Pace, B., Olsen, G.J., Stahl, D.A., Sogin, M.L. and Pace, N.R. 1985. Rapid determination of 16S ribosomal RNA sequence for phylogenetic analyses. *Proc. Natl. Acad. Sci. USA* **82:** 6955-6959.

Lang, J.D. and Haselkorn, R. 1991. A vector for analysis of promoters in the cyanobacterium *Anabaena* sp. strain PCC 7120. *J. Bacteriol*. **173**: 2729-2731.

Lindblad, P., Haselkorn, R., Bergman, B. and Nierzwicki-Bauer, S.A. 1989. Comparison of DNA restriction fragment length polymorphisms of *Nostoc* strains in and from cycads. *Arch. Microbiol*. **152**: 20-24.

Lindstedt, B.A., Heir, E., Vardund, T. and Kapperud, G. 2000. A variation of the amplified fragment length polymorphism (AFLP) technique using three restriction endonucleases and assessment of the enzyme combination *Bgl* II-*Mfe* I for AFLP analysis of *Salmonella enterica* sub sp. *enterica* isolates. *FEMS Microbiol. Lett*. **189**: 19-24.

Masepohl, B., Gorlitz, R. and Bohren, H. 1996. Long tandemly repeated repetitive (LTRR) sequences in the filamentous cyanobacterium *Anabaena* sp. PCC 7120. *Biochim. Biophys. Acta* **1307**: 26-30.

Mazel, D., Houmaird, J., Castets, A.M. and Tandeau de Marsac, N. 1990. Highly repetitive DNA sequences in cyanobacterial genomes. *J. Bacteriol*. **172**: 2755-2761.

Nishihara, H., Niwa, H., Watanabe, M., Nagashima, M., Yagi, O. and Takanura, Y. 1997. Random amplified polymorphic DNA

(RAPD) analyses for discriminating genotypes of *Microcystis* cyanobacteria. *Biosci. Biotech. Biochem.* **61**: 1067-1072.

Nubel, V. , Garcia-Pichel, F. and Muyzer, G. 1997. PCR primers to amplify 16S rRNA genes from cyanobacteria. *Appl. Environ. Microbiol.* **63**: 3327-3332.

Orkwiszewiski, K.G. and Kanaf, A.R. 1974. Genetic transformation of the blue green bacterium, *Anacystis nidulans*. *Arch. Microbiol.* **98**: 31.

Porter, R.D. 1986. Transformation in cyanobacteria. *CRC Crit. Rev. Microbiol.* **13:** 111-132.

Porter, R.D. 1988. DNA transformation. *Meth. Enzymol.* **167**: 703-712.

Porter, R.D., Buzby, J.S., Polon, A., Fields, P.I., Dubbs, J.M. and Stevens, S.E. 1986. Genes from the cyanobacterium *Agmenellum quadruplicatum* isolated by complementation: characterization and production of merodiploids. *Gene* **41**: 249-260.

Prasanna, R. and Kaushik, B.D. 1995. Nitrogen fixation and *nif* gene organization in branched heterocystous cyanobacteria: Variation for the presence of *xis* A. *Folia. Microbiol.*.**40**: 176-180.

Rasmussen, U. and Svenning, M.M. 1998. Fingerprinting of cyanobacteria based on PCR with primers derived from short and long tandemly repeated repetitive sequences. *Appl. Environ. Microbiol.* **64**: 265-272.

Rippka, R., Deruelles, J., Waterbury, M., Herdman, M. and Stanier, R.Y. 1979. Generic assignments, strain histories and properties of pure cultures of cyanobacteria. *J. Gen. Microbiol.* **111**: 1-61.

Rouhiainen, L., Sivonen, K., Buikema, W.J. and Haselkorn, R. 1995. Characterization of toxin producing cyanobacteria by using an oligonucleotide probe containing a tandemly repeated heptamer. *J. Bacteriol.* **177**: 6021-6066.

Schneider, G.I., Lang, J. and Haselkorn, R. 1991. Promoter recognition by the RNA polymerase from vegetative cells of the cyanobacterium *Anabaena* 7120. *Gene* **105**: 51-60.

Sevelkoul, P.H., Aarts, H.J., de Haas, J., Dijkshoorn, L., Duim, B., Osten, M., Rademaker, J.L., Schouls, L. and Lenstra, J.A. 1999. Amplified fragment length polymorphism analysis: the state of an art. *J. Clin. Microbiol.* **37**: 3083-3091.

Shestakov, S.V. and Khuyen, N.T. 1970. Evidence for transformation in the blue green alga *Anacystis nidulans*. *Mol. Gen. Genet.* **107**: 372-375.

Shestakov, S.V. and Reaston, J. 1987. Gene transfer and host-vector systems of cyanobacteria. *Oxford Surveys Plant Mol. Cell Biol.* **4**: 137-166.

Singh, P.K. and Pandey, K.D. 1985. Genetics of blue green algae. In: *Widening Horizons of Plant Sciences*. Edited by Malik, C.P., Cosmo Publ., New Delhi.

Staley, J.T., Bryant, M.P., Pfennig, N. and Holt, J.G., (eds.) 1989. *Bergey's Manual of Systematic Bacteriology*, Vol. 3, Williams and Wilkins, Baltimore. pp. 2298.

Stam, W.T. 1980. Relationships between a number of filamentous blue green algal strains (Cyanophyceae) revealed by DNA-DNA hybridization. *Arch. Hydrobiol. Suppl.* 56, *Algological Studies* **25**: 351-374.

Stam, W.T. and Venema, G. 1975. The use of DNA-DNA hybridization for determination of the relationship between some blue green algae (Cyanophyceae). *Acta Bot. Neerl.* **26**: 327-342.

Stanier, R.Y., Kunisawa, R., Mandel, M. and Cohen-Bazire, G. 1971. Purification and properties of unicellular blue green algae (Order: chroococcales). *Bact. Rev.* **35**: 177-205.

Stevens, S.E. Jr. and Porter, R.D. 1980. Transformation in *Agmenellum quadruplicatum*. *Proc. Natl. Acad. Sci. USA* **77**: 6052-6056.

Stulp, B.K. and Stam, W.T. 1984. Genotypic relationships between strains of *Anabaena* (Cyanophyceae) and their correlation with morphological affinities. *Br. Phycol. J.* **19** : 287-301.

Tandeau de Marsac, N., Borrias, W.E., Kuhlemeir, C.J., Caslets, A.M., von Arkel, G.A. and van den Hondel, C.A.M.J.J. 1982. A new approach for molecular cloning in cyanobacteria cloning of an *Anacystis nidulans met* gene using a Tn901 induced mutant. *Gene* **20**: 111-119.

Thiel, T. 1994. Genetic analysis of cyanobacteria. In: *The Molecular Biology of Cyanobacteria*. Edited by Bryant, D.A., Kluwer Academic Publ.. pp. 581-611.

Thiel, T. and Poo, H. 1989. Transformation of a filamentous cyanobacterium by electroporation. *J. Bacteriol.* **171**: 5743-5748.

Thiel, T. and Wolk, C.P. 1987. Conjugal transfer of plasmids in cyanobacteria. *Meth. Enzymol.* **153**: 232-243.

van Beikum, A., Duim, B., Regelink, A., Moeiler, V., Quint, W. and van Alphen, L. 1994. Genomic DNA fingerprinting of clinical *Haemophilus influenzae* isolates by polymerase chain reaction amplification: comparison with major outer membrane protein and restriction fragment length polymorphism analysis. *J. Med. Microbiol.* **41**: 63-68.

van den Hondel, C.A.M.J.J., Verbeek, S., van der Ende, A., Weisbeck, P.J., Borrias, W.E. and von Arkel, G.A. 1980. Introduction of transposon Tn901 and a plasmid of *Anacystis nidulans:* Preparation for cloning in cyanobacteria. *Proc. Natl. Acad. Sci. USA* **77**: 1570-1574.

Vaneechoutte, M., Rossau, R., De Vos, P., Gillis, M., Janssens, D., Reepe, N., De Rouck, A., Fiers, T., Clacys, G. and Kersters, G. 1992. Rapid identification of bacteria of the Commamonadaceae with amplified ribosomal DNA restriction DNA analysis (ARDRA). *FEMS Microbiol. Lett.* **93**: 227-234.

Versalovic, J., Koeuth, T. and Lapski, J.R. 1991. Distribution of repetitive DNA sequences in eubacteria and application to fingerprinting of bacterial genomes. *Nucl. Acids Res.* **19**: 6823-6831.

Vos, P., Hogers, R., Bleeker, M., Reijans, M., Lee, T. van de, Homes, M., Frijters, A., Pot, J., Peleman, J., Kuiper, M. and Zabeau, M. 1995. AFLP: a new technique for DNA fingerprinting. *Nucl. Acids Res.* **23**: 4407-4414.

Ward, D.M., Bateson, M.M., Weiler, R. and Ruff-Roberts, A.L. 1992. Ribiosomal RNA analysis of microorganisms as they occur in nature. *Adv. Microb. Ecol.* **12**: 219-286.

Wayne, L.G., Brenner, D.J., Calwell, R.R., Gamont, P.A.O., Kandler, O., Kochevsky, M.J., Moore, L.H., Moore, W.E.C., Murray, R.G.E., Stackebrandt, E., Starr, M.P. and Truper, H.G. 1987. Report of the ad hoc committee on reconcilation of approaches to bacterial systematics. *Int. J. Syst. Bacteriol.* **37**: 463-464.

Williams, J.G.K. and Szalay, A.A. 1983. Stable integration of foreign DNA into the chromosome of the cyanobacterium *Synechococcus* R2. *Gene* **24**: 37-51.

Wilmotte, A. 1991. Taxonomic study of marine oscillatorionaceaen strains (Cyanophyceae, Cyanobacteria) with narrow trichomes. I. Morphological variability and autoecological features. *Arch. Hydrobiol. Suppl.* **69**: 215-248.

Wilmotte, A. 1994. Molecular evolution and taxonomy of the cyanobacteria. In: *The Molecular Biology of Cyanobacteria.* Edited by Bryant, D.A., Kluwer Academic Publ., Netherlands. pp. 1-25.

Wilmotte, A. and Stam, W.T. 1984. Genetic relationships among cyanobacterial strains originally designated as *Anacystis nidulans* and some other *Synechococcus* strains. *J. Gen. Microbiol.* **130**: 2737-2740.

Woese, C.R. 1987. Bacterial evolution. *Microbiol. Rev.* **51**: 221-271.

Wolk, C.P. 1991. Genetic analysis of cyanobacterial development. *Curr. Opin. Genet. Dev.* **1**: 336-341.

Wolk, C.P., Cai, Y. and Panoff, J.M. 1991. Use of a transposon with luciferase as a reporter to identify environmentally responsive genes in a cyanobacterium. *Proc. Natl. Acad. Sci. USA* **88**: 5355-5359.

Wolk, C.P., Elhai, J., Kuritz, T. and Holland, D. 1993. Amplified expression of a transcriptional pattern formed during development of *Anabaena. Mol. Microbiol.* **7**: 441-445.

Wolk, C.P., Vonshak, A., Kehoo, P. and Elhai, J. 1984. Construction of shuttle vectors capable of conjugative transfer from *Escherichia coli* to nitrogen fixing cyanobacteria. *Proc. Natl. Acad. Sci. USA* **81**: 1561-1565.

Wood, A.M. and Townsend, D. 1990. DNA polymorphism within the WH7803 serogroup of marine *Synechococcus* spp. (Cyanobacteria). *J. Phycol.* **26**: 576-585.

Worden, A.Z., Chisholam, S.W. and Bindee, B.J. 2000. *In situ* hybridization of *Prochlorococcus* and *Synechococcus* (Marine cyanobacteria) spp. with rRNA targetted peptide nucleic acid probes. *Appl. Environ. Microbiol.* **66**: 284-289.

Chapter 16

PRODUCTION OF TRANSGENIC CYANOBACTERIA AND THEIR APPLICATIONS

Rashmi Tyagi and B.D. Kaushik

Abstract

Cyanobacteria constitute a prokaryotic, autotrophic group of useful experimental organisms. Some of the cyanobacteria do not differentiate heterocysts but achieve temporal and spatial separation of photosynthesis and nitrogen fixation. Study to know complete genomic information of these organisms has been an important and recent aspect of cyanobacterial biotechnology. Genes responsible for regulating various processes in cyanobacteria are being put under a focused study. Information available concerning production of transgenic cyanobacteria and their importance has been discussed.

Key words: Biotechnology, cyanobacteria, genetic engineering, nitrogen fixation, transgenic.

Introduction

Genetic manipulation includes the techniques used for improving the characteristics of organisms by modifying their genetic make up through mutagenesis, selection as well as genetic engineering (gene cloning and recombinant methods). Molecular biology and biomedical research experienced a revolutionary change with the development of gene manipulation. In the United Kingdom and other countries, gene manipulation is defined as "The formation of new combinations of heritable material by the insertion of nucleic acid molecules, produced outside the cell, into any virus, bacterial plasmid or other vector system so as to allow their incorporation into a host organism in which they do not naturally occur but in which they are capable of continued propagation."

Cyanobacteria are a diverse group of gram-negative prokaryotes characterized by oxygenic, plant-like photosynthesis. They vary in structure from unicellular to filamentous, marine to fresh water in habitat and many strains are capable of fixing atmospheric elemental dinitrogen and many are able to alter their light harvesting phycobiliprotein composition to match the quality of incident light. Species of *Nostoc, Anabaena* and several others are capable of reducing atmospheric nitrogen in specialized cells, called heterocysts, which are formed upon starvation for combined nitrogen (Stewart, 1970). Several cyanobacteria, *e.g. Plectonema, Synechococcus, Oscillatoria* etc. do not form heterocysts but achieve temporal and spatial separation of photosynthesis and nitrogen fixation (Weare and Beneman, 1974). The property of nitrogen fixation by cyanobacteria is being exploited to supplement nitrogen in rice cultivation (Venkataraman, 1979) Kumar (1962) for the first time showed sexuality (through transformation) in cyanobacteria. Nucleotide sequence of the entire genome of *Synechocystis* sp.PCC6803 was completed by Kaneko *et. al.* (1996), and this was the first report of the complete genomic information of a photoautotrophic organism. The molecular organization and regulation of expression of *nif* (nitrogen fixation) genes has been studied by many workers. All non-heterocystous cyanobacteria studied have a continuous *nif* HDK cluster, whereas the heterocystous forms show a similar arrangement in the heterocysts (Kallas *et. al.*, 1983), but in vegetative cells *nif* K and nif D genes are separated by 11kb intervening DNA sequence. The rearrangement of these genes occurs by a unique temporally and spatially regulated phenomenon involving site- specific recombination, and an enzyme excisase. This enzyme is encoded by a *xis* A gene present in the 11kb intervening sequence which restores continuity between *nif* K and *nif* D genes.

Prasanna and Kaushik (1994) have described the hybridization with xis A gene (to be used as molecular marker to indicate the possibility of rearrangement) in non-heterocystous cyanobacteria *Plectonema* sp., *Lyngbya* sp. and *Synechocystis* sp. Prasanna and Kaushik (1995) also performed hybridization studies in branched heterocystous cyanobacteria revealing the presence of *nif* gene but the absence of *xis* gene in *Mastigocladus* sp. and *Scytonematopsis* sp., showing the presence of continuous *nif* HDK cluster in *Fischerella* sp., whereas in *Tolypothrix* sp. and *Scytonema* sp. DNA samples hybridized with *xis* A.

Jennifer *et. al.* (2000) reported that the genes controlling circardian rhythm in cyanobacteria are widely distributed, and a clock system is universal among cyanobacteria. The *kai* ABC gene cluster is important for maintaining circardian rhythm in *Synechococcus* PCC 7942. Forty cyanobacterial strains possessed *kai* C related sequences.

Genetic engineering is a powerful tool and can be used elegantly or destructively depending on the direction of use and it is immoral to propose the destructive uses. Relatively little work has been done on the development of gene cloning systems in cyanobacteria. Genetic engineering of cyanobacteria also requires the development of method for introducing DNA into the host and of a cloning vector to enable the subsequent maintenance and expression of the foreign genes. Despite the limited work done, several promising lines of research are evident. The cloning vectors and gene expression systems for commercially interesting cyanobacteria could be developed in the near future. The prokaryotic nature of cyanobacteria makes them technically more amenable for molecular-genetic analysis. In addition, the absence of complicated interactions between nucleus and organelles is an added advantage.

Although by carrying one or more foreign genes, the GM (genetically modified) organisms look the same as their unmodified, naturally occurring counterparts. The basis of identifying a GM organism is to confirm the presence of the new, inserted DNA or the presence of the foreign protein coded by the inserted gene. In case the foreign protein acts as an enzyme, detection of the GM organism is done by chemical analysis, by detecting the product of the reaction which that enzyme catalyzes. Detection of the foreign protein in a GM organism is mainly done by ELISA (Enzyme-linked Immuno Sorbent Assay). For this, the foreign protein produced by GM organism is first extracted and bound to a plastic plate. An antibody molecule which specifically binds to the extracted foreign protein is then added. The antibody molecule is labelled with an enzyme which catalyses a reaction that gives a coloured product. The intensity of the colour produced is proportional to the amount of foreign protein detected. The most promising method for detecting a GM organism is to identify the inserted foreign gene. For this, a highly sensitive DNA amplification technique, PCR (polymerase chain reaction), is utilized. DNA sequences specific to the inserted foreign gene are amplified. In about 25 rounds of PCR, more than a million copies of target DNA are produced that can be identified by running the amplified DNA on an agarose gel.

Criteria for gene cloning systems

The development of a gene cloning system in a new organism requires suitable host strains, construction of vectors and means of introducing recombinant DNA into the host strain. In cyanobacterial gene cloning experiments, great progress is being made with the unicellular cyanobacteria *Anacystis nidulans* (Golden and Sherman, 1984; Daniell and McFadden, 1986), *Synechococcus* PCC 7942 (McCormick *et. al.*, 1999), *Agmenellum quadruplicatum* PR-6 (Stevens and Porter, 1980; Essich and Wolk, 1990), *Synechocystis* PCC 6803 (Sode *et. al.*, 1992; Marracini *et. al.*, 1993; Mernet *et. al.*, 1993; Subramanian and Kaushik, 2001), and the filamentous cyanobacteria *Anabaena* (Golden and Wiest, 1988; Elhai and Wolk, 1990; Xiaoqiang *et. al.*, 1997; Subramanian and Kaushik, 2000; *Nostoc* (Singh *et. al.*, 1987), a chromatically adapting *Fremyella diplosiphon* (Cobley *et. al.*, 1993), and *Calothrix* PCC 7601 (Xiaoqiang *et. al.*, 1997). The biotechnological work using commercially important strains such as *Spirulina* has not been started. Though the expression of desaturase genes in *Spirulina platensis* strain C1 (*Arthrospira* sp. PCC 9438) has been studied by Deshnium *et. al.* (2000), who found that the desaturation of fatty acid in membrane lipids to yield the di-unsaturated fatty acid 18: 2 was not regulated by temperature. The levels of 18: 1 and 18: 2 fatty acids remained almost constant either when the cells were grown at 35° C (normal growth temperature) or 22°C and 40°C, indicating a slow response of fatty acid desaturation to temperature in this cyanobacterium and the growth temperature had little effect on the composition of most of the fatty acids in *S. platensis*. However, in *Synechocystis* sp. PCC 6803 (Los *et. al.*, 1997) and *Synechococcus* sp. PCC 7002 (Sakamoto *et. al.*, 1977) the genes for desaturases have been reported to be temperature-dependent.

For recombinant DNA work, the hosts should lack restriction endonucleases, which, otherwise, would cleave unmodified foreign DNA. *Anacystis nidulans* lacks restriction enzymes,whereas, in

Agmenellum quadruplicatum, the enzyme *Aqu* 1 is thought to be responsible for the lower efficiency of transformation, when the unmodified hybrid plasmid is transferred into the cyanobacterial host. The elimination of the *Aqu* I restriction in the vector site resulted in a 10-fold increase in transformation efficiency. *Anabaena* strains also contain a number of restriction enzymes (Roberts, 1984). In addition to removal of restriction enzyme sites during the construction of the cloning vectors, isolation of mutants lacking restriction enzymes (Piechaczyk *et. al.*, 1980), use of heat treatment to selectively inactivate the restriction but not the modification enzymes (Sanders and Klaenhammer, 1980) or the isolation of the appropriate methylases to modify the foreign DNA before introduction into the host (Hattman, 1981), could be the possible solutions to this problem.

Cloning vectors

Cloning vectors can be constructed from plasmids or viruses, and in cyanobacteria the use of plasmids as vectors is much promising. Many cyanobacterial strains have indigenous plasmids (Subramanian and Kaushik, 1993), ranging in size from 2.3 to over 100 kb and the number of plasmids in a strain varies from 1-7 and the average being 1-3. *Anacystis nidulans* contains two plasmids pUH24 and pUH25. The smaller plasmid pUH24 has been used for the construction of most hybrid vectors. Plasmid should have a number of desirable properties to be employed as cloning vectors. In addition to being small in size for easy acceptance of foreign DNA, the plasmid should be able to replicate autonomously in the host organism as well as in the second host (*E. coli* is a good host because of its faster growth rate), should have genetic markers for detection, and a number of restriction sites in non-essential regions for the insertion of foreign DNA.

Since cyanobacterial plasmids are cryptic, i.e. do not code for any plasmid-borne functions, such as antibiotic resistance, therefore, there is need for the introduction of genetic markers. Antibiotic resistance is a commonly used marker. For the construction of the shuttle vector for *A. nidulans* R2, the transposon Tn 901 was used to insert an ampicillin-resistance (Ap^R) marker into the small indigenous plasmid. Resistance markers are also gained when the indigenous plasmid is fused with another plasmid carrying markers. A Cm'Km' shuttle vector pFCLV7, harbouring Cm^r from plasmid pACYC184 and Km^r from pUF3 and pUF12 plasmids of *E. coli*, was constructed for cloning in *Synechocystis* 6803 by Chauvat *et. al.* (1986).

To obtain unique restriction sites for cloning, the smaller indigenous plasmids are chosen for vector development (Schinozaki *et. al.*, 1982). For maintenance of a vector an origin for replication, which is recognized by the host, is necessary. The failure of transformation of certain *E. coli* plasmids into *Synechococcus* might be due to absence of a suitable origin. For constructing cyanobacterial vectors, *ori* site is obtained from indigenous plasmids either by fusion with another plasmid or by cloning a fragment of indigenous plasmid into another vector (Laudenbach *et. al.*, 1983). Such vectors are capable of replicating in both the cyanobacterial and *E. coli* hosts.

Subramanian and Kaushik (2001) have developed vectors pCY 100, pCY 101 and pCY 102 for cloning DNA fragments into *Synechocystis* sp. ARM 340 and *Anabaena cylindrica* ATCC 29414.

Expression vectors

Expression vectors are cloning vectors which carry regulatory sequences, such as promoter and ribosome-binding site, which allow a high rate of transcription of the cloned genes. Antibiotic resistance genes originating from other bacteria can be expressed in the cyanobacterium *Synechococcus* (Reichelt, 1984), indicating that other prokaryotic promoters are recognized by cyanobacteria. The coding sequence for *Saccharomyces cerevisiae* copper metallothionein (CUP1), the protein responsible for enhanced sequestration of Cu^{2+} in yeast, was placed under the control of an inducible synthetic *E. coli* promoter in the cyanobacterial vector plasmid pTrclK. Strain R2-PIM8 (*smt* A) of *Synechococcus* sp. PCC 7942 was transformed with the resulting construct (plasmid pMcK2), and the yeast CUP1 gene was integrated into its genome. In the transgenic cyanobacteria, the integrated CUP1 gene was transcribed and produced a protein product with the expected metallothionein characteristics (McCormick *et. al.*, 1999). For maximum expression, however, it may be necessary to use a strong cyanobacterial promoter.

Transformation and conjugal transfer of vectors

Of the known natural gene transfer mechanisms in bacteria (conjugation, transduction and transformation), only transformation and conjugation have been successfully demonstrated in cyanobacteria. Kumar (1962) for the first time showed sexuality, through transformation, in cyanobacteria (*Anacystis nidulans*). Many unicellular cyanobacteria are transformable, including *Anacystis nidulans* (Shestakov and Khyen, 1970; Golden and Sherman, 1984; Daniell and McFadden, 1986), *Aphanocapsa* (Astier and Espardellier, 1976), *Syhechocystis* (Grigorieva and Shestakov, 1982; Martinez-Ferez and Vioque, 1992; Ashby and Mullineaux, 1999; McCormick *et. al.*, 1999), *Gloeocapsa alpicola* (Devilly and Houghton, 1977), *Agmenellum quadruplicatum* (Stevens and Porters, 1980; Essich *et. al.*, 1990), *Synechococcus* (Williams and Szalay, 1983). Subramanian and Kaushik (2001) made several attempts to perform transformation experiments in *Anabaena cylindrica* and *Synechocystis* sp., however, they did not get any success. DNA-mediated genetic transfer (transformation) has not been successfully demonstrated in filamentous cyanobacteria and this failure has been attributed to the presence of a variety of restriction endonucleases which attack and degrade homo-and hetero-specific DNA in these cyanobacteria (Thiel and Wolk, 1987). This lack of gene transfer system has been the major block in the development of cloning systems for filamentous cyanobacteria. Trehan and Sinha (1981) have reported the transformation in a filamentous strain, but is unconfirmed. Although, Singh *et. al.* (1987) have confirmed the successful transformation of herbicide resistance genes Mat^r (Machete resistance) and Bas^r (Basalin resistance) from *Gloeocapsa to Nostoc muscorum*.

A number of factors affect transformation, including cell density, DNA concentration, physiological state of the recipient and the conditions during transformation. The reported optimal conditions for transformation are somewhat variable and may depend on the strain. Golden and Sherman (1984) have reported that the efficiency of transformation was enhanced in the absence of light or in the presence of photosystem-II-inhibitor. In *Gloeocapsa*, calcium chloride treatment was required for transformation; whereas in *Agmenellum quadruplicatum* $CaC1_2$ reduced the frequency of transformation.

Many cyanobacteria can receive broad host-range, P-incompatibility group (Inc P) plasmids by conjugation, and can maintain those plasmids. The capacity to receive a conjugative plasmid probably depends only on the structure of the outer cell membrane. If broad-host-range plasmids cannot be maintained in a host, a chimeric plasmid can be constructed by DNA cloning techniques. Such plasmids contain sequences that are necessary for replication in different hosts and can serve as a shuttle vector, which can be able to replicate in both the hosts. Derivatives of pBR322 can be mobilized by several conjugative plasmids (RP4) if the vector contains the *bom* (basis of mobility) region of pBR322 and if trans-acting factors required for mobilization are provided by helper plasmids.

Kumar and Ueda (1984) reported conjugation in the cyanobacterium *Anacystis nidulans*. Wolk *et. al.* (1984) have used the method of conjugation to transfer vectors into *Anabaena* strains. A number of conjugative plasmids are capable of mobilizing other plasmid, if a region of DNA, called *bom* (basis of mobility, also called *ori*T region), is present on the plasmid and a number of trans-acting factors are provided. Second, the plasmid must encode or be provided with a DNA-nicking protein (encoded by a gene called *mob*), that specifically recognizes the *bom* site (Elhai and Wolk, 1988). pBR322 and other widely used derivatives of pMP1, a Col E1-like plasmid, retain the *bom* site but not the *mob* gene of its natural parent. These derivatives can be transferred only when the nicking protein is provided in trans, by a helper plasmid. The donor may, therfore, contain several plasmids to facilitate the transfer of DNA to cyanobacteria:

1. Conjugal plasmid, e.g. RP4
2. The helper plasmid(s) e.g. a plasmid encoding the ColE1-specific nicking protein recognizing the *bom* site or one carrying a gene for a methylase to circumvent restriction by the recipient.
3. The cargo plasmid (shuttle vector or suicide vector) which is really wished to be transferred e.g. pBR322 with *bom* site; pUC 303 for *Synechococcus* R2.

Twig and Sherratt (1980) and Subramanian and Kaushik (2001) utilized the *bom* site to construct vector for cloning in *Anabaena*. For conjugal transfer into *Anabaena*, the broad host-range conjugative plasmid, RP4 which was originally isolated from *Pseudomonas*, was used; helper plasmids pDS 401 or pGJ 28 were used to provide the trans-acting factors (Wolk *et. al.*, 1984) . RP4 is used to transfer DNA in a wide range of unicellular and filamentous cyanobacteria. To obtain stable conjugal transfer it is necessary that (1) conjugal contact be made, (2) transferred DNA escape restriction or degradation, and (3) the DNA replicate autonomously or integrate into one of the replicons of the recipient. Even with transformable unicellular cyanobacteria, conjugation appears to be the preferred route of DNA transfer in certain cases, particularly when a large segment of foreign DNA is to be transferred , because DNA taken up by cyanobacteria appears to be randomly cut early during the transformation process. Thus, the efficiency of transfer of a segment of nonhomologous DNA decreases exponentially with its length. The conjugation proceeds by a different mechanism and transfer of plasmids by conjugation leads to a large number of merodiploid formation. Subramanian and Kaushik (2001) in *Anabaena* and *Synechocystis*, Marracini *et. al.* (1993) in *Synechococcus* and *Synechocystis*, Flores and Wolk (1985) in *Nostoc* and *Fischerella* and many others have transferred the plasmid vectors through conjugation.

Two strategies have been employed for gene cloning in cyanobacteria:

1. One uses shuttle plasmid vectors that carry both a cyanobacterial and an *E. coli* replicon, each of which functions in its indigenous host. Depending on the sequence of the cloned gene, the selection applied, and the cyanobacterial hosts, the cloned gene can be maintained on the autonomous plasmid or can be integrated into the cyanobacterial chromosome.
2. The other strategy exploits the efficient recombination system of *Anacystis nidulans* R2 and *Aphanocapsa* PCC6803 for a direct integration of the cloned gene into the chromosome. By this strategy, a DNA vector that lacks a cyanobacterial replicon, but carries similar chromosomal sequences, is applicable. According to the imposed selective conditions, the cloned gene, either alone or together with the vector sequences, integrates into the chromosome.

Both the strategies have been employed for cloning of reporter genes. Williams and Szalay (1985) cloned small fragments of *Synechococcus* chromosomal DNA into an *E. coli* plasmid that could not replicate in the cyanobacteria. Foreign DNA was inserted into the cyanobacterial DNA segment and the chimeric plasmid introduced into the *Synechococcus*. Since the plasmid could not replicate in the cyanobacterial host, it could only be maintained by integration into the chromosome. Integration occurred through homologous recombination via the *Synechococcus* DNA fragment. This method of introducing foreign DNA into a host has advantage in that it does not require extensive vector development for the new organism and that the DNA is stable even in the absence of selection. The site of integration could be used as a selective marker. Disadvantages of this method include the absence of gene amplification and difficulty of gene recovery from chromosomal DNA, if further gene manipulation is desired.

Genes cloned in Cyanobacteria

A large number of genes have been cloned in cyanobacteria. The genes *cat* (Friedberg, 1988), *lac* Z (Subramanian and Kaushik, 2001), *luc* (Subramanian and Kaushik, 2001), and *lux* A&B (Schmetterer *et. al.*, 1986) have been cloned in cyanobacteria which catalyze the production of chloramphenicol, acetyl transferase, β-galactosidase and luciferase, respectively. The encoded β-galactosidase activity can be easily measured in extracts or in permeabilized cells. Cyanobacterial colonies expressing *lac* Z when treated with the indicator 5-bromo-chloro-3-indolyl- β-D-galactoside (X-Gal) turn from green to turquoise, surrounded by the blue halos. The *lux* operon has been originally derived from the bacterium *Vibrio fischeri* and *V. harveyi*, and the *luc* operon from the American fire fly *Photinus pyralis*. Schmetterer *et. al.* (1986) transferred the *luc* genes from *V. harveyi* and *V. fischeri* into *Anabaena* sp. The contranscribed *lux* A and *lux* B genes from *V. harveyi* and *V. fischeri* have been transferred into *Anabaena*. The measurement of the encoded luciferase in the cyanobacterium is an extremely sensitive assay method. Luciferase catalyzes the oxidation of an aldehyde and of a reduced flavin mononucleotide ($FMNH_2$). Cells containing luciferase produced light on addition of the aldehyde and the resultant light emission is quantitated. Subramanian and Kaushik (2001) have used the technique of cloning of the *lac* Z and *luc* genes in cyanobacteria *Anacystis cylindrica* ATCC 29414 and *Synechocystis* sp. ARM 340 to assess the growth and establishment of inoculated species under field conditions as well as the stability of these reporter genes in soil microcosm.

The coding sequence for *Saccharomyces cerevisiae* copper metallothionein (CUPI), was transformed and integrated into the genome of *Synechococcus* sp. PCC 7942 (McCormick *et. al.*, 1999). The integrated CUPI gene was transcribed and produced a protein product with the metallothionein characteristics in the transgenic cyanobacterium and this property could be used for metal recovery and bioremediation of aquatic environments.

Most insecticides currently used are chemicals, but there are successful biological insecticides also. Resistance to chemical insecticides has occurred in several insect species. Management of such resistance remains a challenge, and new methods of controlling such species are needed. Various proteins have been identified as possessing insecticidal activities, e.g. α-amylase, lectins and insecticidal proteins from *Bacillus* species. The proteins produced by the *Bacillus* species possess the most potent insecticidal activities. Use of transgenic cyanobacteria to control insects is a new technology. *Bacillus thuringiensis* subsp. *israelensis* has been widely used as a biological pesticide to control mosquitoes and black flies, which are vectors for infectious human diseases, without adverse effects on other organisms. The larvicidal activity of this organism is due to the production of an endotoxin. The applications of *B. thuringiensis* sub sp. *israelensis* for mosquito control are limited by the short half-life of existing preparations under field conditions. This problem can be overcome by the cloning of δ-endotoxin genes into cyanobacteria which inhibit the breeding zones of mosquitoes and are used by them as a food source. Several attempts have been made to produce transgenic mosquitocidal cyanobacteria (Angsuthanasombat and Panyim, 1989; Chungjatupornchai, 1990; Murphy and Stevens Jr., 1992; Soltes-Rak *et. al.*, 1993, 1995; Xugong *et. al.*, 1993) and some success has been achieved in expressing single *cry* genes in unicellular strains, but the transformed cells displayed lower levels of mosquito-larvicidal activity than the level required for effective and practical use as bioinsecticides in natural aquatic environment. Various combinations of the genes *cry* IVA, *cry* IVD and p20 from *Bacillus thuringiensis israelensis* were introduced into *Anabaena* sp.strain PCC 7120, which displayed toxicity against third-instar larvae of *Aedes aegypti* at levels greater than any level previously reported for transgenic cyanobacteria (Xiaoqiang *et. al.*, 1997).

The practice of using diazotrophic cyanobacteria as an efficient source of biofertilizer for rice crops has been adapted in many developing countries (Venkataraman, 1972). These mosquitocidal cyanobacterial strains could be efficiently grown in the production tanks as well as in the rice fields without any application of the insecticides, like malathion and thus reducing the pollution effect in the environment. The present biofertilizer technology, however, suffers from a lack of herbicide-resistant cyanobacterial strains, which can grow and add fixed nitrogen in herbicide-treated rice fields. Singh *et. al.* (1987) have reported the natural occurrence of resistance for two herbicides used in rice fields, due to the presence of Machete-resistance (Mat^r) and Basalin-resistance (Bas^r) genes, in a strain of *Gloeocapsa* and have successfully transferred these genes into *Nostoc muscorum*.

Conclusions

Rapid advances in genetic engineering since 1970 have created the potential for large benefits to society through the development of novel organisms and new products. However, these potential benefits are balanced by the introduction of new problems and potential hazards. The regulations governing biotechnology are complex and governmental approval is required for field testing of GM organisms. These regulations have led to high costs and long delays for companies wishing to test new products in the environment.

Growing genetically modified algae or other organisms is still not popular in many parts of the world as it is feared that the artificially engineered organisms could be unsafe and harmful for the environment due to spread of the foreign gene to unmodified organisms. Environmental activists in many countries are supporting the anti-GM movement, which does not approve the genetic modification of organisms. Over the safety of GM food, Food Ethics Council has been formed in many countries, whose role is to decide the issues, such as food safety choice and labelling of GM foods. The European Union has introduced strict rules for labelling GM food stuffs and a Biosafety Protocol is being worked out under the UN's Biodiversity Convention.

Although there are now controversies over the use of GM organisms, it must be realized that GM organisms are specially created for introducing useful traits not existing in their natural counterparts. But many GM organisms have a revolutionary impact on the agricultural sector and their production should not be discouraged. It is necessary to conduct a controlled research on creating value-added GM organisms and thorough testing of the potential benefits or any likely drawbacks of the new life forms on human life, before they are unleashed on the society at large, are also necessary. The cyanobacterial biotechnology is in its infancy, and continues to improve rapidly. More transgenic cyanobacteria can be formed now, and the control of gene expression is becoming better understood.

Acknowledgement

One of the authors , Dr. Mrs. Rashmi Tyagi is thankful to CSIR for financial assistance.

References

Angsuthanasombat, C. and Panyim, S. 1989. Biosynthesis of 130-kilodalton mosquito larvicide in the cyanobacterium *Agmenellum quadruplicatum* PR-6. *Appl. Environ. Microbiol.* **55**: 2428-2430.

Ashby, M.K. and Mullineaux, C. W. 1999. Cyanobacterial *ycf* 27 gene products regulate energy transfer from phycobilisomes to photosystems I and II. *FEMS Microbiol. Lett.* **181**: 253-260.

Astier, C. and Espardiellier, F. 1976. Mise en evidence d'un chez une cyanophycee du genre *Aphanocapsa. C.R. Acad. Sci.* Paris **282**: 795-797.

Chauvat, F., De Vries, L., Van der Ende, A. and Van Arkel, G.A. 1986. A host-vector system for gene cloning in the cyanobacterium *Synechocystis* PCC 6803. *Mol. Gen. Genet*. **204**: 185-191.

Chungjatupornchai, W. 1990. Expression of the mosquitocidal protein genes of *Bacillus thuringiensis* subsp. *israelensis* and the herbicide-resistance gene *bar* in *Synechocystis* PCC 6803. *Curr. Microbiol.* **21**: 283-288.

Daniell, H. and McFadden, B.A. 1986. Characterization of DNA uptake by the cyanobacterium *Anacystis nidulans*. *Mol. Gen.. Genet*. **204**: 243-248.

Deshnium, P., Paithoonrangsarid, K.,Suphatrakul, A., Meesapyodsuk, D., Tanticharoen, M. and Cheevadhanarak , S. 2000. Temperature-independent - and dependent expression of desaturase genes in filamentous cyanobacterium *Spirulina platensis* strain C1 (*Arthrospira* sp. PCC 9438). *FEMS Microbiol. Lett*. **184**: 207-213.

Devilly, C. I. and Houghton, J.A. 1977. A study of genetic transformation in *Gloeocapsa alpicola*. *J. Gen. Microbiol.* **98**: 277-280.

Elhai, J. and Wolk, C.P. 1988. Conjugal transfer of DNA to cyanobacteria. *Meth. Enzymol.* **167**: 747-754.

Elhai, J. and Wolk, C.P. 1990. Developmental regulation and spatial pattern of expression of the structural genes for nitrogenase in the cyanobacterium *Anabaena*. *EMBO. J.* **9**: 3379-3388.

Essich, E., Stevens, S.E.Jr. and Porter, R.D. 1990. Chromosomal transformation in the cyanobacterium *Agmenellum quadruplicatum*. *J. Bacteriol.* **172**: 1916-1922.

Flores, E. and Wolk, C.P. 1985. Identification of facultative heterotrophic nitrogen fixing cyanobacteria able to receive plasmid vectors from *E. coli* by conjugation. *J. Bacteriol.* **162**: 1339-1341.

Friedberg, D. 1988. Use of reporter genes in cyanobacteria. *Meth. Enzymol.***167**: 736-747.

Golden, S.S. and Sherman, L.A. 1984. Optimal conditions for genetic transformation of the cyanobacterium *Anacystis nidulans* R2. *J. Bacteriol.* **158**: 36-42.

Grigorieva, G.R. and Shestakov, S.V. 1982. Transformation in the cyanobacterium *Synechocystis* sp. 6803. *FEMS Microbiol. Lett.* **13**: 367-370.

Hattman, S. 1981. DNA methylation. In: *The Enzymes*. Edited by Boyer, P.D. Academic Press, New York. **14**: 517-548.

Jennifer, L., Scheffer, J., Lee, A., Painter, M.P. and. Miao, V.P.W 2000. Genes controlling circardian rhythm are widely distributed in cyanobacteria. *FEMS Microbiol. Lett.* **189**: 129-133.

Kallas, T., Rebiere, R., Rippler, R. and Tandeau de Marsac, N. 1983. The structural nif genes of the cyanobacteria *Gloeothece* sp. and *Calothrix* sp. share homology with those of *Anabaena* sp. but the *Gloeothece* genes have a different arrangement. *J. Bacteriol.* **155**: 427-431.

Kaneko, T., Sato, S., Kotani, H., Tanaka, A. and Asamizu, E. 1996. Sequence analysis of the genome of the unicellular cyanobacterium *Synechocystis* sp. strain PCC 6803. II: Sequence determination of the entire genome and assignment of potential protein-coding regions. *DNA Res.* **3**: 109-136.

Kreps, S., Ferino, F., Mosrin, C., Gerits, J., Mergeay, M. and Thuriaux, P. 1990. Conjugal transfer and autonomous replication of a promiscuous Inc Q plasmid in the cyanobacterium *Synechocystis* PCC 6803. *Mol. Gen Genet*. **221**: 129-133.

Kumar, H.D. 1962 Apparent genetic recombination in a blue-green alga. *Nature* **196**: 1121-1122.

Kumar, H.D. and Ueda, K. 1984. Conjugation in the cyanobacterium *Anacystis nidulans. Mol. Gen. Genet.* **195**: 356-357.

Laudenbach, D.E., Straus, N.A., Gendel, S. and Williams, J.P. 1983. The large endogenous plasmid of *Anacystis nidulans:* mapping, cloning and localization of the origin of replication. *Mol. Gen. Genet*. **192**: 402-407.

Los, D.A., Ray, M.K. and Murata, N. 1997. Differences in the control of the temperature-dependent expression of four genes for desaturases in *Synechocystis* sp. PCC 6803. *Mol. Microbiol.* **25**: 1167-1175.

Marracini, P., Bulteau, S., Chauvat, C.C., Bouvier, P.M. and Chauvat, F. 1993. A conjugative plasmid vector for promoter analysis in several cyanobacteria of the genera *Synechococcus* and *Synechocystis. Plant Mol. Biol.* **23**: 905-909.

Martinez-Ferez, I.M. and Vioque, A. 1992. Nucleotide sequence of the phytoene-desaturase gene from *Synechocystis* sp. PCC 6803 and characterization of a new mutation which confers resistance to the herbicide norflurazon. *Plant Mol. Biol.* **18**: 981-983.

McCormick, P.M., Cannon, G.C. and Heinhorst,S. 1999. Expression of the copper metallothionein CUPI from *Saccharomyces cerevisiae* in the cyanobacterium *Synechococcus* R2-PIM8 (smt A). *Curr. Microbiol.* **38**: 155-162.

Mermet, B.P., Chauvat, C.C., Marracini, P. and Chauvat, F. 1993. Transfer and replication of RSF 1010 derived plasmids in several cyanobacteria of the genera *Synechocysfis* and *Synechococcus*. *Curr. Microbiol*. **27**: 323-327.

Murphy, R.C. and Stevens, S.E. Jr. 1992. Cloning and expression of the *cry* IV D gene of *Bacillus thuringiensis* subsp. *israelensis* in the cyanobacterium *Agmenellum quardruplicatum* PR-6 and its resulting larvicidal activity. *Appl. Environ. Microbiol.* **58**: 1650-1655.

Piechaczyk, M., Jeanteur, P. and Louarn., J.M. 1980. An easy method for the selection of restriction and modification-deficient mutants of *Escherichia coli* K-12. *Gene.* **11**: 173-175.

Prasanna, R. and Kaushik, B.D. 1994. Physiological and molecular genetic aspects of nitrogen fixation in non-heterocystous cyanobacteria. *Ind. J. Expl. Biol.* **32:** 248-251.

Prasanna, R. and Kaushik, B.D. 1995. Nitrogen fixation and *nif* gene organization in branched heterocystous cyanobacteria: variation in the presence of *xis* A. *Folia Microbiol.* **40**: 176-180.

Reichelt, B.Y. 1984. *Genetic studies on a unicellular cyanobacterium, Synechococcus.* Ph.D. thesis, Univ. New South Wales, Australia.

Roberts, R.J. 1984. Restriction and modification enzymes and their recognition sequences. *Nucleic Acids Res.* **12**: 167-204.

Sakamoto, T., Higashi, S., Wada, H., Murata, N. and Bryant, D.A. 1997. Low-temperature-induced desaturation of fatty acids and expression of desaturase genes in the cyanobacterium *Synechococcus* sp. PCC 7002. *FEMS Microbiol.Lett.* **152**: 313-320.

Sanders, M.E. and Kalenhammer, T.R. 1980. Restriction and modification in group N streptococci: effect of heat on development of modified lytic bacteriophage. *Appl. Environ. Microbiol.* **40**: 500-506.

Schemetterer, G., Wolk, P. and Elhai, J. 1986. Expression of luciferases from *Vibrio harveyi* and *Vibrio fischeri* in filamentous cyanobacteria. *J. Bacteriol.* **167**: 411-414.

Shestakov, S.V. and Khyen, N.T. 1970. Evidence for genetic transformation in blue-green alga *Anacystis nidulans. Mol. Gen. Genet.* **170**: 372-375.

Shinozaki, K., Tomioka, N., Yamada, C. and Sugiura, M. 1982. Cloning and characterization of a plasmid DNA from *Anacystis nidulans* 6301. *Gene* **19**: 221-224.

Singh, D.T., Nirmala, K., Modi,D.R.,Katiyar, S. and Singh, H.N. 1987. Genetic transfer of herbicide resistant genes from *Gloeocapsa to Nostoc muscorum. Mol. Gen. Genet* . **208**: 436-438.

Sode, K., Tatara, M., Takeyama, H., Burgers, J.G. and Matsunaga, G.1992. Conjugative gene transfer in marine cyanobacteria: *Synechococcus sp., Synechocystis* sp. and *Pseudomonas* sp. *Appl. Microbiol. Biotechnol.* **37**: 369-373.

Soltes-Rak, E., Kushner, D.J., Williams, D.D.and Coleman, J.R. 1993. Effect of promoter modification on mosquitocidal cry IVB gene expression in *Synechococcus* sp. strain PCC 7942. *Appl. Environ. Microbiol.* **193**: 2404-2410.

Soltes-Rak, E., Kushner, D.J., Williams, D.D. and Coleman, J.R. 1995. Factors regulating *cry* IV B expression in the cyanobacterium *Synechococcus* PCC 7942. *Mol. Gen. Genet.* **246**: 301-308.

Stevens, S.E. Jr. and Porter, R.D. 1980. Transformation in *Agmenellum quadruplicatum. Proc. Natl. Acad. Sci.* USA. **77**: 6050-6056.

Stewart, W.D.P. 1970. Algal fixation of atmospheric nitrogen. *Plant Soil.*32: 555-558.

Subramanian,S. and Kaushik, B.D. 1993. Isolation and stability of plasmid DNA of cyanobacterium, *Westiellopsis prolifica* ARM 701.*Indian J. Expt. Biol.* **321**: 655-657.

Subramanian, S. and Kaushik, B.D. 2001. Development of genetic markers in cyanobacteria and stability of genetically marked strains in soil. *World J. Microbiol. Biotech.* **17**: 535-544.

Thiel, T. and Wolk, C.P. 1987. Conjugal transfer of plasmids to cyanobacteria. *Meth. Enzymol.* **153**: 232-243.

Trehan, K. and Sinha, U 1981. Genetic transformation in a nitrogen-fixing filamentous cyanobacterium. *J. Gen. Microbiol.,* **124**: 349-352.

Twig, A.J. and Sherratt, D. 1980. Trans complementable copy number mutants of plasmid Col. E1. *Nature* **283**: 216-218.

Venkataraman, G.S. 1972. *Algal Biofertilizers and Rice Cultivation.* Today and Tomorrow Publ., New. Delhi. pp. 75.

Venkataraman, G.S. 1979. Algal inoculation in rice fields. In: *Nitrogen and Rice.* International Rice Research Institute, Los Banos, Philippines. pp. 311-321.

Weare, N.M. and Beneman,J.R. 1974. Nitrogenase activity and photosynthesis in *Plectonema boryanum. J. Bacteriol.* **199**: 258-265.

Williams, J.G.K. and Szalay, A.A. 1983. Stable integration of foreign DNA into the chromosome of the cyanobacterium *Synechococcus* R2. *Gene* **24**: 37-51.

Wolk, C.P., Vonshak, A., Kahoe, P. and Elhai, J. 1984. Construction of shuttle vectors capable of conjugative transfer from *Escherichia coli* to nitrogen-fixing filamentous cyanobacteria. *Proc. Natl. Acad. Sci. USA* **81**: 1561-1565.

Xiaoqiang, W. Vennison, S. J., Huirong, L., Ben-Dov, E. , Zaritsky A. and Boussiba, S. 1997. Mosquito Iarvicidal activity of transgenic *Anabaena* strain PCC 7120 expressing combinations of genes from *Bacillus thuringiensis* subsp. *israelensis*. *Appl. Environ. Microbiol.* **63**: 4971-4975.

Xugong, X., Renqiu, K. and Yuxiang, H. 1993. High Iarvicidal activity of intact cyanobacterium *Anabaena* sp. PCC 7120 expressing gene 51 and gene 42 of *Bacillus sphaericus* 2297. *FEMS Microbiol. Lett.* **107**: 247-250.

Section IV
APPLICATIONS OF ALGAE

Chapter 17

ROLE OF ALGAE IN SUSTAINABLE AQUACULTURE

Amrik S. Ahluwalia and Gagandeep Kaur Khosa

Abstract

'Algae' has been an important component of aquatic ecosystem which also supports the growth and survival of fish and related fauna. Therefore, in addition to algal habitats, culture and nutritional value of algae is of direct significance for the development and economics of fish stock. Importance of native algal inoculum for addition to the fish tanks has been discussed. There should be proper documentation of useful native algae and their proper cultures should be made available to the fish farmers. Ideal algal cultivation system utilizes organic wastes and fertilizers and produce a variety of commercial products while returning little or no waste to the environment. More work is required to evaluate the possibilities of producing algal based fish feed on a commercial level. Suitability of a particular alga or group of algae for feed, and the stage (fish fry, fingerlings or later) or type of fish needs to be thoroughly investigated to make aquaculture sustainable in India. Efforts have also been made at the global level to increase the production of algal products and also their use as fish feed. Excessive growth of algae can sometimes prove harmful for the aquacultural practices, therefore, continuous monitoring of toxic algal species in a fish pond needs to be carried on. Suitable control measures must be taken to keep noxious algae under check, without causing harm to fishes in pisciculture.

Key words: Algae, aquaculture, fish feed, phycocolloids, pisciculture, succession.

Introduction

Algae comprises multitude of heterogenous and polyphyletic assemblage; ranging from minute unicells to gigantic kelps. These ubiquitous organisms are found on rocks, soil, trees as well as predominant in water bodies, both marine and fresh water of tropical, temperate and polar regions (Ahluwalia, 1993).

Culture of aquatic organisms for their commercialization is known as aquaculture. Sustainable aquaculture includes the judicious use of the resources (i.e. algae, fishes etc.) thus ensuring their availability for the present generation as well as preserved for the future. With increasing population, there is an acute pressure on land for the available resources and hence their over exploitation and even depletion. In this context, aquatic resources are being explored as an alternative to sustain the ever increasing number of human race for food.

The importance of algae in aquaculture has been well known since ages. The organisms constitute the basic link of the food cycle of aquatic animals (DeSilva and Anderson, 1995). Without a proper appreciation of the role of these autotrophic organisms (algae), the sustained maximum yield of fish cannot be obtained or maintained (Reitan *et at.*, 1997). Since total global aquacultural production has been rising every year, this industry is projected to grow at the rate of 8% per year for foreseeable future (Tacon, 1998) and hence the importance of algae. Decreasing natural catches and increasing world dependence on the fish as a food source has led to a growing interest in aquaculture. Central to sustainable aquaculture is the need to establish and maintain a food chain to support the animals until they achieve the market size. The annual global aquacultural production of marine algae is currently placed around 6.5×10^6 tonnes (Tacon, 1998).

Moreover, need for aquaculture arises because of the useful products extracted from various algae at the commercial level. Algae are used to extract chemicals that are used as food additives, thickeners and gelling agents. In addition, many algal organisms yield ingredients of culture media, plant growth regulators, anticancer (Mynderse *et. al.*, 1977; Gerwick *et. al.*, 1994; Patternson *et. al.*, 1994), antiviral and antibiotic substances (Mason *et. al.*, 1982; Gleason and Paulson, 1984; Flores and Wolk, 1986). Dried biomass or cell extracts produced from *Chlorella* (Lee, 1997; Yamoguchi, 1997) and *Spirulina* (Emben *et. al.*, 1994; Vonshak, 1997) have dominated the commercial opportunities. These are much better substitutes for food/feed because they have higher contents of proteins, carbohydrates, fats, minerals, vitamins and even trace elements. Their products are marketed as health food and collectively are likely worth many hundred million dollars. Other major products of industrial utilisation include phycocolloids such as agar-agar, alginates, carrageenans and diatomite. Recently, attention is being focussed on production of oils and fatty acids from some unicellular algae (McLachlan *et. al.*, 1999). Algal oils are also used as biofuels (Neenan *et. al.*, 1986). It has been found that algal oils contained long chain polyunsaturated fatty acids (LCPUFAs). These polyunsaturated fatty acids form an important and essential supplement in the human food. Algae are the only vegetable source of these acids (Cohen *et. al.*, 1995; Behrans and Kyle, 1996).

Algae in aquaculture

It has become evident that the algae form an important source of commercial products, health food, chemicals, fish feed etc. Humans depend upon this source directly (as health food) or indirectly (as fish products). Important substances like vitamins A and D, which are commercially obtained from the liver of shark and other fishes, originate their synthesis in the planktonic algae, particularly those forming the food for such fish (Vashishta, 1995).

Fish and fish oils are good sources of LCPUFAs and these are enriched with both DHA (docosahexaenoic acid) and EPA (eicosapentaenoic acid). However, due to accumulation of toxins at contamination level in some fish oils, alternative dietary sources have been sought. Therefore, algae can also directly serve this purpose.

Culture of algae

A water body can support growth of different algal organisms irrespective of their importance as fish food. There is a need to manipulate the growth of those algae that may enrich phytoplankton feed of the fish. The first stage in obtaining culture is the knowledge about the distribution of the required species in nature. Collection and culture of algae from various water bodies in a particular area is of importance to find their specific food value, data on their economics and knowledge of their ecology, including choice of medium pH, temperature of culture etc. Unfortunately, many organisms required in culture, occur only in small numbers in their natural habitats and their direct isolation is thus difficult since individual cells have to be picked out. The cells of algae are picked up in small pipettes drawn out to a fine point and closed by a piece of plugged pressure tubing. Most microscopic algae can be picked up and transferred in this way under a binocular microscope. Surface sterilization of algae can be carried out in the same way by passing the alga through solution containing hypochlorite or by exposure to ultraviolet light. Algae forming zoospores are made bacteria free by subculturing the zoospores immediately after their release. In many instances, the sample from the natural habitats e.g. soil, aquatic plant or plankton, can be stimulated to grow faster by the addition of appropriate nutrient media. Then washing by transfer through a number of watch glasses containing sterilized culture media is carried out in order to reduce the number of contaminant bacteria. Unialgal, axenic and clonal cultures are thus prepared from the forms collected from pond water. Isolated species can be characterized for their growth response, macromolecular components and ill effects, if any (Regnault *et. al.*, 1995).

Cultures are generally made initially in liquid media, which may be relatively simple inorganic media or may contain organic compounds necessary for growth of the organisms *e.g.* many flagellates have been shown to require vitamin B_{12}. Algae may be grown successfully on nutrient media supplemented with gelatin, silica gel or agar, the later is in common use. An agar (Difco bacto grade) gel of 1% concentration is prepared in flasks, petridishes or as slants in test tubes. It is usually best to sterilize the agar in water and equal volume of nutrient solution (double concentrated) separately and mix these after autoclaving to get appropriate concentration. Most important use of agar slants is in maintaining pure stock cultures for occasional subculturing. Petridish cultures are used to investigate algal flora by spreading enriched inoculum containing a range of algae and picking out unialgal colonies. The agar in the dish may be inoculated by spraying a suspension of the algae on the surface of agar or streaking the suspension by means of wire loops. Some algal species will not grow permanently on agar, but the technique can still be used to separate the alga from other algae and from bacteria by planting out a suspension and immediately picking individual cells from the surface with a pipette and transforming them to liquid culture.

Mass culture techniques have been designed to test the use of algae for food production, for inoculating soils and sewage disposal beds, for industrial extraction of chemicals, antibiotics etc. Production of algae in the sea and freshwater is comparable to that of agricultural productivity, on land. Studies of algal biomass production have utilized a wide variety of species from brown algae, red algae, green algae, diatoms and cyanophytes. Aquatic biomass production has some advantages such as availability of area other than agricultural land, buffered temperature variation, easy manipulative nutrition, abundant light availability and a continuous and shorter growth cycle. The algae can be put into two main groups (a) Microalgae (b) Macroalgae.

Microalgae - are microscopic, either unicellular or filamentous or colonial. These include species of *Scenedesmus, Chlorella, Selenastrum, Spirulina, Anabaena, Anacystis, Chaetoceros, Skeletonema, Dunaliella* etc. (Table 17.1). Diatoms are used because of their high fat content. The main aim of these cultures is to achieve a high cell density, and hence a high rate of carbondioxide supply is maintained either by bubbling carbondioxide enriched air through the medium, or by stirring, shaking or circulating the medium. Complicated systems have been designed to measure the rate of uptake and changes in the culture medium and to correct these. The cell density is often photoelectrically measured and nutrient supply is regulated automatically, so that culture maintains a high rate of division.

The productivity of microalgae is higher than that of macroalgae (30-60 dry ash free metric tonnes per hectare per year or DAFMT ha^{-1} yr^{-1}). Microalgae have also been contemplated as a biomass source for the conversion to methane. Because of harvesting cost and nutritional demands, they are considered for direct biosynthetic production of high energy fuels (Glycerol, lipid, hydrogen). Controllable variables that influence the yield include nutrition, design and operation of pond, harvesting procedure etc. Nitrogen and phosphorus are limiting factors and application by addition of sewage (it support luxuriant growth of unicell or filamentous algae e.g. *Chlorella, Chlamydomonas, Scendesmus, Euglena, Oscillatoria, Microactinium* and *Anacystis*) or recycling of processing wastes has been proposed. Unlike macroalgae, carbon availability plays a key role in their cultivation (Manoharan *et. al.*, 1999). To overcome carbon limitation, rapid mixing of culture is required or carbondioxide must be pumped into the system.

Table 17.1: Commonly cultured algal species for aquaculture

Amphora	*Anabaena*
Chaetoceros	*Chlamydomonas*
Chlorella	*Cyclotella*
Dunaliella	*Gomphonema*
Isochrysis	*Nannochloropsis*
Navicula	*Nitzschia*
Pavlova	*Phaeodactylum*
Skeletonema	*Tetraselmis*
Thalassiosira	*Spirulina*

One of the problems with microalgal biomass production is difficulty of harvesting. In most operations, the algal solids will comprise only 0.06-0.015% of the total weight of water (algal slurry, so the microalgae have to be collected and concentrated atleast 1000 times before they can be utilized). They can be concentrated while still in large ponds by inducing aggregation of cells, either through chemical additives (chemical flocculation) or by activation of natural biological mechanisms (bioflocculation). In bioflocculation, aggregation is caused by production of extracellular macromolecules which resulted in cellular agglutination. Once aggregated, they are harvested by either natural sedimentation, dissolved air floatation or filteration by screens with micropores (Venkataraman, 1991; Nagle and Paul, 1999).

The favourable forms can be selected and inoculated in slightly larger containers and then in the open tanks (Table 17.2). They can further be upgraded to larger ponds after careful examination at various levels. After proper scrutiny for these features, an alga can be recommended for inoculation in a fish pond. Regular supply of algal culture encourage pisciculture. Efforts should be made to use only native algal forms i.e. algae collected from areas near their site of collection. It has been known that native forms acclimatize easily and perform better than other forms to changing growth conditions (Ahluwalia *et. al.*, 1993; Davies and Grossman, 1994; Fernandez *et. al.*, 1998; Spalding, 1998). Algae like *Spirulina sp.* are very rich in protein, vitamins and other important substances. Although its harvest and further processing for its use as health food is a costly affair, yet the culture of such algae in fish pond can be a boon for aquaculture. The fish fry can take such algal forms directly and hence overhead expenses of harvesting and processing are kept to the minimum. A flow chart for lab. to land transfer of aquaculturally important algae has been given (Table 17.3).

Table 17.2: Characteristics of selection of an alga for aquaculture

— Unicellular, filamentous or colonial with high protein value
— Faster growth rate
— Forming homogenous suspension
— Broader temperature tolerance
— Simpler nutritional requirements
— Acceptability
— Broader tolerance to salts
— Non-toxic
— Cost effective
— High yield of required products

In aquacultural practices, nitrogen-fixing blue-green algae are preferred because the expenses on supplemental nitrogen in the fish feed can be reduced by such forms. Secondly, nitrogen fixing capability afforded an advantage to these organisms, of growing in a water body without combined nitrogen compounds. These organisms, in turn, reduced or eliminated the growth of competing non-nitrogen-fixing microorganisms. Many blue-green algae with high protein content, suitable amino acid profile, fatty acids, lipids and other substances are essential for the growth of fish (Ahluwalia and Paul, 1987; Dua *et. al.*, 1993). The nature of their cell wall gives much better digestibility to the aquacultural organisms in comparison with eukaryotic algae. Finally, some blue-green algae are tolerant to a wide variety of environments (Tang *et. al.*, 1997) and can grow well. Thus, direct feeding of nitrogen-fixing blue-green algae to cultured organisms have significant advantages over other algal organisms (Mitsui,1979) and can even outcompete other algae, if nutrients are low.

Pigmentation forms an important criterion for culture-grown organisms because colour can affect their commercial acceptability. Artificial diet lacks natural source of pigments that give characteristic coloration to trouts and other fish. Therefore, carotenoid (astaxanthin)

Table 17.3: Flow chart of lab. to land transfer of algae for various products and pisciculture

Collection of algal sample from the ponds

↓

Identification

↓

Isolation to clonal, axenic cultures

↓

Growth studies in the lab.

↓

Labs flask cultures (10 ml - 10 L)
(To know broad parameters of algal growth)

↓

Greenhouse covered ponds (4 m^2)
(Inoculum for large ponds)

↓

Indoor ponds (20 m^2)
(To assess harvesting techniques and rough cost)

↓

Outdoor ponds (200 m^2)
Trials for : Harvesting
: Application
: Growth
: Toxicity

↓

Outdoor ponds (500 m^2)
(Mass cultivation)

↓

Commercial production and use

is added to the feed. In natural food chain, algae are the primary source of astaxanthin and other such pigments. *Haematococcus* is a rich source of astaxanthin (Grunewald *et. al.*, 1997) and is being used as an artificial diet. Thus feeding on such algae improves the colour quality of fish (Reitan *et. al.*, 1994) and hence its market value.

Macroalgae are large multicellular organisms, either called kelps or seaweeds. They are also utilized for both food and extraction of various chemicals. Current yield of kelps ranges from 5-39 dry ash free metric tonnes per hectare per year(DAFMT ha^{-1} yr^{-1}). Macroalgae can be utilized economically as energy alternative by increasing both the area under propagation and biomass yield for unit area. Three crucial factors to control yield can be nutrition, physical constraint and competition. Nitrogen is the key limiting factor as ammonia and nitrate are the only nitrogen forms readily utilized by macroalgae. Additional nitrogen results in higher photosynthetic rate and higher yield. The residue of the biomass to fuel conversion process has been reapplied to algal cultures and it works as nutrient source. The deep waters are enriched in nitrogen and phosphorus and recirculating these waters supplies the algae with the required nutrients. Useful macroalgae comprise Chlorophytes (25 spp.) e.g. *Monostroma, Caulerpa, Enteromorpha, Ulva,* Phaeophytes (54 spp.) e.g. *Laminaria, Undaria, Alaria, Eisenia, Ecklonia,* and Rhodophytes (81 spp.) e.g. *Porphyra, Palmaria, Gracilaria, Gelidium, Hypnea* and *Eucheuma*. Some of the major products obtained from macroalgae are phycocolloids. Important among these are agar-agar, alginates, carrageenans etc.

Agar-agar - It is produced by certain red algae as mucilage and stored alongwith cellulose in the cell wall. Agar is obtained commercially from some members of families Gracilariaceae and Gelidiaceae (e.g. *Gracilaria, Gelidium* and *Pterocladia*). The extract is gel containing galactose and a sulphate. It is produced in the form of powder, cakes and granules. Agar-agar is of great value in the preparation of foodstuffs and is particularly used in the articles of diet for invalids. It is used as base in culture media for bacteria, fungi, algae and other tissues. It is also useful in pharmaceutical industry as preservative in icing, canned meat and fish, as clarifying agent in beer and wine production etc. Total production of agar agar world wide is approximately 7000 tonnes (McLachlan, 1985). Agarose is a highly purified form of agar. It has uses in biomedical applications, electrophoresis, diffusion, filteration, immunological and separation techniques. It is also important for encapsulation and immobilization of microbial cells.

Commercial production of agar was world monopoly of the Japanese for many years. The agarophytes are collected by diving, dragging, or raking them offshore at low tide. In the traditional processing procedure, the plants are cleaned and bleached in sun, with several washings in fresh water used to facilitate bleaching. The material is boiled for several hours, and the extract is acidified. This extract is then frozen and thawed. On thawing, water flows from the agar, carrying impurities with it. The agar that remains is dried and marketed as flakes or cakes. The more modern method extracts the agar under pressure in autoclaves (Cosson *et. al.*, 1995). The agar is decolorized and deodorized with activated charcoal, filtered under pressure, and evaporated under reduced pressure (Lee, 1999).

Alginates - Alginic acid is a complex polyuronic polymer made of â-1, 4-linked D-mannuronic acid and L-glucaronic acid units. Chief source of alginates are brown algae such as *Ascophyllum, Lamanaria, Fucus, Lessonia, Ecklonia, Macrocystis and Eisenia*. Alginates are used as a stabilizer in icecream, milk and fruit squashes, and as gelling or thickening agent for jams, jellies and sauces. It is also useful in paper industry. Other important uses include fire proofing of fabrics, textile and paint industry. Alginates encapsulated islets of Langerhans cells that are capable of producing insulin when needed, have been implanted into animals as a replacement for the endocrine functions of pancreas (Jensen, 1995). World Production of alginates is approximately 17000 tonnes annually worth US$ 100 million.

Carrageenan (caragheen) – It is a cell wall polysaccharide (sulphated glycans) and is a mucilaginous extract. Its chief source is a red alga *Chondrus crispus*, and to a lesser extent species of *Gigartina, Eucheuma, Hypnea, Iridaea* and *Phyllophora*. In the Philippines, *Eucheuma* is extensively cultured as a source of carrageenan (Velasquez, 1972). Carrageenans are used in food, textile, pharmaceutical, leather and brewing industries. These are also used to stabilize emulsions and as remedy for cough. Carrageenans form a part in tooth pastes, deodorants, cosmetics and paints. Their total quantity produced annually is approximately 13000 tonnes worth few hundred million dollars (Firth, 1969; Lembi *et. al.*, 1988). Commercial extraction is similar to that for agar although carrageenans cannot be purified by freezing. The dried alga is washed with freshwater to reduce the salt contents and then boiled with 2-4 parts of the alga to 100 parts of water. The soluble carrageenan is separated from the insoluble residue in a centrifuge. Following filteration and some evaporation under vacuum, the carrageenan is dried on a rotary drier (Lee, 1999).

Chemicals – Many important chemicals and toxins are extracted from various algae. Iodine is obtained from genera such as *Laminaria, Ecklonia, Fucus* and *Eisenia*. Red algae *Rhodomela* and *Polysiphonia* are the source of bromine. Kelps are a source of soda and potash. Algae are known to produce many pharmacologically active compounds and thus are the rich source of biologically active natural products (Nagle and Paul, 1999). Seaweeds are also rich in copper, iron, zinc, cobalt, vanadium, molybdenum, manganese, boron and chromium.

Health food – In many islands, people depend upon algae such as *Porphyra, Laminaria, Undaria, Gracilaria, Alaria, Asparagus* etc. for their food requirements (Van Der Meer and Patwary, 1995). *Ulva, Enteromorpha* and *Monostroma* are other edible forms of algae. Chapman and Chapman (1980) listed 160 sp. of seaweeds eaten by humans (25 chlorophytes, 54 phaeophytes, 8 rhodophytes), because of its food value (as vitamins, proteins and trace elements), flavour, colour and texture. Annual value of food algae from three genera is in excess of US$ 1.5 billions.

Algae in pisciculture

'Algae' comprising planktonic, attached, marine and freshwater forms provides the primary food for fish and other aquatic animals. In sea, marine algae grow as attached forms on rocks near the sea shore, or occur as floating plankton. They manufacture their own food material employing solar energy and are consumed by the fish. Many conventional forms of aquacultural feed and protein production are inefficient in terms of trophic level energy conversion, because they rely on organisms that are far from the ultimate source of energy (solar radiation). However, food and protein production systems based on direct utilization of photosynthetic microorganisms circumvent this efficiency loss by eliminating one or more steps of food chain (Mitsui, 1979). Algae, therefore, presents a very promising application in aquaculture. Based on feed conversion ratios and body weight gains, algal diets have been found to be superior to groundnut cakes and natural phytoplankton and zooplankton. The great fishing grounds of the oceans are formed on the seashore either rich in attached forms of algae or present in very large numbers in the form of plankton. A number of marine fish found in the coral reefs in tropical oceans subsist on marine algae. Some of the common examples are as follows: -

i) *Zebrasoma flavescens* (Sturgeon fish) feeds on the red alga, *Amansia glomerata*, and brown alga, *Ectocarpus* sp. (Randhawa, 1959).

ii) *Forcipiger longirostris*, (common fish of Hawaii) feeds on the blue-green alga *Hydrocoleus cantharidosmum* (Randhawa, 1959).

iii) *Rhinomugil corsula* feeds on unicellular algae such as *Ankistrodesmus*, *Scenedesmus*, desmids (*Cosmarium*, *Closterium*), filamentous algae (*Spirogyra*, *Oedogonium*) and diatoms (*Navicula, Synedra, Pinnularia and Fragilaria*) (Randhawa, 1959).

iv) *Catla catla* feeds on species of *Oscillatoria*, *Anabaena*, *Synedra*, *Fragilaria*, *Navicula*, *Melosira*, *Spirogyra*, *Oedogonium*, *Chlorococcum* and *Cosmarium* directly or indirectly through rotifers, *Daphnia* and crustaceans.

v) *Oscillatoria* is the most favoured blue-green alga consusmed by 56 species of fishes. Others in order of preference are *Spirulina*, *Anabaena*, *Microcystis*, *Lyngbya* and *Merismopedia*.

vi) Common carp (*Cyprinus carpio*) feeds on members of Bacillariophyceae, Chlorophyceae, Cyanophyceae, Crustaceans, Dipteran Larvae, Rotifers, Molluscs and various micro and macro organisms (Sharma, 1994). The gut content examination of this fish revealed the presence of *Melosira sp.* (Bacillariophyceae) throughout the year, closely followed by *Synedra sp.* and *Cymbella sp.* Some of the diatoms were found only for some part of the year such as *Anomoneis sp.*, *Diatoma sp.*, *Fragilaria sp.*, *Navicula sp.*, *Cyclotella sp.* and *Stephanodiscus sp.*, *Ceratium sp.* was the most dominant dinophycean member followed by *Peridinium sp.* Amongst Chlorophyceae, *Botryococcus sp.* was present throughout the year followed by species of *Pandorina*, *Actinastrum*, *Cosmarium* and *Staurastrum*. However, *Microcystis sp.* was recorded as the most dominant blue-green alga in the gut contents followed by *Oscillatoria sp.* and *Merismopedia sp.* (Subla and Dass, 1970).

vii) Lasker (1975) found that populations of *Gymnodinium splendens* from the 15m deep zone were actively grazed by northern anchovy (*Engraulis mordax*) but *Chaetoceros* and

Thalassiosira were generally non-significant. Laboratory studies revealed that the anchovy accepted *Gymnodinium, Gonyaulax, Prorocentrum nicans* and *Peridinium trochoideum* without preference but did not take *Chlamydomonas, Dunaliella, Leptocylindrus, Thalassiosira* or *Dictylium* (Saura and Jerda, 1977). There is evidence that fish can take preferentially the darker coloured cells of *Peridinium*.

viii) The striped mullet (*Mugil cephalus*) derives a major part of its food from the microalgae epipelic in the surface mud but the algal content of the food varies considerably with the habitat of fish (Odum, 1970).

ix) Studies on *Chanos chanos* (milkfish) grown in brackish fishponds showed that planktonic phytoflagellates were more nutritious, followed by diatoms and blue-green algae. Filamentous green algae were less readily digested.

x) Catla catla, Labeo fimbriatus, Chanos chanos and *Etroplus suratensis* mainly phytoplankton feeders feed on *Microcystis* bloom.

All the above examples highlight the direct role of algae in fish food. In freshwater lakes and ponds green filamentous algae such as *Oedogonium, Spirogyra, Microspora, Ulothrix, Cladophora* and *Pithophora* serve directly as fish food, or are consumed by midgefly larvae and other insects, rotifers, *Artemia* or *Daphnia* which in turn are eaten by the fish. Apart from feeding on algae, fish lay eggs in nest formed by the algal filaments.

Algae are fed directly and/or indirectly to the cultured larval organisms. Algal production for feed is used as an intensive monoculture for larval stages of certain fish species. In fish feeding, larvae of Turbot (*Scophthalmus maximus*), Halibut (*Hippoglossus hippoglossus*) and microalgae are used in the production of rotifers (*Bachiomuspecictiles*) for transferring essential nutrients from algae to the live food (Jones *et. al.*, 1981; Witt *et. al.*, 1984). The algae may be given directly to the larvae along with the food. In such cases it serves as food both for the fish larvae and the live food (Howell, 1979; Scott and Middleton, 1979; Jones *et. al.*, 1981; Vasquez-Yeomans *et. al.*, 1990; Naas *et. al.*, 1992). Addition of microalgae to the fish feeding tanks alongwith rotifers improved growth and survival of larvae (Howell, 1979; Scott and Middleton, 1979; Lubzens, 1987), whereas short term enrichment of rotifers with algae did not improve growth and survival of larvae in the tank without addition of algae. The algae in larval tanks tended to modify and stabilize the nutritional quality of rotifers in the period before they were consumed by the larvae.

Many marine fish larvae have high dietary requirement for essential highly unsaturated fatty acids for normal growth and development (Watanabe and Ackman, 1974; Witt *et. al.*, 1984; Kanazawa, 1985; Reitan *et. al.*, 1994). In particular, DHA, EPA and recently arachidonic acid (AA) have been recognized as important nutrients for larvae (Estevez *et. al.*, 1997; Harel *et. al.*, 1998). Microalgae are used for short-term enrichment of the rotifers before they are given to the fish larvae for improving the content of such fatty acids in the rotifers (Choutteau and Sorgeloos, 1997). However, short-term feeding with algae will also affect the protein content and dry weight of the rotifers (Olsen *et. al.*, 1989).

Lipid content and fatty acid composition of the rotifers reflected the composition of the algal diet (Rainuzzo *et. al.*, 1994). Hence, algal species used may be an effective tool to control the fatty acids content especially DHA, EPA and LPA (Licosapentanoic acid). Algal groups producing long chain polyunsaturated fatty acids (LCPUFAs) include diatoms, chrysophytes, cryptophytes, dinoflagellates, blue-green algae and others (Cohen *et. al.*, 1995; Behrans and Kyle, 1996). Dinoflagellates are the most suited for DHA production, especially *Crypthecodinium cohnii* (Behrans and Kyle, 1996). This alga alongwith some others contained high levels of DHA and hence used as a source of DHA for the aquaculture food chain (Kashiwakura *et. al.*, 1994; Barclay and Zeller, 1996). A number of algae have been proposed for production of EPA including species of *Nitzschia* (Boswell *et. al.*, 1992), *Nannochloropsis* (Sukenik, 1991), *Navicula* (Tan and Johns, 1996), *Phaeodactylum* (Molina *et. al.*, 1995) and *Porphyridium* (Cohen *et. al.*, 1995). Ratio of DHA/EPA is critical for larval development (Brett and Muller-Navarra, 1997). Recently, AA is also receiving attention as a potentially important nutrient for larval nutrition and it is possible that the ratio of these 3 fatty acids might be more important than their absolute level (Harel, 1998). Humans have poor capacity of synthesizing DHA. However, it forms nearly 20-25% of total fatty acids present in the grey matter of brain. It has also been abundant in heart muscle tissues, sperm cells and retina rod outer segments. DHA can be obtained from various dietary sources and species of *Chlorella*, *Dunaliella* and *Spirulina* as well (Emben *et. al.*, 1994; Nagle and Paul, 1999).

Although fish oils are rich in LCPUFAs, they are reported to be unsuitable for infants, because of the presence of EPA. It significantly lowered the growth rate and caused other development difficulties in the infants (Carlson *et. al.*, 1994).

The need for reliability to support zooplankton and larvae has led to number of different designs for algal culturing systems, ranging from ponds to tanks and to sophisticated photobioreactors (Chaumont, 1993; Qiang and Richmond, 1994; Borowitzka, 1997; Specktorova *et. al.*, 1997). Photobioreactors are very efficient for growing algae, but the maximum algal cell densities attained are relatively low. Low densities necessitated large volume cultures and resulted in substantial cost of harvesting the algae. Heterotrophic algae can be grown in conventional fermentors for reducing growth costs (Day *et. al.*, 1991; Orus *et. al.*, 1991; Barclay *et. al.*, 1994; Gladue and Maxey, 1994). In some cases, algae are cultured using glucose. The cost of producing heterotrophic algal biomass is estimated to be lesser than other methods of algal biomass production (Gladue and Maxey, 1994). Theoretical cost of producing algae phototrophically is estimated to be of the order of higher magnitude (Wilkinson, 1998). A number of commonly used aquacultural algae, including *Chlorella, Nitzschia, Cyclotella, Tetraselmis* (Day *et. al.*, 1991; Gladue and Maxey, 1994) and *Synechocystis sp.* (Anderson and McIntosh, 1991) have been reported to have the ability to grow heterotrophically (Regnault *et. al.*, 1995).

The yield of protein from algae is much higher as compared to other agricultural crops (Venkataraman, 1991; Dua *et. al.*, 1993). *Spirulina* has high protein content (upto 65%) and its crude lipids vary from 9-15% on dry weight basis (Cifferi, 1983). It is rich in isoleucine, leucine, phenylalanine, tyrosine and valine, but is comparatively deficient in sulphur containing amino acids. *Spirulina* lacks toxicity, has low nucleic acid content, and its lipids are chiefly made of unsaturated fatty acids, so do not produce cholesterol. Therefore, *Spirulina* is used as a super health food for patients of coronary illness and obesity. According to Richmond and Preiss (1989), 20g of dried *Spirulina* can easily be produced per m^2 per day, amounting to 50,000 kg per hectare. World production of *Spirulina* has been over 4500 tonnes per year. First, commercial product of *Spirulina* was reported in 1976 and currently it is grown in more than 40 countries including India.

Seasonal succession in a typical fish pond

It is well known that in natural waters there is a succession or periodicity of algae corresponding to various seasons of the year. Such observations on lakes, rivers and small bodies of water have been reported (Iyengar and Iyengar, 1932; Iyengar, 1933,1939; John, 1936; Phillipose, 1940; Gonzalves and Joshi, 1946; Iyengar and Venkataraman, 1951; Roy, 1954, 1955; Dutta *et. al.*, 1954). Holsinger (1955) has studied the ecology of algae in three small lakes and found that in small confined waters, the succession is more marked than in larger water bodies and rivers. It is known earlier that typical fish ponds for rearing fish remain moderately or highly alkaline throughout the year with varying concentration of nitrogen and phosphorous. Dissolved oxygen also remains fairly high.

The succession of algae in such ponds has been noteworthy. As temperature came down slightly with first showers of monsoons towards end of June, the sky remained partially cloudy and some of the water became slightly turbid. These conditions favoured the development of a number of unicellular volvocales, colonial forms and euglenineae. Some of these formed scum on the surface. Heavier rains in July and August increased the water level and some of the water became quite turbid; the temperature remained moderately high and sky remained overcast. The most common algae during this period were members of colonial volvocales, chlorococcales, and euglenineae, blue-greens, diatoms, yellow green algae, dinoflagellates etc. Most of these algae except the chlorococcales continued till October. *Euglena tuba* also appeared during bright spells. During continuous sunshine (Nov.-Feb.), the air was cooler and there was practically no rainfall and the water remained clear. Most common algae during this period have been enlisted in Table 17.4. During summer months (March-June) of high temperature and long hours of bright sunshine with little rainfall, the blue-green algae became more dominant than the euglenineae. However, E. tuba continued in large numbers. The second phase of chlorococcales appeared and continued till the end of June. Few diatoms and euglenineae also occurred quite often throughout the year.

Information summarized in table 4 indicated that algae showed fluctuations according to climate change and temperature. Therefore, these conditions should be kept in mind while recommending or choosing an alga for aquaculture as these organisms have great influence on the fish stock.

Table 17.4: Annual distributional pattern/seasonal succession of algae in a typical fish pond

Month	*Algal type and class*	*Algal genera*
June	Unicellular Volvocales	*Carteria globosa, Chlorogonium sp., Chlamydomonas* reticulata,
	Colonial forms	Gonium pectorale,
	Euglenophyceae	*Euglena pisciformis, E. proxima, Trachelomonas sp.*
July-Aug.	Colonial Volvocales	*Eudorina, Pandorina and Volvox*
	Chlorococcales	*Chodatella chodati, Actinastrum, Coelastrum microporum, Dictyosphaericum hanizchii, Micractinium pusillum, Oocystis sp., Pachycladon unbrinus, Pediastrum simplex, P. duplex, P. tetras, Scenedesmus denticulatus, S. arenalus. var. bicaudatus, S. opoliensis,•S. quadricauda, Tetraedron minimum, T. planctonium, Closterium and Staurastrum.*
	Euglenophyceae	*Euglena ehrenbergii, E. acus, E. oxyuris* var. *charkowiensis, Lepocinclis salina, Phacus ranula var. trachelomonas, T. hispida, T. playfairii, Strombomonas verrucosa var. conspersa.*
	Blue-greens	*Anabaena, Arthrospira, Oscillatoria, Microcystis flos-aquae*
	Diatoms	*Melosira granulata, Nitszchia palea*
	Yellow green algae	*Botryococcus sp.*
	Dinoflagellates	*Coelastrum scabrum, Dysmorphococcus variabilies, Kirchneriella lurians, Phacotus lenticularis, Peridinium inconspicum, Sphaerellopsis fluviatiles*
October	Dinoflagellates	Most of above algae except *Chlorococcales, Euglena tuba* appear during bright spell.
Nov-Feb.	Chlorophyceae	*Botryococcus braunii, Anabaena spiroides, Euglena tuba, E. viridis, Lepocinclis salina, L. fusiformis, Microcystis flos-aquae, M. aeruginosa, Melosira granulata, Pediastrum duplex, Peridinium inconspicum, Physocytium confericola, Symurarella, Trachelomonas sp.*
March-May	Euglenophyceae	*E. tuba and other Euglenineae are rare*

Contd.........

Month	Algal type and class	Algal genera
March-May	Chlorococcales	*Actinastrum hanizschii, Pediastrum simplex, Scenedesmus quadricauda.*
	Diatoms	*Amphora, Navicula sp., Pinnularia, Stamoneis, Rhopalodia, Euglena acus, E. gaimei, Trachelomonas hispida, T. armoda, Volvocina, Strombomonas verrucosa*
	Blue-greens	Also dominant

Succession in a river

Some clearcut distinctions can be made out on comparing the succession of algae in a pond with that of a river. Rivers are full with water during monsoon and there are rapid currents of turbid water During this period the phytoplanktons are very poor and usually comprised of a few algae. By October, the water clears up and current becomes much less, the river assumes almost lacustrine conditions. A number of algae come up and continue to develop throughout winter. Some other algae become dominant during summer months. Though there is poor development of phytoplanktons during the monsoon season, there is fairly good development during winter and summer. The predominant algae being diatoms, blue-greens and green algae. In most of the Indian rivers, members of Chlorophyceae, Bacillariophyceae and Euglenophyceae have been dominant in summer (March-June) whereas Xanthophycean, Chrysophycean, Dinophycean and Cyanophycean algae dominate in water (Nov. – Feb.). Some Chlorophycean and Bacillariophycean members are also prevalent during this period. Monsoon season supported members of Chlorophyceae, Bacillariophyceae, Euglenophyceae and Cyanophyceae. Seasonal distribution of algae in Indian rivers have been shown in Table 17.5.

Succession in estuaries and brackish water lakes

In tidal rivers and estuaries, the phytoplanktons are usually poor during the monsoon season consisting essentially of fresh water forms (Iyengar and Venkataraman, 1951; Dutta *et. al.*, 1954; Roy, 1954). Brackish water or saline species of diatoms and a few dinoflagellates dominate during winter and summer. The proportion of fresh water and saline forms also depend on their distance from the sea. Species like Dunaliella, Anabaena, Anabaenopsis etc. which

Table 17.5: Seasonal distribution of algae in an Indian river

	Season	Algal types
(i)	Monsoon season (July-Oct)	*Eudorina, Pandorina, Pediastrum simplex, Synedra ulna* and few other diatoms
(ii)	Winter Season (Nov.-Feb.)	*Amphora, Anabaena, Botryococcus, Ceratium, Cylinderospermum, Cymbella, Dictyosphaerium, Dinobryon, Gloeotrichia, Microcystis, Mallomonas tonsurata, Navicula sp., Nitzschia, Oscillatoria, Pediastrum simplex, Peridinium sp., Spirogyra sp.*
(iii)	Summer season (March-June)	*Anabaena, Cymbella, Dictyosphaerium, Euglena granulata, Nitzschia, Oscillatoria, Pediastrum, Spirogyra, Synedra, Trachelomonas sp.*

grow in brackish water can be cultured at commercial level to supplement the fish feed, piggery feed and other protein requirements.

Larval feed

Microalgae are widely used in first feeding of marine fish larvae (Jones *et. al.*, 1981; Lubzens, 1987; Naas *et. al.*, 1992; Reitan *et. al.*, 1993). Nearly 152 million fish larvae were produced commercially in 1990 with the help of algae and other live feed organisms (Liao *et. al.*, 1991). Favoured genera of microalgae for larval feed included species of *Chaetoceros, Thalassiosira, Tetraselmis, Isochrysis, Nannochloropsis, Chlorella, Pavlova, Phaeodactylum, Navicula, Dunaliella, Nitzschia and Skeletonema* (De Roeck *et. al.*, 1993; Renaud *et. al.*, 1994; Takeyamma *et. al.*, 1996; Duerr *et. al.*, 1998). Experience with aquacultural studies has shown that the presence of microalgae and the composition of bacterial flora in water are essential for the growth, survival and viability of marine fish larvae (Naas *et. al.*, 1992; Reitan *et. al.*, 1993; Skjermo *et. al.*, 1997). Ingestion of microalgae by the larvae is essential for their viability (Reitan *et. al.*, 1994; Tytler *et. al.*, 1997) and the microbial colonization of the young larvae (Bergh, 1995; Skjermo and Vadestein, 1993; Skjermo, 1996). In the event of any interruption in the supply of algae, the entire food chain can get broken, resulting in loss of fish larvae and eventually decreased production of the adult fish. Indirect means of providing algae to the larvae are through *Artemia*, Rotifers and *Daphnia*, which in turn either are fed to the target larval organisms or are taken directly by the fish (DeSilva and Anderson, 1995). Therefore, algal feed is of utmost importance for the growth of fish fry larvae as well as adult fish. Unicellular forms or simple filamentous forms are the preferred organisms for fish fry. Presence of large sized algal organisms in the pond are harmful for the fish fry.

Aspects of feeding of fishes on algae

As fishes graze on algae, they are termed as algivorous. They feed both on the planktonic (Kupferberg *et. al.*, 1994) and the benthic habitats of algae. Cushing (1971) estimated that herbivorous fish consumed half or more of their body weight per day and that algae reproduced approximately at 1-1.5 divisions per day. In a study, Ryther *et. al.* (1970) observed that annual fish production must consume more than 1013 g phytoplankton carbon annually and for this an annual rate of 103 g carbon m-2 is required to satisfy the fish production alone. A mean rate of 10g carbon m-2 day-1 of net organic production was estimated to support just 1/3rd of the fish need. According to McAllister *et. al.* (1960) grazing fish appeared to be the restraining factor on algal population and without this depletion, the phytoplankton could theoretically increase upto 50 times to its observed size because nutrient starvation looked unlikely in north pacific areas, as phytoplankton can be stratified in the water column. In normal tropical eutrophic waters, annual yield has been over 1000 lb of fish per acre, but waters containing blooms of one or more species of algae were more productive, yielding over 2000 lb of fish per acre. Still a higher production can be obtained by stocking phytoplankton feeding fish. Temple tanks normally support fish like *Catla catla, Labeo fimbriatus, Chanos chanos* and *Etrophus suratensis* that are basically phytoplankton feeders and feed on permanent blooms of *Microcystis* and therefore show higher fish productivity.

Gunn *et. al.* (1977) found that feeding rate vary according to the algae taken by the fish. Different fishes have preference for various algae. Feeding C_{14}^{-} labelled phytoplankton to *Tilapia* sp. and *Haplochromis sp.*, it was shown that *Tilapia* on an average assimilated about 43% of carbon per day and about 66% by *Haplochromis*. The figures, however, vary according to the algae ingested e.g. upto 70-80% from *Microcystis, Anabaena* and *Nitzschia* and only 50% from *Chlorella* (Moriarty and Moriarty, 1973). Nearly 23% of carbon from *Spirogyra* and 67% from *Anabaena flors-aquae* were ingested by the brown bullhead fish.

Feeding of zooplankton and herbivorous fish on phytoplankton is not synonymous with uptake since a proportion of algal material often passes unharmed through the guts especially when feeding occurred in rich phytoplankton blooms. Some early records failed to take into account the pH changes occurring in fish guts during feeding. Common African fresh water fish (*Tilapia nilotica*) does not feed at night and it commenced to feed only at pH-7, and at this pH the algae are not affected and these may pass into the intestine and out as viable algal cells. However, as feeding proceeds, the acid producing cells of the stomach reduced the pH below 2 and thus breakdown of the algal cells leading to their digestion (Moriarty, 1973).

Harmful algae in aquaculture

Algal blooms

It has now been known that algae constitute the main diet source of many aquatic organisms including fishes. Investigations on marine phytoplankton, the prime synthesizers of all organic matter in the sea, has assumed greater importance since over half a century. Sometimes there has been excessive development of phytoplankton; due to which water assumed distinct colour, as green, yellow green, yellowish or greenish. Thick scum may develop on the surface of water and this is called *'Algal Bloom'*. Although these blooms are suitable for yielding a high weight of fish if stocked with an appropriate number of phytoplankton feeders but formation of a dense scum on the surface of the tank by these blooms would tend to cause fish mortality and cause economic losses to aquaculture in general and fisheries in particular. Ganapati (1950) have reported mortality of this kind in many of the fish forms. Harmful algal blooms (HAB) can be defined as events where the concentrations of one or several harmful algae reach such levels that can cause harm by killing fishes, or cause accumulation of toxins in marine organisms which eventually harm other organisms (Nagle and Paul, 1999). There can be accumulation of algal toxins in fish that in turn get biomagnified and become toxic to consumers including man (Moore, 1977). During recent years, coastal marine waters resulted in killing of invertebrates and cultured fish due to the harmful effect of algal blooms. The killing can be due to physical damage to gill tissue or due to oxygen deficiency. HABs produced species specific toxic substances. Haptophytic alga, *Chrysochromulina polylepis* formed huge bloom which apparently proved highly toxic to fish. Much damage was done to fisheries, especially to recently established fish farms where salmons were being grown in netted enclosures positioned out of sea. Only other member of the Haptophyta that has been reported to cause toxic blooms, namely the brackish water species *Prymnesium parvum*. This can inflict serious damage on fish in breeding ponds as well (Shilo, 1967). Various blue-greens (cyanobacteria) produce one or more toxins belonging to general classes: lipopolysaccharides, endotoxins, hepatotoxins, tumor promoters and neurotoxins. These affect not only zooplanktons and other herbivores, but also humans and aquacultural food organism. Incidents of poisoning are associated with large population or blooms of blue-green algae. Some 50-70% of such blooms generate toxins, and experts believe that all blue-green algal blooms should be regarded as toxic until disapproved (Codd *et. al.*, 1995; Sivonen, 1996; Carmichael, 1997). Most blooms forming dinoflagellates are photosynthetic forms such as *Gonyaulax*, but seasonally high populations of the heterotrophs *Protoperidinium depressum* and *Oxyrrhis maxina* can occur in estuaries or other nearshore waters. Bloom dynamics and physiology of the toxic, bloom forming *Gymnodinium breve* were reviewed by Steidinger *et. al.*(1998).

Microcystis sp. has also been found to be toxic to some fish. Fish fry can get entangled into the branched filaments of algae and die. Therefore, species of *Hydrodictyon*, *Cladophora* and *Oedogonium* should be avoided in ponds being used for fish fry rearing.

Algal species occurring at high concentration so as to discolour the water or produce potent toxins included taxonomic groups comprising – diatoms, dinoflagellates (Plumley, 1997; Baden *et. al.*, 1998) other flagellates and cyanophytes. *Alexandrium tamarense* and *Gyrodinium aureolum* can cause harm even at their low concentration with no discoloration of water. Other harmful species included that of Raphidophytes, *Meterosigma akashiwo, Chaetoceros convolutus* and *C. concavicornes*. These can kill fish at aquaculture sties, but are not harmful to humans (Horner *et. al.*, 1997). In addition, certain microalgae produced undesirable taste/odour metabolites that although not toxic, caused significant losses in aquacultural production and contributed to public concern that water resources containing these compounds may be harmful. Algal derived toxins have been identified as posing significant problem to commercial fisheries and aquaculture (Shumway and Cembella, 1992; Anderson *et. al.*, 1993; E.C.O.H.A.B., 1995, Nagle and Paul, 1999).

Prymnesium parvum (flagellate) has been reported from several parts of Europe. Fish poisoning and then killing generally involved the central nervous system. The stability of toxins has been found to be dependant on a number of factors (pH, suspended soil particles, phosphate, ultraviolet radiations etc.). A sample containing 240 cell mm^3 were found to be far more toxic than another sample with 5 times as many cells. Certain cofactors (divalent ions and organic compounds) appeared necessary before maximal toxic effects were obtained. The toxic material has been isolated (prymnesin) and described as a proteinaceous substance. *Dictyota mertensii* produced a sterol (Dictyol-H), which protected against herbivorous fish (Fleury *et. al.*, 1994). *Uroglena volvox* produces toxic fatty acid that effect fishes. Living cells of *Uroglena* and *Dinobryon* can excrete aldehydes ketones (mainly n-heptanal) into the water, which gives unpleasant taste and odour (Niholls, 1995). It has been suggested that the observed increase in frequency as well as member of species identified as harmful ones, basically were the result of eutrophication of some coastal areas (Smayda, 1990). The need for an increase in the food production calls for intensified exploitation of fresh water and marine living resources, either through fishery or wild stocks, through large scale culture of fish and algae in the coastal areas.

Red tides

Red tides are caused by dinoflagellatges (Graham and Wilcox, 2000) particularly photoautotrophic members of genera *Ceratium, Prymnesium, Olisthodiscus, Skeletonema, Alexandrium, Mesodinium, Gymnodinium, Gonyaulax, Glenodinium, Dinophysis, Prorocentrum* and the heterotrophic genus *Noctiluca*. Red tides are brought about by dense growth of planktonic algae and accumulation of carotenoid pigments which are responsible for its colouration. Ho and Hodgekiss (1993) have discussed the appearance and significance of red tides. In coastal waters of Japan, red tides are caused by members of heterokontophytes (Raphidophyceae) e.g. species of *Chatonella, Fibrocapsa* and *Heterosigma* (Fukuyo, 1992). Red tides are primarily a phenomenon of tropical and subtropical areas, but also occur in the temperate zone in late spring and summer. They are particularly frequent in areas close to the coast. The red tides last for about 15-20 days and can cause the death of enormous number of fish. The correlation between mass mortality of marine life and the occurrence of red tides has been known for a long time. The increasing number of red tides organisms is accompanied by an increase in the variety of species and change in population dominance with dinoflagellates gradually replacing diatoms. Reasons for fish death may be oxygen depletion and/or physical interference with gills or toxins. It is found that secretion of poison from the cells of *Ptichodiscus brevis* killed large number of fish. In laboratory, 0.6 x 2.1 x 10^6 cells l^{-1} were shown to be poisonous to fish, while in nature much higher concentration of cells occurred during blooms. In North sea too, red tides occurred and caused poisoning (Adam *et. al.*, 1968). Red tides caused by *Gonyaulax polygramme* persisted there for more than 3 months. The maximum concentration of *G. polygramme* reached 23,000 cells per ml and several fish kills due to deoxygenation during night hours have been reported (Kumar, 1999). It has been speculated that diverse combination of substances produced by benthic filamentous blue-green algae is essential for survival in a complex grazer community structure (Nagle and Paul, 1999).

Control measures for harmful algae in aquaculture

Algae are normal inhabitants of surface waters and help in maintaining a balance in aquatic environment and oxygen level essential to fish life. However, sometimes they may pose serious problems due to their excessive growth and the formations of water blooms. Some filamentous algae can form extensive accumulation at/or near the surface of the water and appear as floating scums, mats or blankets. Fragments of the algal carpet may detach and form dense floating mats on water surface. The massive growth of algae, in addition to impairing the beauty of ornamental garden ponds and sea, reducing their aesthetic appeal, may also cause death of fish by preventing penetration of oxygen into water. Decomposition of dense algal population itself, with associated organisms and debris, can also give rise to foul odour and bring about oxygen depletion in water, resulting in fish kills. In such situations, the control of super abundance of algae becomes a necessity (Saxena, 1979).

The algal nuisance varied considerably from season to season and from place to place. It may be controlled by mechanical or chemical means, depending upon the type of algae, the nature and extent of the control desired and dimensions of the water body. Mechanical control, however, is purely temporary, uncertain and time-consuming.

Control measures should be undertaken early during the season when the algae are developing rapidly, but have not attained nuisance proportions. The most effective algal control is obtained under calm and sunny conditions during and after the treatment. Ideally, the treatment should be given in the morning on sunny day, to allow several hours of sunlight. Under heavy infestations, more than one treatment may be necessary. During heavy growth of algae, the solid mat of filaments should be broken before the treatment. To avoid suffocation of fish due to lack of oxygen caused by the decaying vegetation, only half of the water area should be treated at a time. The other half should be treated after sufficient interval of time (1-2 weeks). It is not advisable to apply solid or liquid algicides to the ponds directly because such a practice tends to form a heavy concentration of the chemicals near the bottom and can cause destruction of fish and other aquatic organisms. Application of chemicals to the surface of a pond cause mixing and even distribution throughout the area to be treated, resulting in an effective algal control. Utrasonic radiation was first suggested as a potential control technique by Fogg (1969) but field trials by Menday and Buck (1972) led to the death of fish and therefore, this method was not recommended for the fish ponds.

Biological control

Several phycoviruses (LPP-2, SM-1, N-1, AS-1) could conceivably be employed to control the nuisance algal growth since they can cause lysis of the sensitive forms of algae. Recently, the eucaryotic algae *Chlorella* and *Sirodotia* have also been reported to be attacked by the phycoviruses. The addition of virus particles to cultures of susceptible algae resulted in a considerable increase in virus titres and a rapid fall in algal cell numbers (Safferman, 1973).

A better method for algal control consisted in introducing suitable crustacean or fish fingerlings into the water body. The animals feed on algae either directly or indirectly, and may be harvested for food at maturity. *Gymnodinium helveticum* (dinoflagellate) showed nonselective and nonpredatory phagotrophy and these cells contained various algae including blue-greens inside them (Irish, 1979).

Chemical control

The control of various algae by potential algicidal chemicals has been summarized in Table 17.6.

Table 17.6: Control of various algae by potential algicidal chemicals

	Algicide	*Algae controlled*
1.	Copper sulphate	*Anabaena, Anabaenopsis, Cladophora, Chara, Hydrodictyon, Microcystis, Oscillatoria, Spirogyra, Spirulina.*
2.	Cutrine-plus	*Anabaena, Anabaenopsis, Cladophora, Chara Hydrodictyon, Microcystis, Nostoc, Oscillatoria, Phormidium, Plectonema, Spirogyra, Spirulina.*
3.	Dichlorophen (Panacide)	*Anabaena, Anabaenopsis, Aphanothece, Microcystis, Spirulina, Spirogyra, Plectonema, Phormidium, Rhizoclonium.*
4.	Diuron (Karmex) wettable powder	*Anabaena, Anabaenopsis, Aphanothece, Chara, Chorella, Cladophora, Lyngbya, Microcystis, Nostoc, Nitzschia, Oscillatoria, Pithophora, Phormidium, Plectonema, Rhizoclonium Scenedesmus, Spirulina.*
5.	Dichlone (2,3-DNQ) wettable powder	*Anabaena, Anabaenopsis, Aphanothece, Microcystis, Nostoc, Phormidium, Plectonema, Spirulina.*
6.	ZDD wettable powder	*Anabaena, Anabaenopsis, Aulosira, Euglena, Lyngbya, Microcystis, Nostoc, Nitzschia, Oscillatoria, Phormidium, Plectonema..*
7.	Algistat	*Anabaena, Anabaenopsis, Aphanothece, Microcystis, Phormidium, Plectonema, Spirogyra, Spirulina.*
8.	Potassium permagnate (water soluble crystal)	*Anabaena, Anabaenopsis, Chorella, Hydrodictyon, Microcystis, Nitzschia, Oscillatoria, Phormidium, Plectonema, Scenedesmus.*
9.	Simazine (Gesatop, Princep wettable powder)	*Chara, Hydrodictyon, Pithophora, Rhizoclonium, Spirogyra.*
10.	Ansar 529 (liquid herbicide)	*Chlorella, Euglena, Microcystis, Nitzschia, Phormidium, Plectonema, Rhizoclonium, Scenedesmus, Spirogyra.*
11.	Lasso (Alachlor) Herbicide liquid concentrate).	*Euglena, Microcystis, Scenedesmus.*

Chemical treatment is rapid and effective. Algicides must kill noxious algae, without affecting fish or other aquatic life. Algicide copper sulphate selectively kills the algae within a dose range of 0.25-9.5 ppm. The effective dose varied with the chemical character of water as well as concentration of algae. Chlorophenyl dimethyl urea, a photosynthetic inhibitor has also been found to be effective in checking the growth of planktonic algae. Besides, there are other promising algicides as antibiotic, quinones, substituted hydrocarbons and phenols.

Quinone - 2,3 - dichloro-1,4 - nepthoquinone is more selective and toxic to many bloom forming cyanophytes than to other algae. This chemical can be effectively used to control water blooms caused by blue-green algae.

A particularly suitable algicide in widespread use is dichlorophen. It is inexpensive, potent and non toxic to animals and human beings, and can be used for preventing algal growth in open air swimming pools.

Artificial circulation of water in reservoirs under pressure is an effective method of reducing the growth of planktonic algae. The basis underlying this method is not clear but the breakdown of chemical stratification in water bodies appears to be of significance.

Acknowledgement

Thanks are due to the Chairman, Department of Botany, Panjab University, Chandigarh for providing research and other facilities during the course of this work.

References

Adam, J.A., Seaton, D.D., Buchanan, J.B. and Longbottom, M.R. 1968. Biological observations associated with the toxic phytoplankton bloom of the east coast. *Nature* **220**: 24-25.

Ahluwalia, A.S. 1993. Habitats and distribution. In: *Cyanophyta*. Edited by Anand, N., Bishen Singh Mahendra Pal Singh, Dehradun. *Indian Phycol. Rev.* **2**: 41-58.

Ahluwalia, A.S.and Paul, R. 1987. Fatty acid composition of a nitrogen-fixing blue-green alga, *Nostoc ellipsosporum*. *Ann. Biol.* **3**: 56-58.

Ahluwalia, A.S., Reena and Dahuja, S. 1993. Studies on the role of algalization on growth and yield of rice (*Oryza sativa* Linn.) *Res. Bull. Panjab Univ.* **43**: 55-60.

Anderson, D.M. , Galloway, S.B. and Joseph, J.D. 1993. Marine biotoxins and harmful algae: A national plan. Woods Hole Oceanoogr. *Inst. Tech. Report*. WHO 93-02.

Anderson, S.L. and McIntosh, L. 1991. Light-activated heterotrophic growth of the cyanobacterium *Synechocystis* sp. Strain PCC 6803: a blue-light requiring process. *J. Bacteriol.* **173**: 2761-2767.

Baden, D.G., Rein, K.S. and Gawley, R.E. 1998. Marine toxins: How they are studied and what they can tell us? In: *Molecular Approaches to the Study of the ocean*. Edited by Cooksey, K.G. Chapman and Hall, London. pp. 487-514.

Barclay, W. R. and Zeller, S. 1996. Nutritional enhancement of N-3 and N-6 fatty acids in rotifers and *Artemia nauplii* by feeding spraydried *Schizochytrium sp. J. World Aquacult. Soc.* **27**: 314-322.

Barclay, W.R., Meager, K.M. and Abril, I.R. 1994. Hetrophine production of long chain ω-3 fatty acids utilizing algae and algae like microorganisms. *J. Appl. Phycol.* **6**: 123-129.

Behrans, P.W. and Kyle, D.J. 1996. Microalgae as a source of fatty acids. *J. Food Lipids* **3**: 259-272.

Bergh, Q. 1995. Bacteria associated with early life stages of Halibut, *Hippoglossus hippoglossus* L., inhibit growth of a pathogenic *Vibrio* sp. *J. Fish Dis.* **18**: 31-40.

Borowitzka, M.A. 1997. Microalgae for aquaculture: opportunities and constraints. *J. Appl.Phycol.* **9**: 393-401.

Boswell, K.D.B., Gladue, R., Prima, B. and Kyle, D.J. 1992. SCO production by fermentative microalgae. In: *Industrial Applications of Single Cell Oils*. Edited by Kyle, D.J. and Ratledge, C. Amer. Oil Chem. Soc., Champaign, Illinois, pp 274-286.

Brett, M.T. and Muller-Navarra, D.C. 1997. The role of highly unsaturated fatty acids in aquatic food web processes. *Freshwater Biol.* **38**: 483-499.

Carlson, S.E., Werkman, S.H., Peeples, J.M. and Wilson, W.M. 1994. Growth and development of premature infants in relation to ω-3 and ω-6 fatty acids status. *World Rev. Nutr. Diet.* **75**: 63-69.

Carmichael, W.W. 1997. The Cyanotoxins. *Adv. Bot. Res.* **27**: 211-255.

Chapman, V.J. and Chapman, D.J. 1980. *Seaweeds and Their Uses*. Chapman and Hall, New York.

Chaumont, D. 1993. Biotechnology of algal biomass production: a review of systems for outdoor mass culture. *J. Appl. Phycol.* **5**: 593-604.

Choutteau, P. and Sorgeloos, P. 1997. Manipulation of dietary lipids, fatty acids and vitamins in zooplankton cultures. *Freshwater Biol.* **38**: 501-512.

Cifferi, O. 1983. Spirulina: the edible microorganism. *Microbiol. Rev.* **5**: 60-70.

Codd, G.A., Edwards, C., Beattie, K.A., Lawton, L.A., Campbell, D.L. and Bell, S.G. 1995. Toxins from cyanobacteria (blue-green algae). In: *Algae and Environmental Affairs.* Edited by Wiessner, W., Schnepf, E. and Starr, R.C., Bio Press, Bristol. pp. 1-17.

Cohen, Z., Norman, H.A. and Heiman, Y.M. 1995. Microalgae as a source of *w*-3 fatty acids. *World Rev. Nutr. Diet.* **77**: 1-31.

Cosson, J., Deslandes, E., Zinoun, M. and Mouradi, G.A. 1995. Carrageenans and agars, red algal polysaccharides. *Prog. Phycol. Res.* **11**: 269-324.

Cushing, D.H. 1971. A comparison of production in temperate seas and the upwelling areas. *Trans. Roy. Soc. S. Africa.* **40**: 17-33.

Davies, J.P. and Grossman, A.R. 1994. Sequences controlling transcription of the *Chlamydomonas reinhardii* beta 2-tubulin gene after deflagellation and during the cell cycle. *Mol. Cell. Biol.* **14**: 165-174.

Day, J.G., Edwards, A.P. and Rodgers, G.A. 1991. Development of an industrial-scale process for the heterotrophic production of microalgal mollusc feed. *Bioresour. Technol.* **38**: 245-249.

DeRoeck, H., Ylair, Y., Bresdin, C., Amiiel, F. and Derriew, A. 1993. *Vitamins, free A.A. and fatty acid composition, of some marine planktonic microalgae used in aquaculture.* Chapman and Hall, New York. pp.319.

DeSilva, S. and Anderson, T. 1995. *Fish Nutrition in Aquaculture.* Chapman and Hall, New York. pp.319.

Dua, S., Kaur, M. and Ahluwalia, A.S. 1993. Functional properties of two pollutant grown green algae. *J. Food Sci. Technol.* **30**: 25-28.

Duerr, E.O., Molnar, A. and Sato, V. 1998. Cultured microalgae as aquaculture feeds. *J. Mar. Biotechnol.* **7**: 65-70.

Dutta, N., Malhotra, J.C. and Bose, B.B. 1954. *Hydrobiology and seasonal fluctuations of the plankton in the Hoogly estuary.* Symp. Marine and Freshwater plankton in the Indo-Pacific. pp. 1-13.

Ecology and Oceanography of Harmful algal Bloom (E.C.O.H.A.B.) 1995. *A national research agenda Workshop Report*. Woods Hole Oceanogr. Inst. pp. 66.

Emben, E.A., Adolf, R.O. and Gulley, R.M. 1994. Dietary linoleic acid influence desaturation and acylation of deuterium-labelled linoleic and linolenic acid in young adult males. *Biochim. Biophys. Acta* **1213**: 227-288.

Estevez, A., Ishikawa, M. and Kanazawa, A. 1997. Effects of arachidonic acid on pigmentation and fatty acids composition of Japanese flounder, *Paratichythys olivaceus*. *Aquacult. Res.* **28**: 279-289.

Fernandez, E., Galvan, A. and Quesada, A. 1998. Nitrogen deassimilation and its regulation, In: *Molecular Biology of Chlamydomonas: Chloroplasts and Mitochondria*. Edited by Richaix, J.D., Goldschmidt, C.M. and Merchant, S., Kluwer Academic Publ., Dordrecht, pp. 637-659.

Firth, F.E. 1969. *The Encyclopedia of Marine Resources*. Van Nostrand, New York.

Fleury, B.G., Kelecom, A., Pereira, R.C. and Teixeria, V.L. 1994. Polyphenols, terpenes and sterols in Brazilian Dictyotales and Fucales (Phaeophyta). *Bot. Mar.* **37**: 457-462.

Flores, E. and Wolk, C.P. 1986. Production by filamentous nitrogen-fixing cyanobacteria of bacteriocin and of other antibiotics that kill related strains. *Arch. Microbiol.* **145**: 215-219.

Fogg, G.E. 1969. The physiology of an algal nuisance. *Proc. R. Soc. Lond.* **173**: 175-189.

Fukuyo, Y. 1992. *Harmful algal blooms in Japan: Japanese perspective*, IOC-FAO/IPHAB-I/3 Annex. V. *pp* 17-26.

Ganpati, S.V. 1950. Ecology of tropical waters. In: *Proc. Symp. Algol.* Edited by Raghavan, D. and Kachroo, P., ICAR, New Delhi. pp.204-218.

Gerwick, W.H., Robert, M.A., Proteau, P.J. and Chen, J.L. 1994. Screening cultural marine algae for anticancer type activity. *J. Phycol.* **13**: 121-127.

Gladue, R.M. and Maxey, J.E. 1994. Microalgal feeds for aquaculture. *J. Appl. Phycol.* **6**: 131-141.

Gleason, F.K. and Paulson, J.L. 1984. Site of action of the natural algicide cyanobaterin in the blue-green alga, *Synechococcus sp. Arch. Microbiol.* **138**: 273-277.

Gonzalves, E. and Joshi, D.B. 1946. Freshwater algae near Bombay-I, The succession of the algae in a tank of Bandra. *J. Bombay Nat. Hist. Soc.* **46**: 154-176.

Graham, L.E. and Wilcox, L.W. 2000. Dinoflagellates. In: *Algae.* Edited by Ryu, T., Tarabojkia, L., Dellas, K., Prenctice Hall, NJ. pp.220.

Grunewald, K., Hagen, C., Braune, W. 1997. Secondary carotenoid accumulation in flagellates of green alga *Haematococcus lacustris. Eur. J. Phycol.* **32**: 387-392.

Gunn, J.M., Qadri, S.V. and Mortimer, D.C. 1977. Filamentous algae as food source for the brown bullhead (*Ictalurus nebulosus*). *J. Fish. Res. Bd. Can.* **34**: 396-401.

Harel, M., Koven, W., Lund, E., Behrens, P.W. and Place, A. 1998. *High DHA (docosahexaenoic acid) enrichment of live food organisms using novel DHA rich phospholipids.* World Aquacultural Society Annual Meeting, Las Vegas, NV. pp.228.

Ho, K.C and Hodgekiss, I.J. 1993. Macronutrients in the Tolo Harbour and their relations to Red Tides. *EMECS News Letter* **4**: 10-11.

Holsinger, E.C.T. 1955. The distribution and periodicity of the plankton of three Ceylon lakes. *Hydrobiologia* **7**: 25-35.

Horner, R.A., Garrison, D.L., and Plumley, F.G. 1997. Harmful algal blooms and red tide problems on US west coast. *Limnol. Oceanogr.* **42**: 1076-1088.

Howell, B.R. 1979. Experiments on the rearing of larval turbot, *Scophthalmus maximus* L. *Aquaculture* **18**: 215-225.

Irish, A.E. 1979. *Gymnodinium helveticum* Penard (F. Achroum Skiya): a case of phagotrophy. *Br. Phycol.* J. **14**: 11-15.

Iyengar, M.O.P. 1933. Contribution of our knowledge of the colonial volvocales of south India. *J. Linn. Soc. Bot.* **49**: 323-373.

Iyengar, M.O.P. 1939. Algal problems peculiar to the tropics, with special reference to India. *Proc. 25th Indian Sci. Congr.* **4**: 1-8.

Iyengar, M.O.P. and Iyengar, M.O.T. 1932. On a *Characium* growing on *Anopheles* larvae. *New Phytol.* **31**: 66-69.

Iyengar, M.O.T. and Venkataraman, G.S. 1951. The ecology and seasonal succession of the algal flora of the river cooum at Madras with special reference to the Diatomaceae. *J. Madras Univ.* **21**: 140-192.

Jensen, A. 1995. Production of alginates. In: *Algae, Environment and Human Affairs.* Edited by Wiessner, W., Schnepf, E. and Starr, R.C., Bio Press, Bristol. pp.79-92.

John, R.P. 1936. *The seasonal succession and ecology of the phytoplankton of an artificial pond at Madras.* M.Sc. Thesis, University of Madras, India.

Jones, A., Prikett, R.A., Douglas, M.T. 1981. Recent developments in techniques for rearing marine flat fish larvae, particularly turbot (*Scophthalmus maximus* L.) on a pilot commercial scale. *Rapp. P.-V. Reun. Cons. Int. Explor. Mer.* **178**: 522-526.

Kanazawa, A. 1985. Essential fatty acid and lipid requirement of fish. In: *Nutrition and Feeding in Fish.* Edited by Cowery, C.S. Mackie, A.M. and Bell, J.G. Academic Press. London. pp. 281-298.

Kashiwakura, M., Seto, A. and Hasegawa, K. 1994. *Heterotrophic microalgal production of docosahexaenoic acid (DHA) and its use for aquaculture of marine juvenile fish.* Third Int. Mar. Biotechnol.Conf., Baltimore, M.D. pp. 61.

Kumar, H.D. 1999. *Introductory Phycology.* East-West Press Ltd., New Delhi. pp.500.

Kupferberg, S.J., Marks, J.C. and Power, M.E. 1994. Effects of variation in natural algal and detrital diets on larval anuran (*Hyla regilla*) life-history traits. *Copeia.* **1994**: 446-457.

Lasker, R. 1975. Field criteria for survival of anchovy larvae: The relationship between inshore chlorophyll maximum layers and successful first feeding. *Fish. Bull.* **73**: 453-462.

Lee, R.E. 1999. *Phycology.* Cambridge University Press. Cambridge. pp. 614.

Lee, Y.K. 1997. Commercial production of microalgae in the Asia-Pacific rim. *J. Appl. Phycol.* **9**: 403-411.

Lembi, C.A., O'Neal, S.W. and Spencer, D.F. 1988. Algae as weeds: Economic Impact, ecology and management alternatives. In:

Algae and Human Affairs. Edited by Lembi, C.A. and Waaland, J.R., Cambridge University Press, Cambridge, U.K. pp. 455-481.

Liao, I.C., Su, M.S. and Su, H.M. 1991. An overview of live feed production system design in Taiwan. In: *Rotifers and Microalgae Culture Systems.* Edited by Fulks, W. and Main, K.L.,The Oceanic Institute, Honolulu. pp. 135-150.

Lubzens, E. 1987. Raising rotifers for use in aquaculture. *Hydrobiologia* **147**: 245-255.

Manoharan, K., Lee, T.K., Cha, J.M., Kim, J.H. and Lee, W.S. 1999. Acclimation of *Prorocentrum minimum* (Dinophyceae) to prolonged darkness by use of an alternative carbon source from triglycerides and galactolipids. *J. Phycol.* **35**: 287-292.

Mason, C.P., Edwards, K.R., Carlson, R.E., Pignatello, J., Gleason, F.K. and Wools, J.M. 1982. Isolation of chlorine containing antibiotic from the freshwater cyanobacterium, *Scytonema hofmanni. Science.* **215**: 400-402.

McAllister, C.D., Parsons, T.R. and Strickland, J.O.H. 1960. Primary productivity at station 'P' in the north east pacific ocean. *J. Cons. Perm. Int. Explor. Mar.* **25**: 240-259.

McLachlan, J.L. 1985. Macroalgae (seaweeds): Industrial resources and their utilization. *Plant Soil.* **89** : 137-157.

McLachlan, J.L., Curtis, J.M., Boutilier, K., Keusgen, M. and Seoguel, M.R. 1999.*Tetrautrepia pomquetensis* (Euglenophyta), A psychrophilic species. Growth and fatty acid composition. *J. Phycol.* **35**: 280-286.

Menday, A. and Buck, K. 1972. A shokwave technique to collapse the vacuole of blue-green algae. *Water Res.* **6**: 279-285.

Mitsui, A. 1979. The use of photosynthetic marine organisms in food and feed production, Biosaline concept. In: *An Approach to the Utilization of Unexploited Resources.* Edited by Halloaenbar, S.P., Biosaline Plenum Press, New York. pp. 177.

Molina, G.E., Sanchez, P., Garcia, J.A., Fernandez, C.F., Sevilla, J.M., Fernandez. F,.G. and Urda, C.J. 1995. Biomass and Icosapentanoic acid productivity from an outdoor batch culture of *Phaeodactylum tricornutam.* UTEX 640 in an airlift tublar photobioreactor. *Appl. Microbiol. Biotechnol.* **42**: 658-663.

Moore, R.E. 1977. Toxins from blue-green algae. *Bioscience*. **27**: 797-802.

Moriarty, D.J.W. 1973. The physiology of digestion of blue-green algae in cichild fish *Tilapia nilotica*. *J. Zool.* **171**: 25-39.

Moriarty, D.J.W. and Moriarty, C.M. 1973. The assimilation of carbon from phytoplankton by two herbivorous fishes: *Tilapia nilotica* and *Haplochromis nigripinnis*. *J. Zool.* **171**: 41-45.

Mynderse, J.S., Moore R.E., Kashinagi, M. and Nortor, T.R. 1977. Antileukemic activity in the Oscillatorieae: Isolation of debromoaphlysiatoxin from *Lyngbya*. *Science* **196**: 538-540.

Naas, K.E., Naess, T. and Harboe, T. 1992. Enhanced first-feeding of halibut larvae in green water. *Aquaculture* **105**: 143-156.

Nagle, D.G. and Paul, V.J. 1999. Production of secondary metabolites by filamentous tropical marine cyanobacteria: Ecological functions of the compounds. *J. Phycol.* **35**: 1412-1421.

Neenan, B., Feinberz, D., Hill, A., McIntosh, R. and Terry, K. 1986. *Fuels from Microalgae: Technology, Status, Potential and Research Requirements*. Solar energy Research Institute, Golden, Colorado. pp.149.

Niholls, K.H. 1995. Chrysophyte blooms in the plankton and neuston of marine and fresh water systems. In: *Chrysophyte algae: Ecology, Phylogeny and Development*. Edited by Sandgren, C.D., Smol, J.P. and Kristiansen, J. Cambridge University Press, Cambridge, UK. pp. 181-213.

Odum, W.E. 1970. Utilization of the direct grazing and plant detritus food chains by the striped mullet, Mugil cephalus. In: *Marine Food Chains*. Edited by Steele, I.H., Oliver and Boyd., Edinburgh. pp. 222-240.

Olsen, Y., Remuzzo, J.R., Vadetein, O. and Jensen, A. 1989. Kinetics of *w*-3 fatty acids in *Brachionus plicatilis* and changes in the food supply. *Hydrobiologia* **186/187**: 409-413.

Orus, M.L., Marco, E. and Martinez., F. 1991. Suitability of *Chlorella vulgaris* UAM 101 for heterotrophic biomass production. *Bioresour. Technol.* **30**: 179-184.

Patternson, G.M.L., Larsen, L.K. and Moore R.E. 1994. Bioactive natural products from blue-green algae. *J. Appl. Phycol.* **6**: 151-157.

Phillipose, M.T. 1940. *The ecology and seasonal succession in a permanent pool at Madras.* M.Sc. Thesis, Univ. of Madras, India.

Plumley, F.G. 1997. Marine algal toxins: Biochemistry, genetics and molecular biology. *Limnol. Oceangr.* **42**: 1252-1264.

Qiang, H. and Richmond, A. 1994. Optimizing the population density in *Isochrysis galbana* grown outdoors in a glass column photobioreactor. *J. Appl. Phycol.* **6**: 391-396.

Rainuzzo, J.R., Reitan, K.I. and Olsen, Y. 1994. Effect of short and long term lipid enrichment on total lipids, lipid class and fatty acid composition in rotifers. *Aquaculture Int.* **2**: 19-32.

Randhawa, M.S. 1959. Address. In: *Proc. Symp. Algology.* Edited by Raghavan, D. and Kachroo, P., ICAR, New Delhi. pp.4-24.

Regnault, A., Chervin, D., Chamnal, A., Piton, F., Calvayrac, R. and Maziak, P. 1995. Lipid composition of *Euglena gracilis* in relation to carbon-nitrogen balance. *Phytochemistry* **40**: 725-733.

Reitan, K.I., Jose, R., Rainuzzo, J.R., Ole, G. and Olsen, Y. 1997. A review of the nutritional affects of algae in the marine fish larvae. *Aquaculture* **155**: 207-221.

Reitan, K.I., Rainuzzo, J.R., and Olsen, Y. 1994. Influence of lipid composition of live feed on growth, survival and pigmentation of turbot larvae. *Aquaculture Int.* **2**: 33-48.

Reitan, K.I., Jose, R., Rainuzzo, J.R., Oie, G. and Olsen, Y. 1993. Nutritional affects of algal addition in first-feeding of turbot larvae. *Aquaculture* **118**: 257-275.

Renaud, S.M., Parry, D.L. and Thinth, L.V. 1994. Microalgae for use in tropical aquaculture: gross chemical and fatty acid composition of 12 species of microalgae from the Northern Territory, Australia. *J. Appl. Phycol.* **6**: 337-345.

Richmond, A. and Preiss, K. 1989. The biotechnology of algal culture. *Interdiscip. Sci. Rev.* **5**: 60-70.

Roy, H. 1954. Periodicity of the plankton diatoms of Chilka lake for the years 1950 and 1951. *J. Bombay Nat. Hist. Soc.* **52**: 112-23.

Ryther, J.H., Menzel, D.W., Hulbert, E.M., Lorenzen, C.J. and Corwin, N. 1970. The production and utilization of organic matter in the Peru Coastal current. *Anton. Beruv. Rep.* **4**: 403-412.

Safferman, R.S. 1973. Phycoviruses. In: *The Biology of Blue-green Algae.* Edited by Carr, N.G. and Whitton, B.A. Blackwell, Oxford. pp. 214-237.

Saura, E.O. and Jerda, C.W. 1977. Acceptance of various species of phytoplankton as food by the larvae of the northern anchovy *Engraulis mordax* and the relative nutritional value of the dinoflagellate *Gymnodinium splendense* and *Gonyaulaux polyedra. Fish Bull.* **75**: 112.

Saxena, P.N. 1979. Why control algae? In: *Control of Noxious algae in garden tanks and ponds.* Edited by Juneja, B.R. Publ. House, Lucknow. pp.2.

Scott, A.P. and Middleton, C. 1979. Unicellular algae as a food for turbot (*Scophthalmus maximus* L.) larvae - the importance of dietary long-chain polyunsaturated fatty acid. *Aqualculture* **18**: 227-230.

Sharma, V.K. 1994. Food and feeding habits of the common carp, *Cyprinus carpio* Linn. from Gobind Sagar reservoir, H.P. India; *Punj. Fish. Bull.* **18**: 25-33.

Shilo, M.. 1967. Formation and mode of action of algal toxins. *Bacteriol. Rev.* **31**: 180-193.

Shumway, S.E. and Cembella, A.D. 1992. Toxic algal blooms; potential hazards to scallop culture and fisheries. *Bull. Aquacult. Assoc. Can.* **92**: 59-68.

Sivonen, K. 1996. Cyanobacterial toxins and toxin production. *Phycologia* **35**: 12-24.

Skjermo, J. 1996. *Microbial ecology of early stages of cultivated marine fish. Impact of fish-bacterial interactions on growth and survival of larvae.* Ph.D thesis, Norwegian Univ. Sci. Tech., Trondhein, Norway.

Skjermo, J., Salvesen, I., Oie, G., Olsen, Y. and Vadstein, O. 1997. Microbially maturated water: A technique for selection of a non-opportunistic bacterial flora in water that may improve performance of marine larvae. *Aquaculture Int.* **5**: 13-28.

Skjermo, J. and Vadestein, O. 1993. The effect of microalgae in skin and gut bacterial flora of halibut larvae. In: *Proc. Ist Int. Conf. Fish Forming Technol.* Edited by Reinertsen, H., Jorense, L., Trinnereim, K. and Dahle, L.A. Trodheim, Norway. pp. 61-67.

Smayda, T.J. 1990. Novel and nuisance phytoplankton blooms in the sea; Evidence for a global epidemic. *Toxic Marine Phytoplankton* pp. 29-40.

Spalding, M. 1998. Mechanisms of carbon dioxide uptake: adaptations to changing carbon availability. In: *Molecular Biology of Chamydomonas: Chloroplasts and Mitochondria.* Edited by Richaix, J.D., Goldschmidt, Clermont M., and Merchant, S. Kluwer Academic Publ., Dordrecht, pp.529-547.

Specktorova, L., Creswele, R.L. and Vaughan, D. 1997. Closed tubular cultivators: an innovative system for commercial culture of microalgae. *World Aquaculture* **28**: 39-43.

Steidinger, K.A., Vargo, G.A., Tester, P.A. and Tomas, C.A. 1998. Bloom dynamics and physiology of Gymnodinium breve with emphasis on the gulf of Mexico. In: *Physiological ecology of harmful algal blooms.* Edited by Anderson, D.M., Cambella, A.D. and Hallegraeff, G.M. Springer-Verlag, Berlin. pp. 133-153.

Subla, B.A. and Dass, S.M. 1970. Studies on feeding habits, the food and the seasonal fluctuations in feeding in nine Kashmir fishes. *Kashmir Sci.* **7**: 25-44.

Sukenik, A. 1991. Ecophysiological considerations in the optimization of eicosapentaenoic production by *Nannochloropsis* sp. (Eustigmatophyceae). *Bioresour. Technol.* **35**: 263-270.

Tacon, A.G.J. 1998. *Private global trends in aquaculture and aquafeed production (1984-1995).* World Aquaculture Soc. Ann. Meeting, Las Vegas, NV.

Takeyamma, H., Iwamoto, K., Hata, S., Takano, H. and Matsunga, T. 1996. DHA enrichment of rotifers: a simple two step culture using the unicellular algae, *Chlorella vulgaris* and *Isochrysis galbana. J.Mar. Biotechnol.* **3**: 244-247.

Ian, C.K. and Johns, M.R. 1996. Screening of diatoms for heterotropic eicosapentaenoic acid production. *J. Appl. Phycol.* **8**: 59-64.

Tang, E.P.Y., Tremblay, R. and Vincent, W.F. 1997. Cyanobacterial dominance of polar freshwater ecosystems: Are high-latitude mat-formers adapted to low temperature? *J. Phycol.* **33**: 171-181.

Tytler, D., Ireland, J. and Murray, L. 1997. Study of the assimilation of fluorescent pigments of microalgae *Isochrysis galbana* by the early larval stages of turbot and hering. *J. Fish. Biol.* **50**: 999-1009.

Van der Meer, J.P. and Patwary, M.V. 1995. Genetic studies on marine macroalgae: A status report. In: *Algae, Environment and Human affairs.* Edited by Wiessner, W., Schnepf, E. and Starr, R.C. Bio Press, Bristol. pp. 235-258.

Vashishta, B.R. 1995. *Algae.* S. Chand and Co.Ltd., New Delhi. pp. 426.

Vasquez-Yeomans, L., Carrillo-Barrios-Gomez, T. and Sosa-Cordero, E. 1990. The effect of the nanoflagellate *Tetraselmis suecica* on the growth and survival of grunion, *Leuresthes tenuis* larvae. *Environ. Biol. Fish.* **29**: 193-200.

Velasquez, G.T. 1972. *Studies and utilization of the Philippine marine algae.* Proc. 7th Int.Seaweed Symp., Tokyo. pp. 62-65.

Venkataraman, L.V. 1991. What next on Spirulina? *National Symosium on Spirulina,* Shri AMMM Chettier Research Centre, Madras, India. pp. 1-2.

Vonshak, A. (ed.) 1997. *Spirulina platensis (Arthrospira): Physiology, Cell Biology and Biotechnology*. Taylor and Francis, Basing Stoke, Hants. pp. 233.

Watanabe, T. and Ackman, R.G. 1974. Lipids and fatty acids of the American and Europeon flat oysters from a common habitat, and after one feeding with *Dicrateria inovnata* or *Isochrysis galbana. J. Fish Res. Bd. Can.* **31**: 403-409.

Wilcox, H. A. 1982. The ocean as a supplier of food and energy. *Experientia* **38**: 31-35.

Wilkinson, L. 1998. *Criteria for the design and evaluation of photobioreactor for the production of microalgae.* World Aquacutlure Soc. Ann. Meeting, Las Vegas, NV. pp.584.

Witt. U., Quentz, G., Kuhlmann, D. and Kattner, G. 1984. Survival and growth of turbot larvae, *Scophthalmus maximus* L. reared on different food organisms with special regards to long chain polyunsaturated fatty acids. *Aquacult. Eng.* **3**: 177-190.

Yamoguchi, K., 1997. Recent advances in microalgal bioscience in Japan, with special reference to utilization of biomass metabolites: A review. *J. Appl. Phycol.* **8**: 487-502.

Chapter 18

LIPIDS FROM MICROALGAE

M.S. Narayan, N. Bhagyalakshmi and L. V. Venkataraman

Abstract

Most of the algal members contain a good amount of lipids which mainly constitute phospho- and glycolipids. In addition to their function as storage products, metabolites and source of energy, the diversity in the pattern of their fatty acid composition among blue-green algae has much phylogenetic significance. Polar lipids formed the major fraction of the total lipids. Pathway of lipid synthesis and factors affecting, and their importance in oil production has been discussed.

Key words: Algae, fatty acids, non-polar lipids, oils, polar lipids.

Introduction

Micro-algae comprise a large group of photosynthetic, autotrophic organisms, which have an extraordinary biosynthetic potential and hence known as "biochemical factories" of the world. In fact, these organisms have created an environment congenial for higher organisms to evolve on the planet Earth by releasing oxygen to the atmosphere. Microalgae attracted much attention during 1950's as a potential source of commercially interesting products such as fats, oils (also known as single cell oil), sugars and functional bio-active compounds. Lipids and fatty acids are constituents of all plant cells, where they function as membrane components, storage products, metabolites, and as a source of energy (Wada *et. al.*, 1994). Moreover, microalgae produce 30 times more oil per unit of growth area than land plants. Lipids are normally found throughout the micro-algal cell, occurring as storage product inclusions in the cytoplasm and as functional components of various membranes. In some cases, they are excreted as extracellular colony matrix as almost pure oleaginous globules. Lipids serve two major roles in cells and tissues. They are the source of energy and quantitatively exhibit the best caloric value of all foods and also function as important structural constituents of cellular membranes, acting as insulators of delicate internal cell organelles (Paul and Ahluwalia, 1987). Glycolipids act as carriers of messenger molecules across the cell membrane.

In algae, lipids function both as structural elements as well as metabolites in the photosynthetic organelles. Algal chloroplasts possess most of the cellular lipids (Harry, 1951). The lipid content of different microalgae varies depending upon the environmental conditions and the type of water they inhabit such as fresh water, brackish water or sea water (Aaronson *et. al., 1980)*. The variation ranges from 1-53% (Table 18.1).

Table 18.1. Total lipids in different classes of algae (per cent dry weight basis)

Algal classes	*Total lipids Average*	*Range*	*Neutral lipids*	*Glycolipids*	*Phospho lipids*
Bacillariophyceae	16	1-39	14-60	13-44	10-47
Chlorophyceae	19	1-53	21-66	6-26	17-53
Chrysophyceae	29	12-39	-	-	-
Cyanophyceae	8	2-13	11-68	12-41	16-50
Dinophyceae	21	5-36	-	-	-
Haptophyceae	29	5-48	-	-	-
Phaeophyceae	4	1-9	-	-	-
Prasinophyceae	11	3-18	-	-	-
Rhodophyceae	9	1-14	41-58	42-59	-
Xanthophyceae	12	6-16	44	17	39

The present review on algal lipids has been broadly classified into three parts comprising non polar lipids, polar lipids and fatty acids. The nature and availability of these lipids and their nature of action have been restricted mainly to blue-green algae.

Non polar lipids

Non-polar lipids are lipophilic in nature and comprise triglycerides, free fatty acids, hydrocarbons and sterols. The neutral lipids and triglycerides are the major storage products (Opute, 1974), and the major non-polar lipid fractions of most of the algae. The hydrocarbon (HC) content ranges from 0.001-0.48% (on dry weight basis) in marine red, brown and green algae. Many green algae synthesize unsaturated hydrocarbons, whereas the fresh water algae synthesize HC ranging from carbon atoms 15-18 (Opute, 1974). The marine algae synthesize hydrocarbons ranging from 0.2-0.6% in blue-greens (BGA) and 0.4-1.6% in green algae (GA) on dry weight basis (Paoletti *et. al.*, 1976). Sterols are present in the unsaponifiable lipid fraction as in higher plants. Algal sterols, like in higher plants, are not homogeneous, but comprise of a mixture of closely related compounds differing in their melting point and optical rotation. Sterols are reported in green algae to the extent of 0.3 to 1.2% and vary from 0.07 to 0.1% in blue-green algae. However, cholesterol and sitosterol are reported in *Spirulina maxima*. *Spirulina platensis* contains 149 mg cholesterol per kg. dry wt. Blue-green algae have more unsaponifiable lipid than the green algae (Paoletti *et. al.*,1976). The value of such lipids ranges from 9.7% to 14.4% on dry weight basis.

Polar lipids

Polar lipids of algae include compounds with hydrophilic moieties (carboxylic groups, alcohols, sugars etc.). Polar lipids of algae are allyl-acyl lipids containing fatty acids and are saponifiable. They can be divided into glycolipids and phospholipids depending upon the polar group. As the polar lipids are associated with cell membranes, they are synthesized by both marine and fresh water algae (Benson and Shibuya, 1962). The major glycolipids of algae are monogalactosyl diglyceride (MGDG), digalactosyl diglyceride (DGDG) and sulpholipids. The predominant phospholipids of algae are phosphotidyl choline or generally known as lecithin, phosphotidyl ethanolamine and phosphotidyl glycerol.

The major acyl lipids of blue-green algae are monogalactosyl diglycerols, digalactosyl diglyceride, sulpho-quinovosyl-diglyceride and phosphatidyl glycerol. The major fatty acid composition of these glycolipids are palmitic, linoleic and linolenic acids (Kataoka and Misaki, 1983). Chlorosulfolipids are also found in fresh water blue-greens. Though these are known to synthesize only phosphotidyl glycerol, the other polar lipids such as phosphotidyl inositol, lysolecithin and phospholipids have been reported in *Spirulina* sp. (Kates and Volcani, 1966; Hudson and Karris, 1974). During early stages of growth, green algae are known to produce larger amount of polar lipids than the polyunsaturated C: 16, and at stationary phase, neutral lipids are synthesized more predominantly containing 18: 1 and 16: 0 fatty acids (Piorreck *et. al.*,1984).

Fatty acids

Fatty acids are carbon chains with numerous variations in their composition in algae. Algae generally synthesize both saturated and unsaturated straight chain fatty acids, mostly with even number of carbon atoms. Arachidonic (20: 0), and fatty acids with still longer chains are rarely present. Small quantities of odd chain fatty acids, 13: 0, 15.0, 17.0 and 19.0 are reported (Pohl and Zurheide, 1979; Ahluwalia and Paul, 1987). Free fatty acids are not common in algae, as they generally occur in bound form. The double bonds in the unsaturated fatty acids are in *cis*-configuration and only *trans* 3-16: 1 fatty acids are found in phosphotidyl glycerol. In blue-green and green algae, the fatty acids range from 14: 0 to 18: 4 (Schneider *et. al.*, 1970). The unsaturated C: 18 fatty acids are dominant in blue-greens whereas C: 16 fatty acids are dominant in green algae. Linolenic acid (18: 3) has been found to be maximum than other fatty acids in *Nostoc ellipsosporum* (Ahluwalia and Paul, 1987).

A large number of algal isolates from soil and fresh water were found to contain palmitic, palmitoleic, linoleic and linolenic acids (Levin *et. al.*, 1964; Nichols and Wood, 1968a; Kenyon and Stanier, 1970). *Spirulina platensis* contain gamma-linolenic acid (18: 3) (Nichols and Wood, 1968a), whereas presence of alpha-linolenic acid is also reported (Kenyon *et. al.*, 1972). The blue-green algae do not synthesize the higher fatty acids such as arachidonic acid or trans-3-hexadecanoic acid, however, arachidic acid (20: 0) has been found in akinetes of *Nostoc ellipsosporum* (Ahluwalia and Paul, 1987). These are not present even in phosphotidyl glycerol fractions (Nichols and Wood, 1968b).

The freshwater green algae do not synthesize 18: 4 and polyunsaturated C: 20 and C: 22 fatty acids which are mainly found in marine algae. The variation in the synthesis of fatty acids may be due to acyl-lipids of blue-green algae originating from photosynthetic apparatus which shows the variation in the synthesizing capacities of the lamellae (Hitchcock and Nichols, 1971).

In *Spirulina platensis*, the lipid content is 8-10% with a higher level of polar lipids than the non-polar ones. The neutral lipids are mainly dominated by the free fatty acids (23.7%). The triglycerides varied in different species and in some cases, their complete absence has been reported (Piorreck and Pohl, 1984). Glycolipids are found in the bound form. *Spirulina platensis* contains higher amount of palmitic acid and lesser amount of C: 18 unsaturated fatty acids; this may be due to the cultivation in tropical conditions and also distinctive biosynthetic pathway. This alga contains high amount of linolenic acid, concentrated in glycolipids, and in monogalactosides. This confirms that the alpha - linolenate containing galactolipids occur in the photosynthetic lamellar structure of blue-green algae (Erwin and Bloch, 1962).

Biosynthesis

The typical pathway of lipid synthesis in microalgae resemble that of higher plants, though a few reports indicate deviations and inter-conversions of lipids. The main pathway of synthesis of oleic acid (Fig. 18.1) occurs both in aerobic and anaerobic conditions. Stearyl ACP synthesized from acetate produced in TCA cycle, serves as the substrate for the synthesis of oleic acid. This stearyl ACP is readily saturated into oleolyl-ACP for oleic acid in the presence of the stearyl-ACP desaturase and with the input of two electrons either from NADPH or water through photosystem I and photosystem II in the presence of intermediate electron carrier, ferredoxin and ferredoxin reductase to molecular oxygen. The oleolyl ACP produced in chloroplast is converted into free oleic acid and ACP by highly specific acyl-ACP thioesterase.

In the further steps of biosynthesis, desaturation of oleic and linoleic acid occur with involvement of two enzymes, oleoyl desaturase and linoleoyl desaturase. From the experiments conducted so far with the membrane fragments, two substrates *viz.* oleoyl-Co-A and β-oleoyl phosphatidyl choline have been found to be involved in the synthesis. Some results suggest that the oleoyl Co-A is the precursor of β-oleoyl phosphotidyl choline and the rate of desaturation of oleoyl Co-A to linoleoyl Co-A is very fast compared to slower desaturation rate of β-oleoyl phosphotidyl choline(only 30-50%). This suggest that if β-oleoyl phophotidyl choline was the true substrate, a full conversion should be seen. Therefore, some workers suggest that only β-oleoyl phosphotidyl choline which is adjacent to oleoyl and linoloyl desaturase rather than those molecules situated far away, is involved. Most evidences suggest that the site of oleic acid desaturation and desaturases lies in endoplasmic reticulum from where they are transported to specific sites such as chloroplast lamellar membranes.

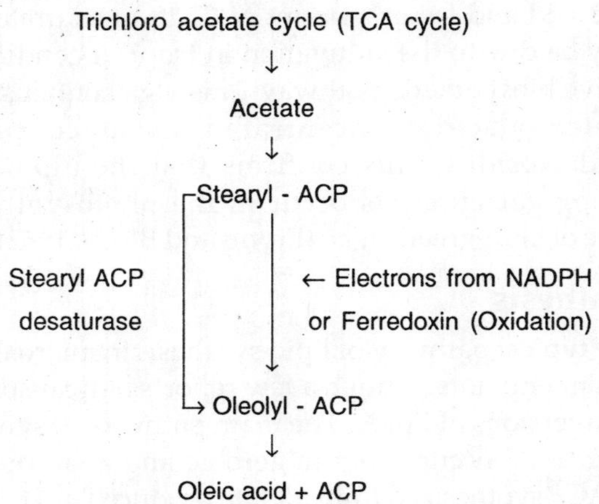

Fig. 18.1: Biosynthesis of oleic acid during photosynthesis

The pathway of elongation and desaturation steps involved in the synthesis of PUFA in microalgae has been highlighted in Fig. 18.2. In chain elongation, two carbon atoms from the donor (either acetyl CoA or melonoyl CoA) are introduced to increase the chain length. The desaturation of fatty acids may proceed in two different routes. Firstly, the omega-3 route where desaturation usually produces linolenic acid (18: 3, n-6) which will yield all cis-eicosapentanoic acid by further chain elongation and two more desaturation reactions. Many microalgae such as species of *Porphyridium*, *Chaetoceros gracilis*, *Nannochloropsis salina*, *Chlorella stigmatophora* (Ben-Amotz *et. al.*, 1985) and species of *Crypthecodinium* (Henderson *et. al.*, 1988) follow this route. In the second route, which is omega-6 route, the additional double bonds are usually introduced towards the carboxyl end to produce arachidonic acid (20: 4, n-6) through desaturation of linoleic acid to linolenic acid (18: 3, n-3) followed by chain elongation as reported in *Porphyridium cruentum* (Nichols, 1973).

Fig. 18.2: Biosynthesis of polyunsaturated fatty acids in algae

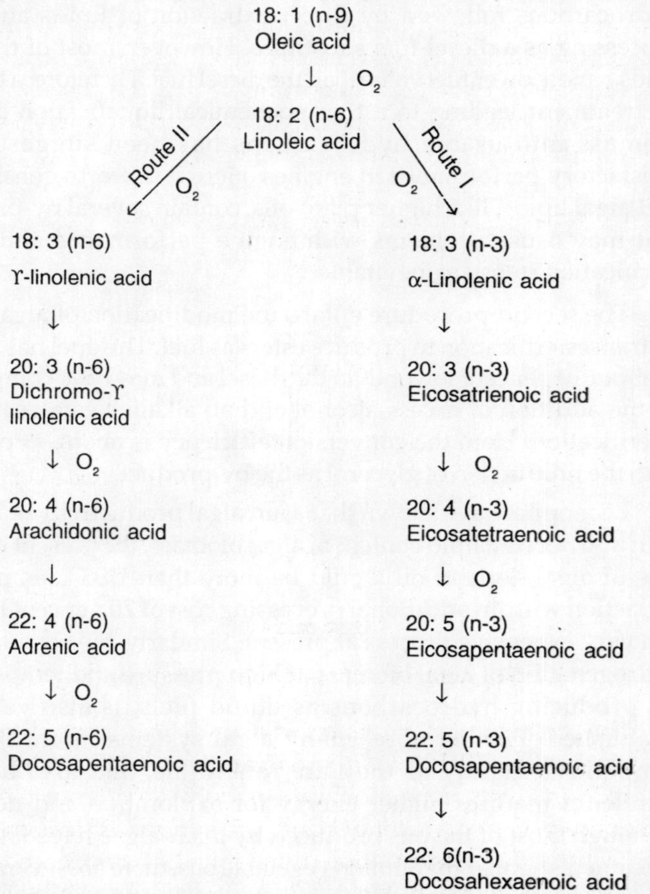

Microalgae as oil producers

Most of the aquatic algae produce lipids or oils as a defense response to stress of either environment or induced which may be exploited for oil or gasoline. These are preferred over land plants for oil production as these algae grow almost five times faster than the most terrestrial plants.

In general, for the production of energy from algal biomass, two basic approaches are employed depending upon the particular alga and the type of lipid-hydrocarbon produced. The first approach involves a simple method of converting nutrients into lipids or hydrocarbons followed by direct extraction of lipids and their processing as a diesel-fuel substitute. However, most of the algal lipids have a lower fuel value than the diesel fuel. Therefore, a thermal pretreatment leading to a thermochemical liquefication of algal biomass into usable hydrocarbons has been suggested for satisfactory performance in engines. Here it is worth considering that algal lipids, like higher plant oils, contain several by-products that may cause problems with engine performance and hence purification steps are inevitable.

The second procedure entails the modification of algal lipids by transesterification to produce esters as fuel. This fuel has a value of about 10 per cent lower than the diesel and may be accomplished by the addition of excess alcohol and an alkaline catalyst during esterification. Here the conversion efficiency is about 98 per cent with the production of glycerol as the by-product.

Economics have shown that at an algal productivity of 75 t/ha with a 30 per cent lipid content of algal biomass, the present cost per litre of algal-derived oil would be more than US$ one, prior to extraction with an additional processing cost of 20 per cent leading to a very expensive process at present. Similarly, the economics of hydrogenation of algal biomass at high pressure and temperature for producing hydrocarbons as liquid fuels, is also yet to be established. Furthermore, many algal systems might become competitive mainly as they are renewable, and also because petroleum requires higher energy for exploration and delivery. Although most of the oils produced by microalgae have fatty acid constituents similar to common vegetable oils, there are several high-value unusual fatty acids that are synthesized exclusively by various algae.

Enhancement and alteration of lipid profile

Total lipids as well as the quality of lipids in algae respond rapidly to fluctuations of the growth conditions or several types of stress and therefore, there are alterations in quality and quantity of lipids. Some of such factors are discussed below:

Carbondioxide

In contrast to higher plants, the lipid accumulating algal cells parallelly perform photosynthesis, indicating that the complete pathway from CO_2-fixation upto triacylglycerol synthesis can be adjusted by the cell. Thus the major raw material required by the alga for the production of lipid is the need for a large amount of carbondioxide during their growth phase, so that under stress conditions, at the expense of already accumulated carbon, lipid synthesis occurs as the photosynthetic activity decreases.

Nitrogen stress

In most of the cases, lipid composition gets altered when there is an addition or restriction of certain components in the medium. For example, nitrogen limitation in culture medium mostly increases lipid content (Borowitzka, 1988) with the exception of *Dunaliella and Tetraselmis suecica* where lipid content increases with nitrogen limitation. The physiological status of the cell and availability of the nutrients, especially the nitrogen, influence the lipid biosynthesis. The cells harvested at the end of linear phase had a high lipid content of 12-22% (Piorreck and Pohl, 1984).

In green algae, more than 70% lipid produced under nitrogen limitation are neutral lipids such as triglycerols, containing C16:0 and C18: 0 fatty acids. When nitrogen is more in the medium, predominant lipids are polar lipids containing polyunsaturated C: 16 and C: 18 fatty acids in most of the organisms whereas in species of *Dunaliella* and *Isochrysis,* C: 18 fatty acids are formed under nitrogen limited condition (Pohl and Zurheide, 1979; Ben Amotz *et. al.,*1985).

Nitrogen starvation doubled the lipid fraction to 40 - 50% in many green algae (Shifrin and Chisholm, 1980). Exhaustion of nitrogen stimulates the fat accumulation. The nitrogen influence on the sterol content of green algae grown autotrophically, photoautotrophically, and heterotrophically indicated that sterol was higher in autotrophic than in heterotrophic cultures by 2 - 20 times. Relative amounts of sterols did not change significantly (Goodwin, 1974).

Other chemical factors

Presence of detergents like Triton X-100, sodium desocycholate (SDC), sodium dodecylsulphate (SDS) or acetyl trimethylammonium bromide in the culture change, the fatty acid composition of the algae (Nyberg and Koskimies, 1984). The addition of 5 ppm of SDS to the culture medium of *Porphyridium purpureum* resulted in higher proportion of eicosapentaenoic acid than control. Silicon limitation shown to enhance lipid production in *Navicula pelliculosa* (Coombs *et. al.*, 1967), and *Cyclotella crytica* (Werner, 1977).

Physical factors

Physical factors such as light, temperature and dissolved oxygen have been reported to alter the array of lipids in microalgae. Low temperature increased the synthesis of polyunsaturated (C: 18) fatty acids in *Monochrysis lutheri* (Ackman *et. al.*, 1980). Temperature has profound effect on the synthesis of linolenic acid in *Cyanidium caldarium* where the initial exposure to a higher temperature denatures the enzyme responsible for the conversion of linoleic acid to linolenic acid by decreasing the degree of unsaturation. *Ochromonas danica* (Aaronson, 1973) and *Chlorella minutissima* (Seto *et. al.*, 1984) have also been studied for lipid composition under different temperature regimes. Decrease in arachidonic acid (a PUFA) occurred in *Porphyridium cruentum* with an increase in temperature from 18°C to 32°C (Cohen *et. al.*, 1987). Temperature plays a crucial role in the degree of alteration of unsaturated fatty acids, particularly in lipids located in the membranes (Deshnium *et. al.*, 2000). In *Spirulina platensis*, the lipids of 18: 1 and 18: 2 increased on shifting the cultures from 35° to 22°C and decreased on shifting the cells to 40°C. This correlated well with the levels of enzymes and patterns of respective transcription and translation indicating that the temperature dependent lipid change is strictly under genetic control (Deshnium *et. al.*, 2000).

Effect of high light intensity on lipid composition of cyanobacteria and microalgae has been reported (Oprcutt and Patterson, 1975). In *Porphyridium cruentum*, an increase in light intensity augmented the polyunsaturated fatty acids (C: 20) (Kost *et. al.*, 1954). Enhanced formation of polyunsaturated C: 16 and C: 18 fatty acids as well as mono-and digalactosyl diglycerides in *Chlorella* was possible under enhanced light conditions (Pohl and Wagner, 1972).

At present even genetic techniques are employed for the large-scale production of lipids in microalgae. However, using non-genetic methods, the scientists have already achieved about 40-60% increase in the lipid content of the algal cell. Some of the PUFA were recently shown to have very high therapeutic value, and hence find great commercial significance. Strains of *Spirulina platensis* under specific environmental and nutritional conditions lead to the enhanced production of GLA (gamma linolenic acid). By the proper choice of the strain and growth conditions, it is possible to increase the GLA content in *Spirulina* from 22% to 32% of the total fatty acids and 1.5% of dry weight (Cohen *et. al.*, 1987). It has also been shown that by using specific herbicides, it is possible to inhibit the specific biosynthetic pathway in blue-green algae (Murphy *et. al.*, 1985). With the increasing health awareness of consumers and selection for particular ingredients in the processed foods, the selection of bioactive fatty acids in the diet will naturally enhance the production of specific oils with essential fatty acids through microalgae.

Conclusively, the available information suggests that microalgae represent a valuable source of a wide spectrum of fats and oils with different application potentials. New developments in the chemical industry will further enhance the range of commercially important products, particularly the replacement of petrochemical feedstock.

References

Aaronson, S. 1973. Effect of incubation temperature on the macromolecular and lipid content of the phytoflagellate *Ochromonas danica*. *J. Phycol.* **9**: 111-113.

Aaronson, S., Berner,T. and Dubinsky, Z. 1980. Microalgae as a source of chemicals and natural products. In: *Algae Biomass*. Edited by Shelef, G. and Soeder, C. J., Elsevier/ North Holland Biomedical Press, Amsterdam. pp. 575-601.

Ackman, R.G., Tocher, C.H. and McLachlan, J. 1980. Marine phytoplankton fatty acids. *J. Fish Res. Board Canada* **25**: 1603-1620.

Ahluwalia, A.S. and Paul, R. 1987. Fatty acid composition of a nitrogen-fixing blue-green alga, *Nostoc ellipsosporum*. *Ann. Biol.* **3**: 56-58.

Ben-Amotz, A., Tornabene, T.G. and Thomas, W.H. 1985. Chemical profile of selected species of microalgae with emphasis on lipids. *J. Phycol.* **21**: 72-81.

Benson, A.A. and Shibuya, I. 1962. In: *Physiology and Biochemistry of Algae*. Edited by Lewin, R.A., Academic Press, New York. pp. 371.

Borowitzka, M.A. 1988. Microalgae as a source of essential fatty acids. *Austr. J. Biotechnol.* **I**: 58-62.

Cohen, Z., Vonshak, A and Richmond, A. 1987. Fatty acid composition of *Spirulina* strain grown under various environmental conditions. *Phytochemistry* **26**: 2255-2258.

Coombs, J., Darley, W.H., Holm-Hansen, O. and Volcani, B.E. 1967. Studies on the biochemistry and fine structure of silica shell formation in diatoms. Chemical composition of *Navicula pelliculosa* during silica shell formation in diatoms saturation synchrony. *Plant Physiol.* **42**: 1601-1606.

Deshnium, P., Paithoonrangsarid, H.A., Phaphugrangkul, T.M. and Cheevadhanaral, A. 2000. Temperature independent and dependent regulation of the desaturase gene expression in *Spirulina platensis. 4th Asia Pacific Conf. Algal Biotechnol.*

Erwin, J. and Bloch, K. 1962. The α-linolenic acid content of some photosynthetic microorganisms. *Biochem. Biophys. Res. Comm.* **9**: 103-108.

Goodwin, T. W. 1974. Sterols. In: *Algal Physiology and Biochemistry*. Edited by Stewart, W. D. P., Blackwell Scientific Publ., Oxford. pp. 266.

Harry,J. D. Jr. 1951. *The Lipids*. Interscience Publ. Inc., New York.

Henderson, R. J., Leftley, J. W. and Sargent, J. R. 1988. Lipid composition and biosynthesis in the marine dinoflagellate *Crypthecodinium cohnii*. *Phytochemistry* **27**: 1679-1683.

Hitchcock, C. and Nichols, B. W. 1971. *Plant Lipid Biochemistry*, Academic Press, New York.

Hudson, J. F. and Karris, G. I. 1974. The lipids of the alga *Spirulina*. *J. Sci. Food Agric.* **25**: 759-763.

Kataoka, N. and Misaki, A. 1983. Glycolipids isolated from *Spirulina maxima*: Structure and fatty acid composition. *Agric. Biol. Chem.* **47**: 3249-2355.

Kates, M. and Volcani, B. E. 1966. Lipid components of diatoms. *Biochim. Biophys. Acta* **16**: 264-278.

Kenyon, C. N., Rippka, R. and Stanier, R. Y. 1972. Fatty acid composition and physiological properties of some filamentous blue-green algae. *Arch. Microbiol.* **83**: 216-236.

Kenyon, C. N. and Stanier, R. Y. 1970. Possible evolutionary significance of polyunsaturated fatty acids in blue-green algae. *Nature* **227**: 1164-1166.

Kost, H. P., Sanser, M. and Manner, G. 1954. Effect of nitrate and sulphate starvation on *Porphyridium cruentum* cells. *Z. Pflanzenphysiol.* **113:** 231-249.

Levin, E., Lennarz, W. J. and Bloch, K. 1964. Occurrence and localization of α-linolenic acid containing galactolipids in the photosynthetic apparatus of *Anabaena variabilis. Biochim. Biophys. Acta* **84:** 471-474.

Murphy, D. J., Harwood, J. L., Lee, K. A., Roberto, F., Stumpf, P. K. and John, J. B. Sr. 1985. Differential responses of a range of photosynthate tissues to a substituted pyridazinone, sandoz 9785: Specific effects on fatty acid desaturation. *Phytochemistry* **24:** 1923-1929.

Nichols, B. W. 1973. Lipid composition and metabolism. In: *The Biology of Blue-green Algae.* Edited by Carr, N. G. and Whitton, B. A. Blackwell Scientific Publ., Oxford.pp. 144-161.

Nichols, B. W. and Wood, B. J. B. 1968a. New glycolipid specific to nitrogen fixing blue-green algae. *Nature* **217:** 767-768.

Nichols, B. W. and Wood, B. J. B. 1968b.Occurrence and biosynthesis of α-linolenic acid in a blue-green alga *Spirulina platensis. Lipids* **3:** 46.

Nyberg, H. and Koskimies-Soininen, K. 1984. The glycolipid fatty acid of *Porphyridium purpureum* cultured in the presence of detergents. *Phytochemistry* **22:** 751-757.

Oprcutt, S. and Patterson, G. W. 1975. Sterol, fatty acid and elemental composition of diatoms grown in chemical defined media. *Comp. Biochem. Physiol.* **50:** 579-583.

Opute, F. I. 1974. Studies on fat accumulation in *Nitzschia palea. Ann. Bot.* **38:** 889-892.

Paoletti, C., Pushparaj, B., Florenzano, G., Capella, P. and Lercker, G. 1976.Unsaponifiable matter of green and blue-green algal lipids as a factor of biochemical differentiation of their biomasses. 1. Total unsaponifiable and hydrocarbon fraction. *Lipids* **11:** 258.

Paul, R. and Ahluwalia, A. S. 1987. Biochemical studies during differentiation in two *Anabaena* spp. *Beitr. Biol. Pflanzen.* **61:** 327-335.

Piorreck, M., Baasch, K. H. and Pohl, P. 1984. Biomass production, total protein, chlorophylls,lipids and fatty acids of fresh water green and blue-green algae under different nitrogen regimes. *Phytochemistry* **23:** 207-216.

Piorreck, M. and Pohl, P. 1984. Formation of biomass, total protein, chlorophylls,lipids and fatty acids in green and blue-green algae during one growth phase. *Phytochemistry* **23:** 217-223.

Pohl, P. and Wagner, H. 1972. Control of fatty acid and lipid biosynthesis in *Euglena gracilis* by ammonia, light and DCMU. Z. *Naturforsch*. **27:** 53-61.

Pohl, P. and Zurheide, F. 1979. Fatty acids and lipids in marine algae and the control of their biosynthesis by environmental factors. In: *Marine algae in Pharmaceutical Science.* Edited by Hope, H. A., Levring, T. and Tanaka, Y. Walter de Gruzter, Berlin. pp. 433-523.

Schneider, H., Gelpi, E., Bennett, E. O. and Oro, J. 1970. Lipids of geochemical significance in microscopic algae.11. Fatty acids of geochemical significance in microscopic algae. *Phytochemistry* **9:** 613-617.

Seto, A., Wang, H. L. and Hesseltine, C. W. 1984. Culture condition affect eicosapentoenoic acid content of *Chlorella minutissima. J. AOCS.* **61:** 892-894.

Shifrin, N. S. and Chisholm, S. W. 1980. Phytoplankton lipids: Environmental influences on production and possible commercial applications. In: *Algae Biomass*. Edited by Shelef, G. and Soeder, C. J., Elsevier/ North Holland Biomedical Press, Amsterdam. pp. 267.

Wada, H., Gombos, Z. and Murata, M. 1994. Contribution of membrane lipids to the ability of the photosynthetic machinery to tolerate temperature stress. *Proc. Natl. Acad. Sci. USA* **91:** 4273-4277.

Werner, D. 1977. Silicate metabolism. In: *The Biology of Diatoms.* Edited by Werner, D., California Univ. Press, Berkeley. pp. 498.

Chapter 19

SEAWEED UTILIZATION: A REVIEW

K. Sivakumar

Abstract

Seaweeds are macroscopic algae which form an important component of the marine living resource. They have been harvested by man for centuries particularly in Japan and China where they form a part of the staple diet. The use of seaweeds as food, fodder and manure are well known in many countries. Marine algae contain more than 60 trace elements in a concentration much higher than in terrestrial plants. They also contain protein, iodine, bromine, vitamin and substances of stimulatory and antibiotic nature. Seaweeds are the only source for the production of agar, alginate and carrageenan. These phytochemicals are extensively used in various industries such as food, confectionary, textile, pharmaceutical, dairy and paper mostly as gelling, stabilizing and thickening agents.

Key words: Agar-agar, algin, carrageenan, manure, pharmaceutical, seaweed, trace elements.

Introduction

The ever increasing human population and the limited resources of land, force the human community to find out an alternate source for food and energy. As on day, the aquatic system appears to be rich in resources which could be exploited beneficially.

Seaweeds are macroscopic marine forms, abundantly growing on rocky seashore regions. Indian resources in seaweeds, relatively speaking, cannot be considered large; but they are sufficiently valuable. An interest has been created in seaweed meal for human consumption, and seaweed supplements to stock food, medicines and manure. Some of the large seaweeds which show considerable tissue differentiation with organs analogous to the root, stem and leaf of the higher plants, may still be termed a thallus for being structurally simple. They are of various colours including green, brown and red etc. but majority of them contain the green chlorophyll pigments and hence are autotrophic. Although the mode of nutrition resembles that of higher plants, yet the method of absorption differs, as they do not possess specialized absorptive organs comparable with root hairs, but food material in solution diffuses through the cell wall over the whole surfaces.

Historical account

The history of algae is as old as that of Botany. The first writings on algae appeared in the ancient Chinese classics, where references to the collection, eating and the medicinal uses of algae are found. There are also references in Roman and Greek literature about the use of algae for cosmetic purposes. The Chinese named them as *Tsao* for having aesthetic significance whereas the Howaiians called them *Limu* as they used algae as a food. In the north coast of France, algae were known to be used as early as twelveth century. The use of brown seaweeds as fertilizers climaxed in France during the seventeenth century. The utilization of certain algae for agar making was introduced in China. It spread to Japan and later, turned out to be a great industry with its foundation in the seventeenth century. Although Linnaeus (1753) first introduced the name Algae but the delimitation of this group was made by A.L. De Jussieu (1789). Jussieu's characterization of algae was not of much value since it was based entirely on macroscopic features. Actually the great pioneer in the study of algae was C.A. Agardh (1824), who first started describing algae considering almost all aspects of the life history. It was Blackman (1990) who postulated a flagellate ancestry for the algae, expressed earlier by Klebs (1893), and suggested that there are three distinct tendencies, the volvocine, tetrasporine and chlorococcine which were recognized in the evolution of green algae.

In India Royle (1839), first reported the occurrence of some members of the lemaniaceae in the mountain streams of the Himalayas. Following him were Decaisse and Montagne who described *Actinotrichea fingillis* and *Calothrix indica* in 1842 and 1849. Griffith (1847) reported a number of species of *Chara* from Bengal. Carter (1858) collected *Eudorina elegans* from the pools and Hobson

(1863) published an account of the desmids from Bengal and Bombay. From 1919 onwards more Indian workers got interested in algae. During 1919 to 1932, Ghose carried on investigations on the blue-green algae of Punjab. M.O.P. Iyengar was another pioneer investigator who published a good number of papers on algae of South India, both fresh water and marine.

Seaweed used as food

Large number of seaweeds have entered into the diets of human beings from ancient times. The earliest records of the Chinese, mentioned such food plants as *Laminaria* and *Gracilaria* in their 'Materia Medica' several thousand years ago. The ancient inhabitants of Japan ate *Porphyra* as a health supplement to their rice diet. Its use became wide spread not only in Japan, but in China as well, in course of time. Kombu, a Japanese food is prepared from stipes of species of *Laminaria*. The Hawaiians considered algae a necessary staple of their daily diet. The most widely used food alga in western Europe in recent centuries was Irish moss, or carragheen (*Chondrus crispus*), which was cooked with milk, seasoned with vanilla or fruit, and made into a highly palatable dish known as blancmanges. The jellying qualities of Irish moss gave the alga an early food use.

Man, thus obtained carbohydrates, vitamins, (algae are especially rich in vitamins A and E, and they contain some C and D), and inorganic substances, e.g., Iodine (goiter is unknown among the people who eat seaweeds), not to mention the benefits of the mild laxative action of the ingested algae, Witsch (1959) stated that vitamin B value of young cultures of *Chlorella* equals that of lemon juice. In Japan, powdered *Chlorella ellipsoidea* has been used successfully after mixing with green tea. In Germany and in the United States considerable work is being carried out on the suitability of mass cultures of *Chlorella* and *Spirulina* as alternative source of animal feed and of human vegetable food.

Algae used as fodder

The orientals developed wide human uses for marine algae, but Europeans profitted by extensive use of these plants for stock feed. In Iceland and Scandinavia in the British Isles, and among the coast of France, stock has been allowed to wander to the seashore at low tide to feed on seaweeds. Some algae, such as *Rhodymenia palmata* and *Alaria esculenta*, are favourable food of goats, cows and sheep. The milk does not have any taste of algae, nor is the meat inferior because of the seaweed diet. Animals that have been nourished for several generations on algae, show better ability to digest it than those not so habituated.

The storage of grains in many parts of the Europe during world war first led to considerable experimentation with the use of seaweed as food for cow and horses. Stock feed factories were established in France, Norway, Denmark and Germany and various methods of treating and reducing seaweeds to meal or powder were developed.

The favourable results in animal husbandry in Europe led to the industrial processing of the great pacific – coast kelp (*Macrocystis*) for animal rations. Seaweed – meal factories have been operating in the United States for decades, providing supplementary foods for poultry, cattle and dogs. The high mineral and vitamin content of the kelps meal, has made possible its use in various poultry and other animal rations.

Algae used as fertilizers

The value of seaweeds in fertilizing the soil was discovered early in the history of agriculture in coastal Asia, and by the ancient colonizers of the coasts and Islands of North – Western Europe. In some areas of Britain, and along the coast of North-West France, the cutting of rock weeds for manure has been so intensively practiced that it became necessary to regulate it by laws that have now been in effect for nearly 100 years.

In the United States, seaweeds were employed for fertilizers by the thrifty farmers. Not only the fertilization, but also the water holding capacity of the algal fragments in the soil proved effective. These provided valuable small reservoirs of water in close contact with the roots of the cultivated plants. Furthermore, the bulky organic substances decay slowly in the soil and form humus. Again yield of paddy increased substantially when paddy field is inoculated with nitrogen fixing blue-green algae (species of *Tolypothrix, Aulosira, Anabaena, Anabaenopsis, Calothrix, Nostoc* and *Cylindrospermum*).

Algae used as medicine

Medicinal applications of plants are almost as old as their food uses. From the earliest times, laminariales were used for treatment of goiter and other glandular troubles. *Gelidium* was employed for treatment of stomach disorders and for heat induced illness. The gentle swelling of dried *lamanaria* stipes upon exposure to moisture make them surgical tool in the opening of wounds. Similarly, the orientals have employed the same technique in child-birth for expansion of the cervix.

Most widely used algae for medicinal purposes are the agarophytes including *Gelidium, Pterocladia, Gracilaria,* and *Ahnfeltia.* The name agar-agar is of Malaya origin and means 'jelly'. This jelly was obtained by boiling up seaweeds and cooling the resulting liquid. Agar has been useful for stomach disorders and as a laxative. It was originally produced and marketed in China, but the Japanese took over production in about 1662 and maintained a world monopoly till 1940. The most significant date in the utilization of agarophytes was 1881, when Robert Koch proved the value of agar in the cultivation of bacteria. Since that time it has become essential to the work of hospitals and medical research laboratories throughout the world. Besides these, *Chlorella* is used for preparation of antibiotic chlorellin.

Marine algae have been in use as food materials as well as curative agents. Interest in marine organisms as potential source of pharmaceutical agents has been growing in recent years. Sodium alginate, on the basis of old Chinese literature and as folk medicine among the people, has been known as herbal medicine. It can help in treatment of goiter and scrofula as antipyretic and cool refreshing liquid, treatment of sunstroke, urinary diseases, dropsy, cancer, stomach ailments, haemorrhoid anal fistulas, cough, bronchitis, tonsillitis, asthma and as laxative and tonic for lying women (Zeng and Zang, 1984).

Heparin is a sulfated amino polysaccharide and used in cardiovascular surgery and treatment of atherosclerosis (Engelberg, 1984). The Indian coastal waters are enriched with seaweeds, and the red alga *Grateloupia filicina* had shown promising heparin activity. The potency was 30 units/mg and the yield per kg dry seaweed was 5.6 x 10^6 units. Subramnian *et. al.* (1997) have shown that the red algae, *Acanthophora spicifera* and *Hypnea valentiae* yielded heparin 10-20 times higher than the most animal tissues.

Industrial utilization of algae

Kelp Industry

Industrial utilization of seaweeds in Europe had its principal early development in the production of 'Kelp' – a name that originally referred to the ash, rich in soda and potash, derived from burning marine plants. Kelp production began sometimes in the seventeenth century by French peasants and spread to other parts of North-West Europe. Drift weeds were first used, but cutting was later resorted to *Laminaria* and *Saccorhiza* in North Britain. In addition, *Fucus, Ascophyllum, Himanthalia* and *Chorda* were also used. The kelp ash from these plants was widely bought by early industrialists for use in manufacture of soap, glass and alum. Kelp extract contains a number of chemical elements, notably potassium and iodine. About 25% of the dry weight of kelp is potassium chloride. Many species of kelp are used as food for man, especially in the orient. In Northern Europe, these also serve as food for domestic animals such as sheep and cattle.

Algin industry

Algin is the general term designating the hydrophilic, or water loving derivatives of alginic acid. The most commonly known algin is sodium alginate, but other commercially important compounds are the potassium, ammonium, calcium and propylene glycol alignates, as well as alginic acid itself. With the exception of alginic acid and calcium alginate, the algin products offered, commercially, are soluble in water to form viscous solutions. Algin occurs generally throughout the brown algae (*Laminaria, Macrocystis, Sargassum* and *Fucus* spp.) as a cell wall constituent. It has remarkable water absorbing qualities that make it useful in numerous industries in which a thickening, suspending, stabilizing, emulsifying, gel forming, or film forming colloid is required. Thus, algin provides ice-cream with a smooth texture by preventing the formation of ice crystals, Also used in automobile polishes, paints, pigments, drugs and antibiotics. As a stabilizing agent, it serves in the processing of rubber latex and in the printing of textiles. As an emulsifier, it is widely used in water based paints, dressings and cosmetics.

The algin industry has become so important to a wide variety of industries that extensive survey of kelp-bed ecology is an effort to guard against loss of this important resource. Harvesting methods are now carefully regulated, and a huge amount of money is being spent on kelp-bed research throughout the world. Experimental studies are concentrating on the relation of pollution to kelp survival and on kelp-bed grazing organization.

Agar Industry

The outstanding use of the red algae is in the production of agar. This is a dried and bleached gelatinous extract obtained from red algae *Gelidium nudifrons, G. pusillum, G. robustum,* and *Gracilaria verrucosa*. Agar is used extensively in medicine, chiefly as laxative, since it is not digested and increases greatly in bulk with the absorption of water. More important than this medicinal utilization is its use as an essential ingredient in the preparation of medium for the growth of algae, bacteria and fungi. As such, it is indespensible in the bacteriological laboratories because no adequate substitute for agar is known.

Since the introduction of agar into Bacteriology in 1881, the agarophytes have become increasingly industrialized and the technical uses of agar enormously expanded. Modern industry has developed a multi trade of applications from agar. Agar serves widely as a substitute for gelatin, as an antidrying agent in breads and pastry, in improving the slicing quality of cheese, in the preparation of rapidly setting jellies and desserts, and in the manufacture of frozen dairy products, The use of agar in meat and fishing canning has greatly expanded, and hundreds of tons are utilized annually. Agar has proved effective as a temporary preventive for meat and fish in tropical regions, due to the inability of most putrifying bacteria to attack it.

Early industrial use of agar in the orient included sizing fabric, water-proofing paper and cloth, and making rice paper more durable. Modern industry has refined and expanded these uses to meet new needs in the manufacturing of such items as photographic film, shoe polish, dental impression molds, shaving soaps and hand lotions. In the tanning industry, agar imparts stiffness to finished leather. In the manufacture of electric lamps, a lubricant of graphite and agar is used in drawing the hot tungsten wire.

The increasing applications have called for wide expansion of the collection of agarophytes. Since Japan supplied most of the world's markets before second world war, a great amount of research was conducted in an attempt to develop domestic agar supplies in the United States, South Africa, Australia, Newzealand and Russia as the supplies were cut off during the war.

Commercial exploitation of seaweeds in India has started in 1966 (Silas and Kalimuthu, 1987). Important seaweeds harvested for commercial utilization in the Gulf of Mannar include *Gelidiella acerosa, Gracilaria edulis, Gracilaria crassa* and species of *Sargassum* and *Turbinaria*. Species commonly utilized for the extraction of agar-agar are *Gelidiella acerosa, Gracilaria edulis* and *G. crassa*. Species of *Sargassum* and *Turbinaria* are used for the extraction of alginic acid. In India, agar and alginate are manufactured from seaweeds exploited from the natural seaweed beds particularly from Tamil Nadu coast (Kaliyaperumal *et. al.*, 1998). In certain areas of Gulf of Mannar, commercial exploitation of seaweeds is the livelihood of fisherfolk during the seaweed collection season. Resources of agar yielding seaweeds in the shallow waters of Indian coasts are inadequate to meet the demand of industries. Under such circumstances seaweed cultivation is the only alternative to ensure steady supply of raw material to the industry and also in the economic upliftment of coastal population by way of providing employment opportunity to them. Since seaweeds have a lot of essential trace elements and several bioactive substances, they are considered as medical food of the 21st century. In a country like India where about one third of the population live along the coastal zone, people should find seaweed as one of the food sources from the ocean.

Since there is no awareness on the nutritive value and potential of seaweed production in India, seaweed cultivation is lagging behind though India holds a big market for seaweeds.

Inadequate technology

Though the Central Salt Marine Chemical Research Institute (CSMCRI) and Central Marine Fisheries Research Institute (CMFRI) developed technology on an experimental basis long ago, its application in real commercial production did not see the light of the day. Their experiments were pertaining to farming in the sea. When pond culture in easily manageable conditions give better results there is no reason why R & D work for pond culture was not given adequate importance in India. Grazing problem which was reported to be very severe in sea farming could be easily controlled in ponds. Fertilization for a higher production can also be done for a higher turnover and return.

Seaweed cultivation in other countries

The seaweed production in Philippines has touched nearly 40,000 tonnes worth more than US$ 50 million. In China seaweed accounts for 24% of total aquacultural production. Taiwan has successfully demonstrated *Gracilaria* production in milk-fish ponds through manipulation of water depth, transparency, temperature, fertilizer application and polyculture with fish and crab.

Gracilaria culture in Taiwan has become popular since it is used as feed for abalone culture which fetch high price (US$ 20 / kg) in the international market. Taiwan produced around 6000t of *Gracilaria* in 1990 from 330 hectare of water spread area, which is often cultivated with milk-fish, tiger shrimp and mudcrab . It is found unattached on the bottom of ponds which are used for milk-fish culture.

Seaweed is identified as one of the marine plants to treat the shrimp farm effluents effectively. They are planted in ponds, drainage canals and resources in Kung Krabben Bay of Thailand. In Philippines, *G. tenuispitata* and *G. fisheri* are found to grow well especially in drainage reservoir of shrimp farms, yielding 900 kg in 3 months. The water quality of the farm drainage system also shows dramatic improvement. For the sustainable development of shrimp farming, crop rotation is strongly recommended all over the world. The seaweeds offer unique opportunity in not only treating the farm waste but also as a cultiviable species. The farmer could follow water resource / recycle system and minimize the risk of disease outbreak from external sources. Thus the seaweeds serve twin purpose of promoting eco-friendly shrimp culture on one hand and augmenting the farmer's profit on the other. They are, therefore, gaining a lot of importance nowadays in coastal aquaculture.

Species of *Hypnea* have many applications in medicine and as a food item in various countries (Zaneveld, 1959). It is also used for preparation of domestic agar-agar and as a vermifuge in Indonesia and Philippines. Various species of *Hypnea* are boiled to get the gel and this is taken as a food item after mixing with palm sugar, chili, lime juice etc. (Reed, 1906; Edwards *et. al.*, 1982). *Hypnea* is a rich source (53% on dry weight basis) of phycocolloid carrageenan with strong gelling abilities. (Humm, 1944, 1948, 1951; Hamilton and Carroll, 1962). It can be easily modified to the desired gel strength (Humm and Williams, 1948). In Brazil, the yield of carrageenan varied between 48-66% for *H. musciformis* while it was 25-42% in Senegal and Hawaii (Mollion, 1973; Mshigeni, 1979). In India, a yield up to 48% carrageenan has been reported in *H. musciformis* (Rama Rao, 1970).

It is well known that seaweeds are represented in the daily food of the people. Some processed seaweeds in Japan are :

Japanese Name	Common Name and Explanation
Dashi kompu	Dried kelp – boiled in water to make Japanese soup
Natto kompu	Dried kelp – boiled in water to make used in soup
Oboro kompu	Dried kelp – boiled in water used in soup
Wakame kompu	Dried seaweed – *Undaria pinnatifida*
Tororo kompu	Dried kelp – with vinegar, and shredded, used in soup
Oyatsu kompu	Dried kelp – Eaten as it is
Kashi kompu	Dried kelp – Eaten as it is
Su kompu	Dried kelp – Eaten as it is
Hijiki kompu	Dried kelp – *Hijikia fusiformis*. Eaten after boiling in Japanese sauce
Kobu cha	Kelp tea – Drunk like tea
Nori matsuba	Sandwich of squid and *Porphyra tenera*
Kobu matsuba	Sandwich of squid and kelp
Kompu tsukudani	Kelp seasoned in a concentrated solution of Japanese sauce sugar etc. Eaten with rice
Nori	*Porphyra tenera seasoned. Eaten with rice*
Hoshi nori	Dried *P. tenera;* eaten with rice after a little warming. One of the famous Japanese dishes "Sushi" is made by rolling cooked rice in it and cutting the roll into pieces.
Momi nori	Dried *P. tenera* (powdered); eaten with rice.
Chazuke nori	P. tenera; mixed with salt, ajinomoto etc. Eaten with cold rice.

In Japan, wheat flour is currently fortified with flour made out of various kelps. Nevertheless, as Chapman and Chapman (1980) stated, seaweeds form only 0.5% of the vegetable diet of the people.

Philippines has a wealth of green, red and brown seaweeds, and a programme for the utilization of these resources since 1937, *Gracilaria verrucosa* has been increasingly used as a supplementary or substitute feed for large scale farming of milk-fish. The fresh seaweeds are employed from November to June and the dry meal during the dormant season of the alga.

Reports from Utrecht Veterinary University showed a 10 to 20% increase in the blood haemoglobin of animals which had received a 3% supplement of Norwegian brand of seaweed meal for three months. In the Netherlands, it was found that production of milk and the concentration of vitamin-A in milk increased on feeding the animals on this meal.

In Norway, seaweed meals were found to be substitute for a grass meal to avoid deficiency in riboflavin of chick rations, and also to raise the iodine content of eggs. Further when seaweed meals from *Ascophyllum* and *Laminaria* were used as supplements to a low based diet, satisfactory hatching of eggs in the same manner as with grass meal or skimmed milk was observed.

Exact biological investigations with rats in Denmark showed that a dose of 0.05 to 0.20g daily of finely pulverized seaweed was insufficient and caused rickets; although there was some growth and no xerophthalmia. However, a daily dose of 0.30g and over, caused normal development. Rats that were not given algae or were given dose of only 0.025g., showed arrest of growth, osteoporosis and xerophthalmia. A dose of 200g., daily given to milk producing cows, increased the yield, fat and vitamin-A content of milk.

In United Kingdom, dairy cows were fed at the daily rate of 8 oz. of *Ascophyllum nodosum* meal and the response was positive. 3% of the seaweed meal was found to be a substitute for the same amount of dried grass in feed for laying hens to obtain the standard yolk colour. Feeding trials with rats showed that mannitol, fucoidin, calcium alginate and laminarin at the 8% level were equal to starch, but at 16% they were inferior to starch, fucoidin having a strong laxative effect. These meals were made from the blends of several types of brown seaweeds of the North of Scotland and were used successfully in the rations of cattle, pigs, poultry and sheep. It was observed that this meal rendered them less susceptible to certain diseases.

The Turkey Breeders Association declared that there was check on mortality when seaweed supplement was given in the rations. Seaweed supplement are in use in the Kennel for animals and in trout nurseries. The Race Horse owners Association of British

Columbia found that after the horses received a dosage of 'algit' meal (for 14 days), they broke their individual records.

Algin and mannitol

Algin occurs as an intercellular substance partly as calcium alginate and partly as alginic acid; but on the whole in a complex combination in the cell wall of brown algae. In *Laminaria*, iodine and mannitol are minimum in spring and maximum in autumn, whereas protein and fat show the reverse trend. Ash, potassium and alginic acid vary in quality only slightly throughout the year, and therefore harvesting should be done in autumn to secure the maximum by-products i.e. potassium, iodine and mannitol. Alginic acid can form 31% of the dry matter in summer and 20% in winter, whereas mannitol varies from 20% in the summer to 4-6% in winter. In Japanese laminariaceae and fucaceae, the alginic acid content varies form 20% to 30% and 15% to 20% of the dry matter respectively; mannitol in both varying from 20% to 30% during favourable period and only 6% or less in the spring; the potassium content in members of laminariaceae are generally higher than those of fucaceae.

Fucacean members are the chief source of algin in India, but various species of the dictyotaceae are also used , the laminariaceae being unrepresented in tropical waters. Algin yield was estimated to be 15.6-19.2% on dry matter basis in fucaceae and 10.4% for *Padina*.

Agar

China was the cradle of agar manufacture some time before the beginning of the 4th century A.D. and later the industry was borrowed by Japan. It was discovered by China that when agar gel freezes and thaws, it is transformed into a sponge with a pearl like lustre, and this gave a lead to Japanese production which currently is about 3.5 million lb. (1543 metric tons) per annum. Europe imported agar from 1860, and in 1881, Hesse, a contemporary of Koch, used it in culture medium as it proved to be more suitable for routine cultures than gelatin. Shortly thereafter Koch used it in his renowned work in Bacteriology, and published the first note on the new culture medium. Agar is the complex of polysaccharide which is present in considerable amounts along with cellulose in the cell walls of agarophytes. Gelose is the chief component of agar and it is in the sulphate form and contains calcium and magnesium. No substitute is known for agar in its role as the translucent soil of culture gel that lends itself to the mixing in of varied nutrients and chemicals and to a wide range of incubation temperatures. In India, agar has been produced from *Gracilaria* sp. since 1948. About 30 agarophytic species are now known from our waters, and studies on their taxonomy, cytology, ecology and commercial value are in progress in the country.

Seaweeds are storehouse of the important substances like potash, ionic sulphate, trace elements and growth substances beside having every other element and radical required by plants. Seaweed manure seems to increase resistance to disease, though this is yet only a hypothesis. Most of the nutrients including nitrogen compounds are in ionic form and a quick absorption by crop takes place and relatively little in left to be broken down by soil microflora, thus preventing acidic condition of the soil arising from fermentation. In general, the minerals diffuse out from the seaweed thallus rapidly. Seaweed manure holds water and air at the same time and improves the soil in both respects.

The enterprise of Irish farmers in cutting *Fucus* for manure is a speaking evidence of its value especially for root crops because of their requirements of potash. Compared with "green manure" in general, seaweed manure has a little less of nitrogen but a little more of phosphate. The ash content of marine algae is on an average considerably higher than the usual amount of ash in most land plants. The calcareous algae can have up to 90% ash calculated on the living matter basis. Most of the ash is water soluble, the soluble part consisting of chlorides, sulphates, phosphates and carbonates, and alkaline-earth metals. The nitrogen content generally varies from 1-3% of the dry matter with the exception of *Porphyra* which has up to 7% nitrogen.

Red and brown algae have appreciable content of phosphate on dry matter basis (0.2% to 0.3%); brown algae in general have high potash (about 1% to 4%); both red and green algae are characterized by high sulphate content, about 1% to 8% (all dry matter basis). Calcium and trace elements occur in high percentage in almost all the marine algae.

The evidence seem to show that seaweeds would complement in valuable chemical fertilizer by adding trace elements and growth substances and would ensure all round enrichment of the soil. As a fortifier of leaf mold and farmyard manure, seaweeds have a similar role but also contribute towards the required potassium, sulphur and calcium. Our limited resources in seaweeds would stand to gain if seaweeds were incorporated in composts near coastal tracts, or if seaweed manure were used alongwith chemical fertilizer. Macro- and microelements from three species of *Sargassum* have been reported by Selvaraj and Sivakumar (1998) as follows: -

Name of the species	*Macro – elements mg/kg*			*Micro – elements mg/kg*		
	N	*P*	*K*	*Ca*	*Mg*	*Fe*
Sargassum ilicifolium	9383	533	8960	280	88	1176
Sargassum liniarifolium	1953	1507	9631	238	121	793
Sargassum polycystum	1553	434	8765	205	90	751

Carrageenan

Another important polysaccharide from red algae is carrageenan. It is a cell wall polysaccharide and is mucilaginous extract. The source of carrageenan is *Chondrus crispus, Gigartina stellata*, and *Hypnea*. It is used in food, textile, pharmaceutical, leather and brewing industries. It is used as a component of tooth paste, deodorants, cosmetics and paints. It is also used to stabilize emulsions and as remedy for cough.

In spite of our knowledge that the seaweeds can be a source of many commercial products, much work still has to be done to optimize and to improve culture of seaweeds, develop more cost effective methods of its culture, harvesting, extraction, and purification. It is hoped that the genetic engineering methods will be used in future to optimize the techniques for better production and utilization.

References

Chapman, V.J. and Chapman, D.J. 1980. *Seaweeds and their Uses*. Chapman and Hall, London. pp. 334.

Edwards, P., Boromthanarat, S. and Tam, D.M. 1982. Seaweeds of economic importance in Thailand.- Part I. Field Survey, The Government Statistics and Future Prospects. *Bot .Mar.* **25**: 237-246.

Engelberg, H. 1984. Heparin and the atherosclerotic process. *Pharmacol. Rev.* **36**: 91-110.

Hall, J.G. 1986. Sulphated polysaccharides, corticosteroid and lymphocyte recirculation. *Immunology* **57**: 275-279.

Hamilton, R.D. and Carroll, J.J. 1962. A comparison of polysaccharides of *Hypnea musciformis* and *Chondrus crispus*. *Nature* **196**: 1200-1201.

Humm, H.J. 1944. Agar resources of the south atlantic and east gulf coasts. *Science*. **100**: 209-212.

Humm, H.J. 1948. Methods of extracting gelose from seaweeds such as *Hypnea musciformis* and other species of the genus, *Hypnea*. U.S. Pat. 2, **4461**: 991-1009.

Humm, H.J. 1951. The red algae of economic importance agar and related phycocolloids, In: *Marine Products of Commerce*. Edited by Iressler, D.K. and Lemonleds, J.M., Reinhold, New York. pp. 47-93.

Humm, H.J. and Williams, L.G. 1948. A study of agar from two Brazilian seaweeds. *Amer. J. Bot*. **35**: 287-292.

Kaliaperumal, N., Chennubhotla, V.S.K., Kalimuthu, S., Ramalingam, J.R., Krishna Pillai, S., Muniyandi, K., Rama Rao, K., Thomas, P.C., Zaidi, S.H. and Subbaramaiah, K. 1998. Seaweed resources and distribution in deepwaters from Dhanuskodi to Kanyakumari, Tamil Nadu. *Seaweed Res. Utiln*. **20**: 141-151.

Mayer, A.M.S. and Panick, B. 1984. Antitumor evaluation of marine algae in Argentina. *Proc. Int. Seaweed Symp*. **11**: 529-533.

Mollion, M.J. 1973. Elude preliminaire des *Hypnea* an senegeal comme source de phycocolloids. *Bot. Mar*. **16**: 221-223.

Mshigeni, E.K. 1979. Contributions on the content, seasonality and gelling properties of the phycocolloid from three *Hypnea* spp. from Hawaii. In: *Marine Algae in Pharmaceutical Science*. Edited by Hoppe, H.A. Walter de Gruyter, Berlin. pp.721-746.

Oliver, A.M. and Wieir, D.M. 1985. The effect of *Pseudomonas* alginate on rat alveolar macrophage phagocytosis and bacterial opsonisation. *Clin. Exp. Immunol*. **59**: 190-196.

Quinn, M., Parthasarathy, S. and Steinberg, D. 1985. Endothelial cell derived chemotactic activity for mouse peritoneal macrophages and the effects of modified forms of low density lipoprotein. *Proc. Natl. Acad. Sci. U.S.A*. **82**: 5949-5953.

Rama Rao, K. 1970. Studies on the growth cycle and phycocolloid content in *Hypnea musciformis* (Wulf.) Lam. *Bot. Mar.* **13**: 163-165.

Randall, J.E. 1967. Food habits of reef fishes of the West Indies. *Stud. Trop. Ocean* . **5**: 665-847.

Reed, M.S. 1906. The economic seaweeds of Hawaii and their food value. *Ann. Rep. Hawaii Agric. Exp. Stn.* **1906**: 61-68.

Sasikala, S.L. and Makesh, S.V. 1999. Haematological studies and immune response of algal extractives from *Sargassum swaratzii* administered to wistan albino rat, *Rattus norvegius. Seaweed Res. Utiln.* **21**: 155-180.

Schenkman, R.P.F. 1989. *Hypnea musciformis* (Rhodophyta): ecological influence on growth . *J. Phycol.* **25**: 192-196.

Selvaraj, R. and Sivakumar, K. 1998. Biochemical studies on three species of *Sargassum. Seaweed Res. Utiln.* **20**: 59-62.

Silas, E.G. and Kalimuthu, S. 1987. Commercial exploitation of seaweeds in India .*Cent. Mar. Fish. Res. Inst. Bull.*, **41**: 55-59.

Speert, D.P., Lawton, D. and Mutharia, L.M. 1984. Antibody to *Pseudomonas aeruginosa* mucoid exopolysaccharide and the sodium alignate in cystic fibrosis serum. *Pediat. Res.* **18**: 431-433.

Subramanian, A., Kannan, L., Pugalendhi, T. and Murugananthan, A. 1999. Heparin from a seaweed *Grateloupia filicina. Seaweed Res. Utiln.*, **21**: 21-23.

Sugarwa, I. and Lee, K.C. 1984. Fucoidan blocks macrophage activation in an inductive phase but promotes macrophage activation in an effector phase. *Microbiol. Cell Immunol.* **28**: 371-377.

Woods, D.E. and Bryan, L.E. 1985. Studies on the ability of alginate to act as a protective immunogen against infection with *Pseudomonas aeruginosa* in animals. *J. Infect. Dis.* **151**: 582-588.

Zaneveld, J.S. 1959. The utilization of marine algae in tropical south and east Asia. *Econ. Bot.* **13**: 89-131.

Zeng, C. and Zang, J. 1984. Chinese seaweeds in herbal medicine. *Hydrobiologia* **116**: 152-154.

Chapter 20

SPIRULINA IN MODERN INDUSTRIES FOR MANUFACTURING VALUE ADDED DIETARY PACKAGES

B.K. Behera and Manjeet Kaur

Abstract

Spirulina platensis, a blue-green alga (cyanobacterium), has been known as a super health food on this planet due to the presence of a spectrum of useful macromolecules in its tiny filaments. Scientists have been exploring this organism for useful proteins, vitamins, phycocyanins carotenes, and long chain polyunsaturated fatty acids. It has contributed immensely for food, feed, medicine and value-added biochemicals. Because of a reasonably low level of nucleic acids in their cells, this organism is all set to revolutionise the scene for microbial /biomass protein.

Key words: Algal lipids, Cyanobacteria, culture, economics, health food, *Spirulina*.

Introduction

In view of the appearance of new viruses, drug-resistant bacteria, and ineffective antibiotics, scientists are discovering the possibility of certain foods and/or food additives to increase the immune system and confront problems associated with early aging. Scientific documents presently available explain that *Spirulina, a* blue-green alga, that has been existing on earth surface for more than 3-6 billion years (Skulacher, 1994; Brown, 1998), may be one of the disease preventing and anti-aging foods. Beyond anti-aging benefits, scientists are discovering unusual phytonutrients in *Spirulina.*

Utilization of *Spirulina, as* food from Lake Chad in Central Africa has long been well known (Brandily, 1959). The Mexicans use algae in cosmetics. Ciferri (1983), in his excellent review, explained the use of *Spirulina* by athletes from Mexico for a period 30 to 45 days, with good results. In trials taken on Mexican children suffering from severe malnutrition, a significant weight gain was noticed. It has been identified that compounds in blue-green algae are remarkably active against the AIDS virus. A Russian patent was awarded to *Spirulina as* a medical food for improving the immunity of the 'Children of Chernobyl' suffering from radiation illness. Animal researchers are also realizing the power of *Spirulina.* Scientists in North Carolina, Japan and China have discovered that small amounts of *Spirulina* added to animal feed greatly help animals resist infections. Widely used for aquaculture fish, shrimp, pet birds and race horses (Pauw and Pruder, 1986). *Spirulina* will soon find its use in feeds for healthier chickens, turkeys and other animals (Anusuya *et. al.,* 1981; Kay, 1991; Becker, 1994).

As one of the oldest living plants on the planet, *Spirulina* has been a source of food for some cultures for centuries. It has been consumed by millions of people of all ages in the U.S. and 40 more countries for over 15 years. Sold in tablets, capsules and powder, the dark green powder can be blended into fruit and milk-drinks or added to recipes to boost nutritional value.

Keeping the aforesaid important facts in view, *Spirulina has* gained international demand for its high value added biochemical constituents. It has found applications in health foods, therapeutics and specialized feed. As a result, *Spirulina* is cultured on a large scale (both in open and in close bioreactors or tubular-bioreactors), as market for algal products like β- carotene and phycocyanin (called 'lina blue' A). Besides this, *Spirulina* species are rich in value-added lipid and fatty acids. This has been discussed elsewhere in this review.

The present article is designed to understand and review the present state-of-art of *Spirulina* production at industrial level and its increased awareness in using it as feed, food and food additive.

Spirulina production and processing technology

The nutrient medium used for the cultivation of *Spirulina* is simple and highly alkaline ($NaHCO_3$) with pH range between 9 to 11. There has been a wealth of information available on the parameters related to the cultivation and harvesting of biomass of *Spirulina*. On the basis of utility, *Spirulina* can be cultured under variable natural, artificial and laboratory conditions.

Like other microorganisms, *Spirulina* has high specific growth rate (0.3 d^{-1}) under optimal laboratory conditions, 0.2 d^{-1} in natural conditions during summer and 0.1 d^{-1} in winter. Besides this, it has a very short life cycle ranging from 1 day under optimal laboratory conditions and 3 to 5 days under natural conditions (Claudio, 1982).

Occurrence of *Spirulina* under natural conditions

This tiny blue-green alga grows in alkaline water in Africa, Asia, North and South America (Busson, 1971) in latitudes between 35^0S and 35^0N, area of incident solar irradiation from 600 to 850 KJ/ cm^2 / year and total insolation from 3000 to 4000 h/ year (Landsberg, 1966).

Lake Texcoco is an excellent example for large-scale production of *Spirulina*. This lake is situated in the valley of Mexico and located 2200m above sea level but in a semitropical climate. Lake Chad is another example of large-scale occurrence of *Spirulina* under natural habitat. This lake was earlier in French Equatorial Africa and now in the Republic of Chad. Lonar Lake in the Buldhana District of Maharashtra, India is also rich in *Spirulina*.

Culture in open outdoor ponds

One can have open outdoor culture of *Spirulina* of family size units or hyper size *Spirulina* farm on the basis of requirement.

Family-size unit

About a square metre of surface area is sufficient to grow *Spirulina* for daily dose for one person. As suggested by Ripley (1996) small plastic tray or bag or concrete basins and earthern pots, can be used for such purpose. The nutrients required for growing *Spirulina* in batches or as discontinuous culture are supplied in prepared doses. The harvest can be carried out every day or so until the culture comes to senescence phase for lack of nutrients.

Generally, the nutrient medium contains (gm/litre) $NaHCO_3$ 15.0, K_2HPO_4 0.5, $NaNO_3$ 2.5, K_2SO_4 0.6, NaCl 0.2, $MgSO_4$ 0.7, H_2O 0.09, $CaCl_2.2H_2O$ 0.008, Fe-EDTA 0.1ml and the pH of the medium is maintained at 9.0 to 11.0.

The algal biomass is harvested by filtering the suspension through filter cloth having mesh size of 10-100µ. Then it is repeatedly washed with fresh water to remove salts and dried under sunlight by spreading over polyethylene sheets. The dry algal flakes are collected and stored for use as feed.

Hyper-size unit

Seawater is also used for hyper-sized *Spirulina* farm. Presently, this is being carried out on Hainan Island by the South China Sea Institute of Oceanology. During such cultivation, the exhausted sea water medium is usually not recycled into the cultures. The water is enriched with $NaHCO_3$, KNO_3 or Urea, K_2HPO_4 or H_3PO_4 and Fe-EDTA. Addition of such salts is made daily so as to maintain nutrient concentration (mg/l) as 800-1000 (HCO_3), 6-10 (PO_4), 30-50 (Urea), and 1000-1500 (NO_3)

Spirulina culture pond is usually less than 0.5m in depth. Wheel type paddle agitator makes the provision for continuous mixing of water. The mixing by flow is usually more practical and economical than paddle agitator.

After harvesting, the biomass is processed for powder preparation (Fig. 20.1). Before powdering, the algal biomass is ruptured by homogenization or sonication. The fluid is pasteurized to eliminate any microbial contamination and then spray - dried to the final product.

As described above, out-door pond system to produce *Spirulina* is restricted for certain limited purposes like food supplements, food coloring agents, phycobiliprotein isolation (used as fluorescent markers in diagnostics), and as aquaculture feeds for juvenile shellfish and fish fry. However, for certain value added metabolites like deuterated hydrocarbons used as high temperature lubricants, potential algal pharmaceuticals, xanthophyll pigments and the inocula needed for large-scale outdoor production systems. It is essential to go for closed photobioreactors, which allow greater control over the physical, chemical and biological environment and microbial contamination. Plenty of literature is available on application and development of closed photobioreactors. Cook (1950) had taken first credit to develop vertical tubular reactor for algal mass culture system. Davies *et. al.* (1953) developed fermenter with inserted lights. Besides cost factor, this system did not work for significant operating limitations. Various workers incorporated

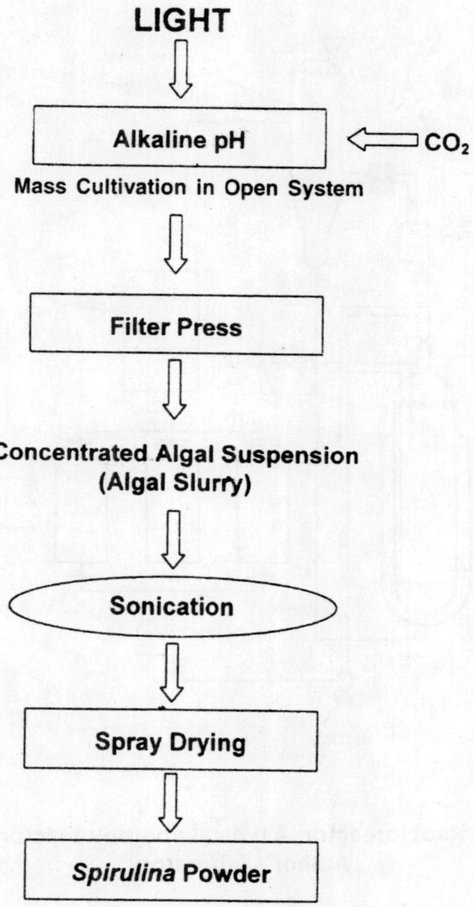

Fig. 20.1: Industrial cultivation and processing of *Spirulina*

subsequently many more modifications. These include tubular photobioreactor made of very narrow tubes; outdoor photobioreactors with either collapsible or rigid tubes (Pirt *et al.*, 1983), horizontally arranged photobioreactor (Miyamoto *et. al.*, 1988), helical tubular photobioreactor (Borowitzka, 1989), light transfer and growth kinetics in photobioreactor (Cornet *et. al.*, 1992), and two plane tubular photobioreactor (Torzillo *et. al.*, 1993).

Schematic diagram of a typical bioreactor is given in the Fig. 20.2. Natural sunlight or artificial light (mostly for indoor

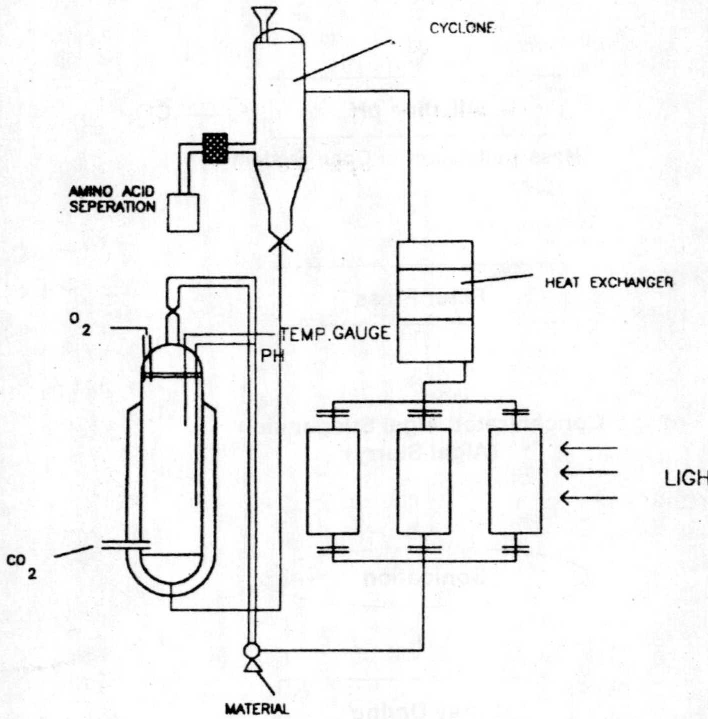

Fig. 20.2: Photobioreactor: A typical photobioreactor present in Author's laboratory

bioreactors) is provided for algal cultures. Sunlight is the cheapest source of light for such purpose as artificial light is too expensive. The fabrication of closed algal systems requires a very short light path to the algae. This has been achieved mainly by using flat transparent panels, tubes introducing light *via* fiber optics and other means. Sunlight receiving type of bioreactor is designed towards orientation of maximum sunlight. The tubular reactor orientation is fixed. In such reactor, provision of light is given so that around the tube a uniform light intensity is being maintained. In flat inclined plate reactor such problem is overcome as the plate reactor mounted over a panel, which can change orientation during daylight.

For better productivity, algal culture inside the bioreactor is to be mixed properly. The culture has to be circulated by using peristaltic diaphragm, tube or centrifugal pumps for this purpose.

Temperature control is achieved by either incorporating a heat exchanger into the design, and/or by evaporative cooling by spraying water on to the surface of the reactor. Evaporative cooling is supposed to be economical as compared to heat exchanger. Heat exchanger system is required for colder climates. In close photobioreactors, provisions for removal of O_2 and perfusion of CO_2 are provided for better growth and development of algae.

During fabrication of photobioreactor, selection of materials to be used in reactor fabrication is important and compatibility of materials with the algae should be maintained. Leaching of toxic chemicals, heavy metals and corrosion of photobioreactors are some of the major avoidable factors while designing a photobioreactor. Glass is an excellent material for large commercial scale tubular reactors but it is too expensive. Costly silicone or teflon tubing can be used in photobioreactor designing. The PVC tubing is the most economical.

Value-added biochemicals

Not much differences have been found in the nutritional qualities of the *Spirulina* grown in the traditional way or in an artificial medium in the laboratory. Dried alga contains 65-68% protein (Clement *et. al.*, 1967), 6 to 7% lipid and 4% nucleic acids, 0.8% chlorophyll, 0.23% β-carotene, 0.12- o.15% xanthophylls and 12-15% phycobilins. Plenty of data are available related to beneficial effect of various biochemicals and minerals present in *Spirulina* (Table 20.1).

Amino acid composition

The amino acid spectrum of *Spirulina* protein is similar to that of other microorganisms (Jones, 1974; Waslien, 1975). The microalgae are typically low in concentration of the sulfur containing amino acids, cysteine, methionine and the basic amino acid lysine (Clement *et. al.*, 1967, Aaronson and Dubinsky, 1982; Becker and Venkataraman, 1982; Ciferri, 1983; Ciferri and Tiboni, 1985; Becker, 1986; Richmond, 1988). As per the specification with protein standards specified by Food and Agriculture Organization (FAO) for egg albumin, milk and meat protein, *Spirulina* protein is inferior.

Table 20.1: Composition of value added biochemicals in *Spirulina* and their beneficial effects

Type of Chemical	*Qty. in Percentage/dry Weight*	*Benefits*	*Reference*
Protein	65 – 68 (Clement *et.al.*,1967)	For combatting malnutrition treatment of protein energy malnutrition	Narsimha *et al.*, 1982
		Treatment of Kwashiorkor and Marasmus disease	Ripley, 1996
Polysaccharides	10-18 (Ciferri,1983)	Enhancement of endonuclease activity and repair DNA synthesis	Becker and Venkatraman. 1984
Lipids	6-7(Clement *et.al.*, 1967)	Lowers serum cholesterol, triglyceride and LDL (undesirable fat)	Nayaka, 1988
		Plays role in the improvement of AIDS conditions	Begin and Das , 1986
γ–linolenic acid	1 – 1.5 (Mahajan and Kamat, 1995)	Controls cholesterol level in human body	Mahajan and Kamat, 1995
		Relieving human pain of tropical pancreatitis and rheumatoid arthritis	Chaturvedi and Habib, 1999
Ash	4 –5 (Ciferri, 1983)	Provide minerals that regulate various metabolic activities of human body	Narsimha *et al.*, 1982
Crude fibre	3.0 – 8.0 (Ciferri, 1983)	Treatment of obesity	Becker , 1986
β-carotene	0.23 (Santillan, 1982)	Good source of vitamin A, Increase serum retinal level. Prevent blindness and eye disease. Increase serum retinal level.	Robert, 1991 Annapurna, 1991 Seshadri, 1993
B-12	0.001(Santillan, 1982)	To help support body's defense, Essential for healthy nerves and tissues	www.spirulina.com
Chlorophyll	0.8 (Santillan, 1982)	Natural cleanser and referred as nature's green magic	www.spirulina.com

However, it is superior than other vegetable sources e.g. wheat, rice and legumes. The total protein content of *Spirulina* is higher than vegetable protein, milk, meat and egg (Belitz and Grosch, 1999).

Fatty acids and lipids

Considerable work has been done to determine the lipid composition of cyanobacteria and to characterize these lipids (Murata, 1987). Significant variations have been reported in the fatty acid content. In *Spirulina platensis* and *Spirulina maxima*, free fatty acids account for 70 to 80% of the total lipids, the remaining being chiefly mono- and digalactosyl glycerides and phosphotidyl glycerol (Hudson and Karis, 1974). The presence of high concentration of γ-linolenic acid, synthesized in *Spirulina platensis* by direct destruction of linoleic acid has been known (Nichols and Wood, 1968). The high content of these two essential fatty acids are of considerable importance for human nutrition. These two acids have tremendous therapeutic value for treatment of hypertension, pre-menstrual tension, various atopic disorders and number of other conditions (Horribin and Manku, 1983).

Among the hydrocarbons, n-heptadeane is the most abundant (65 to 70%) in *Spirulina platensis* and *Spirulina maxima* (Gelpi *et. al.*, 1970; Tulliez *et. al.*, 1975; Fedeli and Farini, 1980). Cholesterol and β-sitosterol are the main sterols present in *Spirulina maxima* and *Spirulina platensis* (Pacletti *et. al.*, 1976; Claudio, 1982). Poly-β-hydroxy-butyrate (PHB) granules, a reserve of carbon and energy, are present in *Spirulina* (Campbell *et. al.*, 1982). This compound accumulates during exponential growth phase to a concentration corresponding to 6% of the dry weight.

Chlorophyll-a is the most abundant in *Spirulina*. Myxoxanthophyll and β-carotene are the major carotenoids (0.2 to 0.4 % of the dry weight, Claudio, 1982). *Spirulina* contains a branched polysaccharide forms 15-20% of the dry weight, composed of only glucose and structurally similar to glycogen (Casu *et. al.*, 1980). The phycobiliproteins have made the most impact in the market place among cyanobacterial pigments. They have been used both as tools for basic research and medical diagnostics as well as food grade dyes.

Spirulina has been reported to contain 10 sterols, with clinasterol as the major constituent. An amount of 20mg/kg cholesterol in

Spirulina platensis and 149mg/kg in *Spirulina maxima* have been reported (Materassi *et. al.*, 1980).

Spirulina as food

Spirulina has been grown either photoautotrophically or heterotrophically with a high demand for food/feed source of biochemicals. Many companies are now engaged in manufacturing *Spirulina* on a large scale.

Spirulina biomass is sold mainly to the health food market in the form of either powder or pills or tablets. *Spirulina* powder is generally mixed with variety of food products including ice cream, cake etc.

Cereals like rice or wheat are mostly deficient in lysine and threonine. The quality of such food can be upgraded by supplementation of *Spirulina* powder. Currently gamma linobenic acid (GLA) is extracted commercially from higher plant oils, especially from *Oenothera piennis* (evening primrose). Although *Spirulina* biomass has been suggested as an economic alternative to GLA. Production of GLA from *Mucor rouxii* is about fifteen times higher than *Spirulina*. The only benefit of *Spirulina* appears to be its suitability for direct consumption, rather than as a source for extraction and purification of GLA.

Animal feed

It has been very successfully demonstrated that *Spirulina* can safely be used as conventional protein source in livestock feed. Becker and Venkataraman (1984) noticed that poultry, pigs and ruminants can safely consume *Spirulina* upto 10% of their daily diet. Higher than this may adversely reduce the growth rate.

Future scope

The future of utilization of *Spirulina* depends on expansion of the range of products to be manufactured from this tiny alga under economic budget. Molecular biology and genetics of *Spirulina* hold promise for new products and improved yield. At present, it is very difficult to say whether cyanobacteria would be in a position to compete *E.coli* or yeast strains in heterotrophic expression of proteins. Many new ideal concepts are yet to be developed to exploit maximum benefit from cyanobacteria as presently being taken from *E.coli* and yeast. The most important in this field is to think for production of secondary metabolites having role in pharmaceutics, food processing , textile and cosmetic industries.

Cyanobacteria have very short life cycle. This quality gives cyanobacteria a competitive status with reference to life cycle of higher plants. Besides this, cyanobacteria are oxygenic autotrophs that are better suited to indoor culture systems.

The extensive coastal belt of India and brackish lakes provide an access to cultivation of various marine algae, especially cyanobacteria to compete world market for single cell protein and various secondary metabolites of commercial value.

References

Aaronson, S. and Dubinsky, Z. 1982. Mass production of microalgae. *Experientia* **38**: 6-10.

Annapurna, V . 1991. Bioavailability of *Spirulina* carötene in pre school children. *J.Clin. Biochem. Nut.* **10**: 145-151.

Anusuya, D.M., Sabbulakshmi, G., Madhavi, K. and Venkataraman, L.V. 1981. Studies on the proteins of mass cultivated, blue green alga (*Spirulina platensis*). *Agri. Food Chem.* **29**: 522-525.

Becker, E.W. 1986. Nutritional properties of microalgae: potentials and constraints, In: *Handbook of Microalgal Mass Culture*, Edited by Richmond, A., CRC Press. Boca Ratan, FL. pp. 339.

Becker, E.W. 1994. *Microalgae: Biotechnology and Microbiology*. Cambridge University Press, Cambridge.

Becker, E.W. and Venkataraman, L.V. 1982. *Biotechnology and Exploitation of Algae – the Indian Approach*. Deutsche Gesellschaft for Technische. pp.126.

Becker, E. W. and Venkataraman, L.V. 1984. Production and utilization of the blue-green alga *Spirulina* in India. *Biomass* **4**: 105-125.

Begin, M.E. and Das, U.N. 1986. A deficiency in dietary gamma linolenic acid or eicosapentaenoic acid may determine individual susceptibility to AIDS. *Med. Hypoth.* **20**: 1-8.

Belitz, H.D. and Grosch, W. 1999. *Food Chemistry*. Springer-Verlag, Berlin. pp. 473-693.

Borowitzka, M.A. 1998. Tubular photobioreactors for large scale algal culture. In: *Cyanboacterial Biotechnology*. Edited by Subramanian, G., Kaushik, B.D. and Venkataraman, G.S. Oxford & IBH Publ. Co., New Delhi. pp. 249-257.

Brandily, M.Y. 1959. Depuis des lustres une tribu primitive du. Tchad exploite la nourriture de I'an 2000. *Sci. Aren.* **152**: 516-19.

Brown, I.I. 1998. The alternative bioenergetic patterns in cyanobacteria as factors of biochemical adaptation. In: *Cyanobacterial Biotechnology* . Edited by Subramanian, G., Kaushik, B.D. and Venkataraman, G.S. Oxford & IBH Publ. Co., New Delhi. pp 21-34.

Busson, F. 1971. *Spirulina platensis* (Gomi) Geitleret Spirulina Geitler J. de Toni. Cyanophycees Alimentaires. Service de Sante. Parc du Pharo, Marseille.

Campbell. J., Stevens, S.E.J.R. and Balkwill, D.L. 1982. Accumulation of poly-β-hydroxybutyrate in *Spirulina platensis. J. Bacteriol.* **149**: 361-363.

Casu, B., Naggi, A. and Vercellitti, J.R. 1980. Polysaccaridi di riserva dela *Spirulina platensis*, estrazione e charatterizziove. In: Prospective della coltura di *Spirulina* in Italia. Edited by Materassi, R., Rome. pp. 145-153.

Chaturvedi, U.K. and Habib, I. 1999. Cyanobacteria as a source of food. *Proc. Acad. Environ. Biol.* **8**: 241-245.

Ciferri, O. 1983. *Spirulina*, the edible microorganism. *Microbiol. Rev.* **47**: 551-578.

Ciferri, O. and Tiboni, O. 1985. The biochemistry and industrial potential of *Spirulina. Ann. Rev Microbiol.* **39**: 503-526.

Claudio, S. 1982. Mass production of *Spirulina*. *Experientia* **38**: 40-43.

Clement, G., Giddey, C.and Menzi, R. 1967. Amino acid composition and nutritive value of the algae *Spirulina maxima. J. Sci. Food Agric.* **18**: 497-501.

Cook, P.M. 1950. *Culturing of Algae.* Edited by Prescott, G.W. and Tiffany, L.H., Charles F. Kettering Foundation. pp. 53-114.

Cornet, J.F., Dussap, C.G., Cluzel, P. and Dubertret, G. 1992. Structural model for simulation of cultures of the cyanobacterium *Spirulina platensis* in photobioreactors. II Identification of kinetic parameters under light and mineral limitations. *Biotech. Bioeng.* **40**: 826-834.

Davis, E.A., Dedrick, J., French, C.S., Milner, H.W., Myers, J., Smith, J.H.C.and Spoehr, H.A. 1953. Laboratory experiments on *Chlorella* culture at the Carnegie Institution of Washington, Department of Plant Biology. In: *Algal Culture from Laboratory to Pilot Plant*. Edited by Burlew, J.S., Carnegie Institution of Washington Publ. 600, Washington. pp. 105-153.

FAO, WHO 1973. *Energy and Protein Requirement*. Joint FAO/WHO Nutrition Report Series NC 52, FAO.

Fedeli, E. and Favini, G. 1980. Caratterizzazione e significato biologico dei componenti lipidici e contaminanti ambientali delle biomasse di *Spirulina* pp. 171-178.

Gelpi, E., Schneider, H., Mann, J. and Oro, J. 1970. Hydrocarbons of geochemical significance in microscopic algae. *Phytochemistry* **9**: 603-612.

Horribin, F. and Manku, M.S. 1983. The seed oil of evening primrose (*Oenothera piennis*): a new oil seed crop. In: *Fats for the Future.* Edited by Brooker, S.G., Renmick, A., Hannan, S.F. and Eyres, L. Duromark Publ. Auckland. pp 99-101.

Hudson, B.J.F. and Karis, I.G. 1974. The lipids of the alga *Spirulina. J. Sci. Food Agric.* **25**: 759-763.

Jones, A. 1974. *World Protein Resource.* MTP Publishers, Lancaster, U.K. pp. 305.

Kay, R.A. 1991. Microalgae as food and supplement. *Crit. Rev. Food Sci. Nut.* **30**: 555-573.

Landsberg, H.E. 1966. *Global Distribution of Solar and Sky Radiation in World Maps of Climatology.* Edited by Rondenwalt, E. and Jusatz, H.J. Springer, Berlin.

Mahajan, G. and Kamat, M. 1995. Linolenic acid production from *Spirulina platensis. Appl.Microbiol. Biotechnol.* **43**: 466-469.

Maio, J.R. 1987. *Spirulina in Jiangxi China.* Academy of Agricultural Science. Soc. Appl. Algology, Lille, France.

Materassi, R., Paoletti, C., Balloni, W. and Florenzano, G. 1980. Some considerations on the production of lipid substances by microalgae and cyanobacteria. In: *Algal Biomass*. Edited by Shelef, G. and Soeder, C.J., Elsevier/North–Holland Biomedical Press. pp. 169.

Miyamoto, K., Wable, O. and Benemann, J.R. 1988. Vertical tubular reactor for microalgae cultivation. *Biotech. Lett.* **10**: 703-708.

Murata, N. 1987. Unique characteristics of cyanobacterial glycerolipids. In: *The Metabolism, Structure and Function of Plant Lipids*. Edited by Stumph, P.K., Mudd, J.B. and Nes, W.D., Plenum Publ.Corp. pp. 603.

Narasimha, D.L.R., Venkataraman, G.S., Duggal, S.K, and Bjorn, O. E 1982. Nutritional quality of the blue-green alga *Spirulina platensis* Geitler. *J. Sci. Food Agric.* **33**: 456-460.

Nayaka, N. 1988. Cholesterol lowering effect of *Spirulina. Nutr. Rep.Int.* **37**: 1329-1337.

Nichols, B.W. and Wood, B.J.B. 1968. The occurrence and biosynthesis of gamma-linolenic acid in a blue-green alga *Spirulina platensis. Lipids* **3**: 46-50.

Paoletti, C., Pushparaj, B., Florenzano, G., Capella, P. and Lercker, G. 1976. Unsaponifiable matter of green and blue green algal lipids as a factor of biochemical differentiation of their biomass. II Terpenic alcohol and sterol fraction. *Lipids* **11**: 266-271.

Pauw, D. and Pruder, G. 1986. Use and production of microalgae as food in Aquaculture. In: *Realism in aquaculture: Achievements, Constraints, Perspective*. Edited by Bilo, M., Rosental, H. and Sindermann, G.J., Bredeve, European Aquaculture Society. pp. 77-106.

Pirt, S.J., Lee, Y.K., Walach, M.R., Pirt, W.M., Balyuzi, H.H.M. and Bazin, M.J. 1983. A tubular bioreactor for photosynthetic production of biomass from carbondioxide: design and performance. *J. Chem. Tech. Biotechnol.* **3313**: 35-58.

Richmond, A. 1988. *Spirulina*. In: *Microalgal Biotechnology*. Edited by Borowitzka, M.A. and Borowitzka, L.J., Cambridge University Press, Cambridge.

Riple, D. F. 1996. *Spirulina Production and Potential*. Edisud, La Calade R.N. 7, 13090 Aix-en-Province. France.

Seshadri, C.V. 1993. *Large scale Nutritional Supplementation with Spirulina alga.* All India Coordinated Project on *Spirulina*, MCRC, Madras, India.

Skulacher, V.P. 1994. Bioenergetics: the evolution of molecular mechanisms and the development of bioenergetic concepts. *Antonie van Leevwenhoek* **65**: 271-284.

Torzillo, G., Carlozzi, P., Pushparaj, B., Montaini, E. and Materassi, R. 1993. A 2-plane tubular photobioreactor for outdoor culture of *Spirulina. Biotechnol. Bioeng*. **42**: 891-898.

Tulliez,J.G., Borries, B.C. and Fevrier, C. 1975. Les hydrocarbures des algues spirulines: nature, etude du devenir del'heptadecane Chez le rat et le pore. *Ann. Nutr. Ailment* **29**: 563-572.

Waslien, C.I. 1975. Unusual sources of proteins for man. *Crit. Rev Food Sci.* **6**: 77-151.

Chapter 21

NITROGEN FIXING CAPACITY OF SOME SELECTED BGA ISOLATED FROM THE RICE FIELD SOILS OF NORTH EASTERN INDIA

N. Irabanta Singh, H. Dorycanta and G. A. Devi

Abstract

Twenty blue-green algal (BGA) strains commonly isolated from the rice field soils of North Eastern India were studied for their nitrogenase. The highest activity was recorded in *Anabaena spiroides* (2.15 mg N/g alga/hr) whereas the lowest activity was recorded in *Dichothrix* sp. (0.30 mg N/g alga/hr). Further, wide variations in N_2 fixing capacity were observed among the species of *Anabaena* (0.36-1.87 mgN/g alga/hr) and *Aulosira* (1.05-1.98 mg N/g alga/hr) whereas the isolates of *Calothrix* (0.43-0.46 mg N/g alga/hr.), *Cylindrospermum* (0.53 mg N/g alga/hr) and T*olypothrix* (0.64-0.72 mg N/g alga/hr) were more or less consistant in their performance.

Key words: Acetylene reduction, bio-fertilizers, blue-green algae, nitrogen fixation, rice fields

Introduction

The states of Arunachal Pradesh, Assam, Manipur, Meghalaya, Mizoram, Nagaland and Tripura occupy a total geographical area of 2.55 lakh Km^2. The area is represented by wide variations in altitude, topography and agroclimate. As such, the summer and winter temperature in the region ranges from 28-30 ^{0}C and –1 to 5°C respectively. The region receives the highest annual rainfall in the world. Thus, heavy rainfall, high humidity, moderate range of temperature, variation in altitude and physiography make the region one of the favourite dwelling place for a variety of fauna and flora. Among these, BGA are regarded as renewable sources of biological nitrogen for their ability to fix atmosphere nitrogen and sustaining soil fertility (Singh, 2001).

In BGA, the energy source for nitrogen fixation is the sun. The enzyme nitrogenase can reduce a variety of compounds like hydrogen cyanide, hydrogen azide, nitrous oxide and acetylene. Earlier, the occurrence and distribution of different BGA from the rice field soils of North Eastern India had been worked out (Annonymous, 1996; 1997). The present paper deals with nitrogenase activity of some selected BGA isolated from the rice field soils of North Eastern India for their possible use as renewable nitrogen source.

Materials and Methods

The method used in the study for measuring nitrogen fixation was a slight modification of the technique adopted by Hardy *et. al.* (1973). In the original method, N_2 fixation was measured in terms of ethylene (C_2H_4) produced by the reduction of acetylene (C_2H_2) and expressed in n mole C_2H_4/mg chl/hr . Standardization of the peak was done by using C_2H_4 formed during the process. However, in the present investigation, standardization of C_2H_2 peak, in which the gas C_2H_2 used was of the same grade as that of the gas used for reduction during N_2 fixation. Thus, N_2 fixation was measured directly in terms of acetylene (C_2H_2) reduced, rather than in terms of ethylene (C_2H_4) formed, and expressed in mg C_2H_2/g alga/hr (1g alga = dry weight of alga used).

Fifteen days old algal cultures were used for the experiment. A known volume of algal sample was introduced in a glass serum bottle and plugged. The gas phase was removed equivalent to 10% of the remaining volume of the bottle and injected an equivalent volume of acetylene. The samples were incubated for 30 minutes in continuous fluorescent light (2-3K lux) and a temperature of 28+1°C. The sample was inactivated after 30 minutes by injecting 0.5 ml of 5N H_2SO_4. A gas sample of 1.0 ml was withdrawn from the serum bottle with an air tight syringe and injected in the chromatograph.

In order to remove more than one sample of gas phase for analysis, positive pressure in the vial was built up by injecting equal volume of water into the vial. Amount of N_2 fixed by a particular BGA was measured by calculating the area of the curve drawn with the help of a recorder.

Results and Discussion

Twenty blue-green algal strains were studied for their nitrogenase activity. Out of these, the highest activity was recorded by *Anabaena spiroides* (Table-1) followed by *Anabaena. oryzae, Anabaena. doliolum, Aulosira fertilissima, Nostoc muscorum, Nostoc. linckia, Anabaena torulosa, Anabaena oscillarioides*, etc. whereas the lowest activity among BGA strains tested was shown by *Dichothrix* sp. (Table 21.1).

Nitrogen fixing capacity in different species of the same genus was found to be variable. Wide variations were observed among the species of *Anabaena* (0.93-2.15 mg N/1g alga/hr), *Nostoc* (0.36-1.87 mg N/1g alga/hr) and *Aulosira* (1.05-1.98 mg N/1g alga/hr), while the isolates of *Calothrix, Cylindrospermum* and *Tolypothrix* spp. were fairly consistant in their performance (Table 21.1).

Kolte and Goyal (1986) also observed wide variations in N_2 fixing capacity among 62 cyanobacterial forms belonging to 9 genera ranging from 0.33-2.45 mg/50 ml algae/ 28 days in *Anabaena*, 0.77-5.83 mg/ 50 ml alga/28 days in *Calothrix*, 0.31-1.15 mg/50 ml alga/ 28 days in *Hapalosiphon*, 0.44-4.04 mg N/50 ml alga/28 days in *Nostoc* , 0.30-3.41 mg N/50 ml alga/28 days in *Scytonema*, 2.55-5.6 mg N/50 ml alga/28 days in *Tolypothrix* and 0.6-2.34 mgN/50ml alga/28 days in *Westiellopsis*.

Among 35 strains of BGA studied from Manipur, *Anabaena spiroides* was found to be the most efficient nitrogen fixer (2.15 mg N/1g alga/hr; Devi, 1997). The same finding was also observed from 27 isolates of Mizoram (2.11 mg N /1g alga/hr; Dorycanta, 1996). In Sweden, *Anabaena* and *Nostoc* were found to be the most potential N_2 fixing algae (Henriksson, 1972) whereas in Japan, *Tolypothrix* was observed as the most efficient nitrogen fixer (5.2 mg N/100 ml of the medium; Watanabe and Yamamoto, 1971). However, Suseela and Goyal (1995) found *Calothrix* as better nitrogen fixer than *Nostoc* and *Anabaena*.

In all these reports, the maximum nitrogen fixing BGA type was different and in the present finding too, the maximum nitrogen fixing BGA type was different, though of the same genus.

Table 21.1: Nitrogenase activity of some selected BGA strains isolated from rice field soils of North Eastern Region of India

Sl.No.	BGA type	Nitrogen fixed (NF) in term of C_2H_2 reduction as mg N/g alga/hr					
		Arunachal Pradesh	Manipur	Mizoram	Meghalaya	Nagaland	Tripura
		NF	NF	NF	NF	NF	NF
1.	*Anabaena spiroides*		–	2.15	2.11	2.11	2.10
2.	*A.oryzae*	2.06	2.06	2.00	2.09	2.06	2.06
3.	*A.variabilis*	1.33	–	1.33	1.03	1.33	1.33
4.	*A.torulosa*	1.43	1.43	1.43	1.43	1.43	1.43
5.	*A.doliolum*	2.01	–	2.05	2.01	2.01	–
6.	*A.oscillarioides*	1.25	1.26	–	–	–	–
7.	*A.cylindrica*	0.93	0.93	0.93	0.96	0.93	–
8.	*Aulosira fertilissima*	1.11	1.98	1.05	1.05	1.11	1.11
9.	*Calothrix marchica*	0.46	0.43	0.46	0.46	0.46	0.46
10.	*Cylindro–spermum sp.*	0.53	0.53	0.53	0.53	0.53	–
11.	*Dichothrix sp.*	0.30	0.31	–	–	0.30	0.31
12.	*Hapalosiphon welwitschia*	0.99	0.99	0.99	0.99	0.99	0.99
13.	*Nostoc punctiforme*	0.36	0.36	0.36	0.36	0.36	0.36
14.	*N.linckia*	1.41	1.45	1.40	1.40	1.41	1.41
15.	*N.muscorum*	1.81	1.87	1.83	1.82	1.80	1.81
16.	*N.calcicola*	0.82	0.82	–	0.78	0.82	–
17.	*N.paludosum*	0.76	0.76	–	0.76	0.76	0.76
18.	*Tolypothrix tenuis*	0.72	0.72	0.64	0.64	0.72	0.72
19.	*Westiellopsis prolifica*	0.77	0.77	0.77	0.76	0.77	0.76
20.	*Scytonema sp.*	0.90	1.01	0.89	0.89	0.90	0.89

Acknowledgement

The senior author records with thanks the financial assistance accorded to him by DBT and CSIR.

References

Anonymous 1996. *Studies on blue green algae from rice field soils of North-Eastern India for biofertilizer technology*. Department of life Sciences, Manipur Univ. Project completion report submitted to DBT, Government of India, New Delhi-3.

Annonymous 1997. *Technology development and demonstration project on biofertilizers (BGA and Azolla) for North East India. Department of Life Sciences,* Manipur University. Project completion report submitted to CSIR, Government of India, New Delhi-12.

Devi, G.A. 1997. *Nitrogen fixing blue green algae from rice field soils of Manipur and their possible use as biofertilizer.* Ph.D. Thesis, Department of Life Sciences, Manipur University, Imphal, 795003. India.

Dorycanta, H. 1996. *Nitrogen fixing blue green algae from rice field soils of Mizoram and Nagaland for biofertilizer technology*. Ph.D. Thesis Department of Life Sciences, Manipur University, Imphal-795003. India.

Hardy, R.W.F., Burns, R.C. and Hotsten, R.D. 1973. Application of acetylene ethylene assay for measurement of nitrogen fixation. *Soil Biol. Biochem.* **5**: 47-81.

Henriksson, E. 1972. Nitrogen fixation in Swedish soil by blue green algae.In : *Taxonomy and Biology of Blue-green Algae.* Edited by Desikachary,T.V. University of Madras (India) pp. 269-273.

Kolte, S.O. and Goyal, S.K. 1983. Effect of ammonium nitrogen on growth and uptake and fixation of nitrogen by blue green algae. *Acta Bot. Ind.* **17**: 172-176.

Singh, N.I. 2001. Studies on cyanobacterial microflora isolated from the rice field soils of North Eastern India. In: *Three decades of Research in Bio-fertilizers and organic farming in North East India.* Edited by Yadava, A.K. and Raychoudhury, S. Regional Bio-fertilizer Development Centre, Imphal , Manipur. pp. 72-83.

Suseela, M.R. and Goyal, S.K. 1995. Growth and nitrogen fixing potential of cyanobacteria from the east coast of India. *Phykos* **34**: 131-134.

Watanabe, A. and Yamamoto, Y. 1971. Algal nitrogen fixation in tropics. *Plant Soil. Special Vol.* pp.403-413.

Section V
CULTURE METHODS & TECHNIQUES

Chapter 22

CULTURE AND CULTIVATION OF MARINE ALGAE

V. Krishnamurthy

Abstract

This review deals with the cultural requirements for laboratory cultivation of algae in general and marine algae in particular and the possible culture media that can be used for marine algae. Also dealt with are the principles involved, and the problems pertinent to cultivation of marine algae on a large scale under field conditions. The cultivation methods adopted for different marine algae are described.

Key words: Algae, culture media, cultivation methods, *Porphyra*, seaweeds

Introduction

The successful culture of any photosynthetic organism such as the algae depends on (1) temperature, (2) nature, intensity and duration of illumination, (3) a source of carbon, (4) major nutrients, *viz.*, potassium, nitrogen and phosphorus, (5) microelements provided, (6) growth factors and vitamins and (7) an atmosphere free of contamination of any kind.

The literature on the general methods of laboratory culture has been reviewed by Bold (1942), Pringsheim (1949), Lewin (1959) and Myers (1962). A review of laboratory culture of seaweeds was given by Subbaramaiah and Krishnamurthy (1967).

Cultural requirements

A prerequisite for nutritional studies is to obtain the organism in bacteria-free (axenic) culture. The bacterial contamination is usually carried along with the alga, and in some cases the mucilaginous envelope of the alga harbours bacteria. Elimination of this bacterial contamination is affected by treatment with ultraviolet radiation (Gerloff *et. al.*, 1950), antibiotics (Spencer, 1952), Jadopax solution (Fries, 1963) and a mixture of detergent and phenol (McDaniel *et. al.*, 1962). Ultrasonication combined with antibiotics, *viz.* penicillin G, chloramphenicol, polymyxin and neomycin, yielded bacteria-free cultures of *Ulva* (Provasoli, 1958b). Fries (1963) obtained bacteria-free cultures of some red algae by treatment with a solution of Jadopax (active part is 5 per cent iodine present in it) alone or in combination with antibiotics.

A special study of the temperature requirements for the culture of most seaweeds has not been attempted. Cultures have been carried out under various temperature conditions, mostly between 13°C and 20°C. It has been ascertained for the Conchocelis phase of *Porphyra* that germination of spores takes place between 6°C and 14°C (Yendo, 1919), vegetative growth is active at 19-20°C (Iwasaki, 1965) and sporulation is active between 18°C and 21°C, and stops at 25°C (Kurogi and Hirano, 1956). The leafy phase of *Porphyra* shows good growth at 14-15°C (Iwasaki, 1965). Similar studies do not appear to have been made for other larger seaweeds and it would be desirable to obtain information on temperature control of vegetative growth and spore formation. An illumination of 800 – 10,000 lux supported good growth of several algae (Fries, 1963).

The nutritional requirements of the marine algae have been reviewed by Provasoli *et. al.* (1957), Provasoli (1963) and Hutner and Provasoli (1964). Culture media which have a composition similar to the natural environment have the best chances of success in the laboratory (Provasoli, 1958a). Most intertidal algae survive salinity ranging from 0.1 to 3.0 times that of sea water (Biebl, 1962) and pH ranging from 6.8 to 9.8 (Kylin, 1927). The red alga *Porphyridium cruentum* has been grown in axenic culture in a completely artificial sea water medium with salinities ranging from 4.5 to 35 ppt and pH range from 5.2 to 8.3 (Jones *et. al.*, 1963). The response of some euryhaline algae to varying range of ionic concentration is rather one of sodium tolerance than that of salinity tolerance (Droop, 1958).

Apart from the major nutrient elements, the micronutrients Fe, Zn, Mn, B, Mo, Co, Cu, V and I are essential for seaweed growth. Some brown seaweeds concentrate heavy metals, the concentration factor ranging from 100 to 30,000 (Black and Mitchell, 1952). The uptake of elements is facilitated by waves and currents. Further, plant hormones and vitamins (B_{12}, thiamine and biotin) are important for the growth of some seaweeds.

A chelator is generally used in synthetic medium containing trace metals. This is usually in the form of ethylenediamine tetra-acetic acid (EDTA). Other chelators used are citric acid and N-N-di(2-hydroxyl ether) glycine. During the growth of an alga in culture, pH changes are likely owing to differential intake of ions by the alga and due to exudates from the alga itself. Such changes in pH may inhibit growth and hence a buffer system should be built into the culture medium. The most common buffer system used by earlier workers consisted of a proportionate mixture of mono and di-hydrogen phosphates and such phosphate buffer systems are still used in a number of laboratories. In more recent practices, the buffer used is either tris –(hydroxymethyl) aminomethane (TRIS) and glycyl glycine (Droop, 1961).

The most used culture medium for larger seaweeds is Erdschreiber solution developed by Foyn (1934) for *Cladophora* and *Ulva*. The variability in the inorganic and organic micronutrients of sea water and the soil extract, and the precipitation during sterilization are the drawbacks of the medium. Various modifications over this medium such as addition of trace metals with or without a chelator (Barker, 1935; Sweeney, 1951; Spencer, 1954), pH buffering, enrichment with vitamins and other organic nutrients were made (Sweeney, 1954). Precipitation was sought to be eliminated by resorting to filter sterilization of the sea water.

In all these media, unknown entities were involved in the use of natural sea water , and hence repeatability of results was doubtful. It was, therefore, necessary to design purely synthetic media. The precipitation of the medium and the soil extract were dispensed with in synthetic media, and the control of trace metals was affected by the use of EDTA, citric acid and N-N-di(2-hydroxyl ether) glycine as chelators, and the control of pH by the use of buffers such as tris-(hydroxymethyl) aminomethane and glycyl glycine (Droop, 1961).

Provasoli (1958) found that synthetic media permit slow callus-like growth of *Ulva lactuca* and for near normal morphology, sea water enriched with N, P, trace metals and vitamins, to which adenine and kinetin were added, was required. Here the morphogenetic substances are not B vitamins. Vitamin B_{12} was, however, required by certain red algae, *Goniotrichum elegans, Nemalion multifidum* (Fries, 1960), *Antithamnion* spp. (Tatewaki and Provasoli, 1964), *Bangia fuscopurpurea* (Provasoli, 1963), Conchocelis of *Porphyra tenera* (lwasaki, 1965). The latter found that different B_{12} analogues, plant hormones (kinetin, adenine, indole acetic acid and gibberellin), purines (xanthine, hypoxanthine and guanine) and pyrimidines (uracil, methyl cytosine and thymine) gave positive results, and all the analogues containing benzimidazole could replace B_{12}.

Commonly used media for marine algae

A number of culture media have been developed by various authors and used with varying success. The media used are of two kinds: (1) Those using natural seawater as the base, to which enrichments in the form of major nutrients, trace elements, vitamins and growth factors were added. The earliest such medium is that of Foyn (1934) and a more recent one is the enriched seawater of Provasoli, PES medium (Provasoli, 1963, 1968), (2) Synthetic or artificial seawater media, where a seawater substitute containing the major salts present in seawater in definite proportions simulating that found in natural seawater is prepared, to which are added trace elements, vitamins and growth factors. In both approaches, chelators and buffers are used. Provasoli *et. al.* (1957) have given an account of the development of culture media for marine organisms and have given a number of formulae for both seawater based and artificial seawater media (Provasoli, 1963).

If it is intended to keep larger marine algae growing in the laboratory for short periods. They can be kept in filtered seawater to which enrichments in the form of potassium nitrate and potassium monohydrogen phosphate are added. For prolonged storage in the laboratory, the seawater may have to be changed periodically.

Following culture media (Provasoli's enriched seawater (PES) can be used for marine algae :

Provasoli's enriched seawater (PES) medium (Prevasoli, 1963)

Stock solutions (each in 100 ml distilled water)		*ml of stock solutions to be used*
Na NO_3	35 g	10
Na_2 glycerophosphate	5 g	10
Vitamins B_{12}	1 mg	10
Thaimine	50 mg	10
Biotin	0.5 mg	10
TRIS for pH 8		
(Mix 5.32 g Trizma HCl and 1.97g Trizma base)		
Fe EDTA (1:1 molar) ($Fe(NH_4)_2$ $SO_4.6H_2O$, 351 mg + Na_2 EDTA 300 mg / 500 ml		250
P II Trace metals		250
(Make total volume to 1250 ml with distilled water)		

PII Trace metal solutions:

Dissolve the following in 1 litre of distilled water

H_3 BO_3	1.14 g
Fe Cl_3. $6H_2$ O	49 mg
Mn SO_4. $4H_2O$	164 mg
Zn SO_4 $7H_2O$	22 mg
Co SO_4. $7H_2O$	4.8 mg
Na_2 EDTA	1 g

Add 20 ml of stock solution given above to 1000 ml of filtered seawater to prepare full strength PES medium.

F/2 medium (Guillard and Ryther, 1962)

Macronutrients	
Na NO_3	75 mg
Na H_2 PO_4. H_2O	5 mg
FeEDTA	5 mg
Na_2 Si O_3. $9H_2O$	30 mg
Vitamins	
Thiamine hydrochloride	0.1 mg
Biotin	0.5 mg
B_{12}	0.5 mg
Trace metals	
Cu SO_4. $5H_2O$	0.0098 mg

Zn $SO_4 . 7H_2O$	0.022 mg
$CoCl_2 . 6H_2O$	0.01 mg
$MnCl_2 . 4H_2O$	0.18 mg
$Na_2MoO_4 . 2H_2O$	0.006 mg
Sea water (of 30% salt concentration)	1000 ml

PH of the medium to be adjusted to 7.9. This medium was primarily intended for marine diatoms, but can also be used for macroalgae.

ASP-2 Artificial Seawater (Provasoli) medium (Provasoli *et. al.*, 1957)

Na Cl	1800 mg
Mg $SO_4 . 7H_2O$	500 mg
KCl	60 mg
Ca Cl_2	10 mg
$NaNO_3$	5 mg
K_2HPO_4	0.5 mg
Na_2 $SiO_3 . 9H_2O$	15 mg
TRIS buffer	100 mg
B_{12}	0.2 μg
Vitamin mix S3	1.0ml
Na_2 EDTA	3.0 mg
$FeCl_2$	0.08mg
$ZnCl_2$	15.0 mg
$MnCl_2$	0.12 mg
$CoCl_2$	0.3 μg
$CuCl_2$	0.12 μg
H_3BO_3	0.6 mg
Glass distilled water	100 ml
pH to be adjusted to	7.6 to 7.8

* 1 ml of vitamin mix S3 contains: thiamine HCl, 0.05 mg; nicotinic acid, 0.01 mg; Ca pantothenate, 0.01 mg; p-amino benzoic acid, 1.0 μg; biotin 0.1 μg; inositol, 0.5 mg; folic acid, 0.2 μg; thymine, 0.3 mg.

APS – 6 medium (Provasoli *et. al.*, 1957)

NaCl	2400 mg
$MgSO_4 . 7H_2O$	800mg
KCl	70 mg
Ca Cl_2	30 mg
K_2 glycerophosphate	10 mg
Na_2 $SiO_3 . 9H_2O$	7 mg

TRIS buffer	100 mg
B_{12}	0.05 µg
*Vitamin mix 8	0.1 ml
$FeCl_2$	0.2 mg
$ZnCl_2$	0.05 mg
$MnCl_2$	0.1 mg
$CoCl_2$	1.0 µg
$CuCl_2$	2.0 µg
H_3BO_3	0.2 mg
Glass distilled water 100.0 ml, pH to be adjusted to 7.4 to 7.6.	

* 1.0 ml of vitamin mix 8 contains; thiamine HCl, 0.2 mg; nicotinic acid, 0.1 mg; putrescine, 0.04 mg; Ca pantothenate, 0.1 mg; riboflavin, 5.0 mg; pyridoxine 2 HCl, 0.04 mg; pyridoxamine 2 HCl, 0.02 mg; p-amino benzoic acid, 0.01 mg; biotin, 0.5 µg; Choline-H_2-citrate, 0.5 mg; inositol, 1.0 mg; thymine, 0.8 mg; orotic acid, 0.26 mg; B_{12} ; 0.05 µg; folinic acid, 0.2 µg; folic acid, 2.5 µg.

Large scale cultivation of marine algae

Cultivation of marine algae for commercial exploitation has been practiced for a very long time, although scientific studies on methods of cultivation and refinement are of comparatively recent origin. From very early times, many seaweeds which were considered delicacies were kept growing in the Royal "limu" gardens in the Hawaiian islands. This practice was perhaps the forerunner of cultivation of seaweeds in other parts of the world. Japan and adjacent regions have been cultivating seaweeds for a long time and particular mention may be made of the cultivation of *Porphyra* spp. which constitute the commercial "nori" . The story of "nori" culture is one of steady progress from a primitive kind of recruitment on bamboo "hibi" to a modern, planned, scientific cultivation using the comparatively recently acquired knowledge of the biology of *Porphyra*.

Krishnamurthy (1967) discussed the necessity for cultivation of seaweeds and set down the principles underlying such cultivation as well as the problems that may develop during the process. It was shown earlier by Humm (1944) that vegetative plants of *Gracilaria confervoides* (= *G. verrucosa)* can continue to grow in a free floating state, provided environmental conditions are favourable. Krishnamurthy (1954) found the species growing vigorously under the reduced salinity conditions in the estuary of river Adyar at

Madras. Isaac (1956) cultured two inch fragments of *G. confervoides* in filtered seawater in the laboratory and showed that the fragments themselves did not grow in length but produced new branches in a characteristic manner. Results similar to these were obtained by the writer at Mandapam for *G. corticata*. Oza (1971) showed that fragments of *G. corticata* as small as 2 mm in length could regenerate and produce proliferations. Further, it was found that the fragments exhibited definite polarity, branches being produced in greater number from the morphological apical end of the fragments, while at the basal end, haptera like structures were formed. Similar results were obtained by the writer for *G. edulis*. However, growth under laboratory conditions was slow. When such fragments were attached to ropes and planted at a suitable depth in the sea, the fragments lengthened rapidly, and produced numerous branches. Some of the branches at the base adhering firmly to the rope and thus anchoring the plants. Raju and Thomas (1971) and Krishnamurthy *et. al.* (1977) described a series of experiments on the rope culture of *G. edulis* in South India.

Krishnamurthy (1967) also emphasized that a knowledge of the seaweed's biology and reproduction is fundamental to successful cultivation and enumerated six major constraints governing seaweed cultivation: 1) selection of a suitable environment for growth, 2) engineering of the seashore for farming, 3) protection of the crop from physical and biotic effects, 4) probable pests and parasites, 5) proper transplantation of young germlings to the field of cultivation and 6) harvesting.

Regarding mass culture of seaweeds, there appear to be four basic approaches (Huguenin, 1975, 1976, 1977; Mathieson, 1981): (1) cultivation in lagoons or lagoon like or pond-like enclosures; 2) cultivation in open ocean using artificial upwelling; 3) culture in intensive raceways and 4) culture in shallow rotating trays with a sprinkler system of irrigation with seawater. The lagoon culture method may use natural areas such as estuaries, bays and ponds which are dependant upon natural flushing by tidal waters. Such methods are used for a variety of seaweeds in Japan , China, Taiwan, and the South West Pacific (Saito, 1976; Cheng, 1969; Lin *et.al.*, 1979; Doty, 1973a, b). The open-ocean upwelling system has been developed for the giant kelp, *Macrocystis pyrifera* (Beleau *et. al.*, 1975; Wilcox, 1975a, b, 1976a, b, 1977). Intensive raceways have been devised for growing unattached seaweeds with aeration by pumping or paddle agitation. Tanks of various shapes have also been used with the water being circulated by either pumps or paddles. These have been used largely in the United States. The sprinkler type of irrigation of seaweeds spread on shallow perforated trays is one of the recent developments in seaweed culture (Huguenin, 1977) and holds forth much promise.

Seaweed culture systems

Culture in natural seawater

The simplest system of seaweed cultivation makes use of natural seawater and of ropes or rope nets as substrate for growing the seaweeds. Sheltered lagoons, estuaries or quite bays are ideally suited for this system since natural flushing by seawater is available and this eliminates the necessity for providing additional nutrients, as well as other physical parameters essential for growth of the seaweeds. A careful selection of the site in the vicinity of natural growth of the seaweed concerned will ensure the proper growth of the crop.

A prerequisite of this system of cultivation is a thorough knowledge of the biology, reproduction and growth of the seaweed and the environmental parameters required for optimum growth. The procedure mainly consists of recruitment of either vegetative fragments or spores of the seaweed on to the substratum, *viz.* ropes or nets. In what may be termed as a single stage culture (Krishnamurthy, 1983), vegetative fragments are inserted into twists of ropes at suitable intervals and these are then planted in the sea (Raju and Thomas, 1971; Krishnamurthy *et. al.*, 1977). Here again two methods of planting the ropes or nets may be recognized: 1) constant depth planting, where the nets are fixed to buoys or barges which keep the nets floating at a fixed depth with reference to the sea surface and 2) constant level planting, where the nets are tied to posts fixed to the sea bottom at a particular level, and tidal rise and fall of the seawater does not affect the level of the ropes but does influence the exposure of the ropes for specified periods, thus simulating conditions in the intertidal regions. This type of planting has been adopted by Raju and Thomas (1971) and by Krishnamurthy *et. al.* (1977). Either method of planting can be adopted accordingly as the seaweed cultivated is normally subtidal or intertidal in its occurrence.

Seawater ponds

Seaweed cultivation in large ponds constructed near the seashore and lined with heavy plastic gauge has been carried out by Marine Colloids Inc. in Nova Scotia (Mathieson, 1977). The seawater is fed into the ponds at a steady rate and agitation is provided by blowing air through the pond. This is a process by which natural seawater is utilized in an artificial pond system. The nearness to the sea takes care of the problem of drainage and circulation of the seawater.

A more sophisticated method involves the use of a land based network of reinforced concrete tanks placed at descending heights (Moeller *et. al.*, 1973). The tanks are enclosed in a greenhouse to use solar energy to heat the tanks and to ensure year round use of the system. The seawater is pumped to the highest tank and the water then flows through the network of tanks by gravity. Aeration of the cultivated seaweeds in these systems (simplest ones) may be facilitated by the mere movement on gravity flow of the seawater. However, in reinforced concrete tanks, aeration by using paddles or a system of aerating tubes connected to a compressor may be used. In these systems, natural seawater is used either in lagoons, creeks and estuaries or in earthern ponds, concrete tanks either in the open or in greenhouses and a continuous flow of seawater through the system and drainage back into the sea are maintained.

Closed seawater circulation systems

Closed seawater circulation systems are quite different in concept and operation. In these, natural seawater is used but is kept circulating in a closed system. In such systems, a fairly large quantity of seawater is kept circulating in outdoor greenhouse aquaria or in indoor cultivation rooms, having controlled light intensity and photoperiod (Neushul *et. al.*, 1976, 1977; Charters and Neushul, 1979). In such systems, light and nutrients are important and should be controlled carefully within prescribed limits. In such a system, operated at the University of California, Santa Barbara, large kelps have been grown in a healthy condition for more than a year.

Culture of spores and propagules

As pointed out by Krishnamurthy (1967), using spores of marine algae to start a mass culture would be of great value because of the large output of spores or other reproductive cells by various marine algae (Table 22.1).

The number of viable vegetative fragments from these algae would be limited. Hence, it is quite logical to assume that cultivation of these algae by spores would be the most advantageous, provided all spores produced by the seaweed could be collected and cultured and the viability of the spores and the survival rate of the sporelings are high. Such a type of cultivation involves two stages: (1) the collection and culture of spores to give sporelings in a healthy state of growth and (2) the transplanting of the sporelings to natural seawater areas for further growth and maturation. The first stage has, necessarily, to be carried out under controlled conditions in a laboratory or "nursery" using culture medium, light, photoperiod and temperature suited to the needs of the seaweed concerned. However, the second stage involves the preparation of a cultivation ground such as a natural seawater area or artificial ponds or tanks, where the further growth of the seaweed can be ensured.

This two stage culture method has been successfully adopted to cultivate *Porphyra* in Japan (Kurogi, 1953a, b; Kurogi and Saito, 1962) where six species of *Porphyra* including *P. tenera* and *P. yezoensis* are grown (Miura, 1974) . This method of cultivation has been adopted in India for *Gracilaria edulis* (Raju and Thomas, 1971; Krishnamurthy *et. al.*, 1977) and recommended for *Enteromorpha, Monostroma, Sargassum, Asparagopsis* and *Hypnea* (Krishnamurthy, 1983).

Table 22.1: Spore output of some marine algae (after Krishnamurthy, 1967)

Name of alga	*Type of Spore*	*approximate no. per plant*
Ulva fasciata	Swarmers (gametes)	1,15,34,400
Sargassum swartzi	Oospores	5,53,300
Gracilaria corticata	Carpospores	42,780
	tetraspores	68,500

Open ocean cultivation

The use of open ocean areas for cultivation of seaweeds was first proposed by Howard A. Wilcox. Based on his proposal, young plants of the giant kelp, *Macrocystis pyrifera*, were attached to a huge net, about 50 to 100 feet below the surface. At the centre of the net was a pipe reaching down to 1000 feet below the surface. A pump drew the nutrient rich seawater from this depth to the surface and radiating hoses distributed the water among the plants attached to the net. This method, using "artificial upwelling" ensured the growth of the seaweed using the nutrient rich deep seawater. This system has been developed to produce the giant kelp on a wide area (Beleau *et. al.*, 1975; Wilcox, 1975a, 1976a, 1977).

Sprinkler system of cultivation

This system , also termed as intensive rotary culture, was suggested by Huguenin (1977). Young seaweed plants or fragments were distributed uniformly in large shallow circular basins or trays containing only a small quantity of water. The seaweeds are rotated and wetted periodically by spraying of seawater. Another method of growing seaweeds based on this principle is to have the seaweeds spread in perforated trays placed one over another and sprinkling seawater from above. The seawater wets the seaweeds on the top tray and percolates down through several trays, thus wetting the seaweeds in succession. This system necessarily involves pumping of seawater and its storage in overhead tanks and the use of sprinklers over a system of stacked trays.

The cultivation of some of the major seaweeds, which can be taken up in India, is discussed as under : -

Porphyra

Commercial cultivation of *Porphyra* originated in Japan about 300 years ago using a very simple and primitive method. It was common knowledge that *Porphyra* plants make their appearance during autumn. Just prior to this season or at the beginning of autumn, bundles of twigs, often bamboo, known as "hibi" were inserted into the sea bottom in shallow water, to collect spores. Spores of *Porphyra* (now known to be the spores of the Conchocelis phase) get attached to these twigs and begin to grow into young thalli which are allowed to develop into mature plants before harvesting. A somewhat similar but still primitive custom was the "rock cleaning" method of the Chinese, developed about 200 years ago. In both the Chinese and the Japanese practices, settlement of spores on cleaned rocks or on "hibi" was left to the vagaries of nature, so the production was highly fluctuating and comparatively low. Net cultivation was introduced around 1920 and this resulted in some increase in production. However, it was the fundamental observation of Drew (1949, 1954, 1956) on the Conchocelis phase of *Porphyra* which led to a revolutionary change in the process of cultivation of *Porphyra*. Drew's observations were followed by researches by a number of other workers (Kurogi, 1953a, b; Krishnamurthy, 1959; Kurogi and Saito, 1962). These researches resulted in the development of a two stage cultivation of *Porphyra*, the first stage being the culture of the *Conchocelis* phase on shells in a laboratory or nursery and the second, the transfer of the sporulating *Conchocelis* phase to the rope nets planted in the sea for the cultivation of the leafy phase of *Porphyra*. The shells are tied to the nets and positioned below them, so that liberated spores can settle on the nets.

According to Miura (1974) six species of *Porphyra* are cultivated in Japan including *P. tenera* and *P. yezoensis*. In China, a seventh species, *P. haitanensis* is also used (Tseng, 1981). Interest in the cultivation of *Porphyra* has also been evinced in the United States, where extensive studies on the biology and culture of *P. nereocystis* have been made (Woessner, 1974, 1979; Woessner *et. al.*, 1977). An account of mass cultivation of *Porphyra* in North America is given by Waaland (1977). In India, studies on the biology (Umamaheswara Rao and Sreeramulu, 1963) and chemical composition (Tewari *et. al.*, 1968) of *P. vietnamensis*, suggest the feasibility of commercial cultivation of this species (Krishnamurthy, 1963).

Gracilaria

Gracilaria is valuable for the agar, yielded by several of its species. The yield of agar by different species of *Gracilaria* from Indian shores has been studied and it has been stated that *Gracilaria* spp. yield from 25% to 46% agar on dry weight basis.

Experimental cultivation of *Gracilaria edulis*, *G. corticata* and *G. verrucosa* was carried out in India by Krishnamurthy and his co-workers at the Marine Algal Research Station, Mandapam, from 1969 onwards and Krishnamurthy *et. al.* (1977) described a series of experiments on rope culture of *G. edulis*.

Gracilaria cultivation has been practiced in China from the late fifties and four approaches have been adopted (Tseng, 1981): (1) Branches of *Gracilaria* were inserted in bamboo splits and these were planted in the littoral region. (2) 5-6 cm long juvenile plants were gathered from nature or from cultures on pebbles and stones and these were scattered in shallow water farms protected from tidal or wave action. (3) net cultivation and (4) polyculture with milkfish, shrimp or crab. *Gracilaria* culture has also been studied and adopted in the United States and Canada. *Gracilaria tikvahiae* was studied in greater detail. Goldstein (1977) studied artificial recruitment of *G. tikvahiae* by seeding tetraspores on shells and after 1-2 weeks' incubation, transferring these to the sea. 50,000 sporelings were outplanted at 2 sites and monitored for 13 months. Neish and Fox (1971) studied growth of *G. tikvahiae* under green house conditions and showed that warm water conditions were necessary for the growth of this alga. *Gracilaria folifera* has been studied for cultivation at Florida and Beaufort and *G. verrucosa* at Florida and British Columbia.

Gracilaria edulis has been successfully cultivated in India and the first report on such cultivation was made by Raju and Thomas (1971). A fuller account of the process was given by Krishnamurthy *et. al.* (1977). Prior to cultivation of the species, environmental data pertaining to the alga were gathered, particularly the depth of occurrence and light requirements for optimum growth. Healthy plants of the alga were selected, thoroughly washed in sea water and 2.5 cm fragments were removed from the apices of branches. These branches were then inserted in the twists of a coir rope at intervals of 10 cm along the length of the rope. Several ropes were prepared in this manner. The ropes were then transferred to the lagoon where bamboo posts were planted at intervals of ten meters. The ropes were tied to the posts and their level adjusted so as to correspond to the level at which the alga grows normally.

The ropes were cleaned at regular intervals in the first month of culture so that no fouling occurred and then left without much attention, except for periodic observations. The plants attained lengths of about 30 cm in about five months. At this stage, the plants were harvested by clipping them close to the rope. After harvesting, the ropes with the remnants of the plants were put back in position. Two more harvests were made at intervals of ten weeks each, giving a total of three harvests in ten months. An average yield of about 4 kg of seaweed per meter of rope was obtained. In subsequent experiments, it was found advantageous to make the first harvest at the end of three months and subsequent harvests each month up to ten months. This practice resulted in greater yield of the seaweed.

Gelidiella

Gelidiella acerosa is a major agar yielding red alga from the Indian coasts. Its cultivation from fragments was first attempted by Krishnamurthy *et. al.* (1977) by the rope culture method adopted for *Gracilaria edulis*. Subsequently, recruitment on coral stones in the subtidal fringe at Yerwadi in South India was successfully carried out by Patel *et. al.* (1977). Only fragments of the thallus have been used as seed material as spore output in this alga is low and seasonally restricted.

Hypnea

Mshigeni (1976) showed the cultivation of *Hypnea curvicornis* from both carpospores and tetraspores. As many as four crops can be raised in a year since the alga attains maturity and sporulates within three months after spore germination. Beneficial effect of nitrate fertilization on three Hawaiian species, *H. curvicornis, H. chordacea* and *H. nidifica* was also studied (Mshigeni, 1978). A different approach was made by Haines (1975) who cultured *H. musciformis* in an artificial upwelling system and found that growth of this species could be correlated with ammonia concentration contrary to the observations of Mshigeni (1978). In India, Rama Rao *et. al.*, (1983) cultivated *Hypnea valentiae* from stellate bulbils produced by this species in profusion.

Enteromorpha

This genus is well represented by about eleven species in India, distributed all over the coast at the uppermost level of the algal zone. Some species are found in estuarine conditions and a few can tolerate highly saline water. *E. prolifera* is harvested for food in Japan. Recently, *Enteromorpha* has been shown to contain antibacterial substances (Sreenivasa Rao *et. al.*, 1979).

Mass cultivation on nets was carried out on *E. flexuosa* and *E. compressa*, growing commonly on the coast of Port Okha (Ohno *et. al.*, 1981). A suspension of swarmers was obtained by keeping 20 g of the thalli in the dark during the night and exposing them to light next day. The suspension was then poured into a tank containing the net, and was agitated several times during the day. The nets were then transplanted into tidal pools. First harvest was made after 4-6 weeks of planting and then repeated every week from successive areas.

Eucheuma

Eucheuma spp. are a source of specific carrageenan and are in great demand by industries. Successful cultivation of this alga has been achieved in the Philippines. The method adopted is to recruit fragments of the alga in rope nets, suspended at a suitable depth in comparatively quite lagoon waters and harvesting the growth after full development of the vegetative thallus.

In India, *Eucheuma* cultivation has been attempted on a fairly large scale by the Central Salt and Marine Chemicals Research Institute, Bhavnagar, at their Marine Algal Research Station, Mandapam. The growing thalli are protected against predators by bagging them in perforated plastic gunnies. A more ambitious programme of *Eucheuma* cultivation is being pursued.

Caulerpa

Caulerpa lentellifera is used as a fresh vegetable and in salads. The alga, which is also named "sea-grapes" was once collected directly from reef flats and after washing, used for edible purposes. *Caulerpa* can be cultivated in ponds in which pieces of the thallus are planted in the bottom at fixed intervals and the crop is flooded with slow movement of seawater. Alternately, the alga can be cultivated in open lagoons with a muddy bottom.

In Okinawa (Japan), a different type of cultivation, known as the cage culture, is adopted. Multistage cages are prepared and at each level of the cage, small fragments of the alga are fixed in the middle and the cage is hung vertically in the sea. Each fragment grows radially and branches soon protrude out of the cage all around. The protruding portions are harvested at periodical intervals.

Along Indian coasts, there are as many as 40 taxa of *Caulerpa*, including *C. lentellifera*, reported as *C. longistipitata* Svedelius (Duraiswamy, 1984). It is possible that some of these can be used for edible purposes. Therefore, there is good scope for cultivation of *Caulerpa* in this country.

Cultivation methods adopted in India

Cultivation methods adopted in India may be classified into one stage or direct culture and two stage culture. The first is based purely on the capacity for vegetative propagation of the seaweed concerned and may be resorted to in two ways. (1) The vegetative fragments can be scattered in a shallow lagoon specially enclosed by means of some kind of barrier which will ensure that the growing seaweeds are not washed away into the sea. The barrier may conveniently be constructed of nylon netting tied to posts planted at intervals in the lagoon and weighted down to keep them in position. (2) Rope culture method (already described above) may be adopted. The ropes can simply form long parallel lines spaced so as to allow movement of a boat for harvesting or may be made into nets of a particular width so that growth of the alga is spread over an area. Either method can be adopted depending upon the alga to be cultivated. Again the ropes may be tied to poles at a particular level or kept down at the bottom of the lagoon resulting in what is known as constant level cultivation. Alternately, the ropes or nets may be attached to barges or floats in such a way that, despite tidal changes, the plants are kept at the same depth below the water surface, a practice in constant depth cultivation.

In two stage culture, spores or other propagules of the alga are used and as a first step, these are made to shed on the suitable substrates kept in aquaria in a nursery on the shore. This, for instance, is done in the cultivation of *Enteromorpha, Monostroma* etc. A similar method can also be adopted for *Gracilaria, Gelidiella, Sargassum, Hypnea,* and in fact, for any marine alga which normally reproduces by means of spores. In *Hypnea valentiae,* stellate bulbils are produced in profusion at certain stages of development and these can be recruited on to shells or stones in the nursery. After the sporelings have got attached firmly to the substrate, they are transferred to the sea and allowed to grow into adult plants.

General considerations

The most suitable environment for cultivation of seaweeds is shallow sheltered waters of creeks and lagoons. The locations should not be subject to turbulence of waves or strong surf action as otherwise the crop may be washed away before it is ready for harvesting. The water may show fluctuating salinity, but this is not a great problem as many seaweeds can tolerate lower salinity. Preparation of the seashore is necessary for cultivation as natural substrates cannot give the same hold for the propagules of the seaweed proposed to be planted and an artificial substratum is usually necessary for the purpose. If the propagules are to be allowed to grow into free floating individuals, the portion of the seawater which is proposed to be farmed has to be enclosed in some fashion. If propagules are to be attached, then some substrates like ropes or nets or stones have to be used and secured in the seawater. Moreover, some method of attaching the propagules to the substrate has to be adopted..

A study of the autecology of the species, proposed to be cultured, would provide clue as to the type of propagule to be used, the stages of growth of the propagule fit for planting and the time or season best suited for planting so that maximised growth and rich harvests are assured.

The crop once established needs to be protected from the physical forces such as waves, surf and tidal changes. The last may sometimes be crucial as uncommonly high tides may disturb optimum conditions of light required for growth of the seaweed and an uncommonly low tide may expose the alga to high isolation which may sometimes be harmful. An important factor which may be detrimental to the crop in the sea is the presence of predators in the form of large herbivorous animals especially fishes. Some protection against these predators will have to be devised to save the crop. Many seaweeds are affected by pests and parasites. Pests in the form of epiphytic animals, algae, parasitic animals and fungi can cause considerable loss of crop, especially when the alga is grown gregariously in an artificial culture bed. A constant vigil has to be maintained to save the crop from such afflictions.

Finally, harvesting will have to be carried out at the appropriate stage of growth to obtain maximum benefits. In many algae, a number of successive harvests can be made at judicious intervals and these have to be determined by careful study.

References

Barker, H.A. 1935. The culture and physiology of the marine dinoflagellates. *Arch.. Mikrobiol.* **6**: 157-181.

Beleau, M.H., Heidelbaugh, N.D. and Van Dyke, D. 1975. Open ocean farming of kelp for animal and human food. *Food Technol.* **29**: 27-30.

Black, W.A.P. and Mitchell, R.L. 1952. Trace elements in the common brown algae and seawater. *J.Mar. Biol. Assn.*, U.K. **30**: 575-584

Biebl, R. 1962. Seaweeds. In: *Physiology and Biochemistry of Algae.* Edited by Lewin, R.A., Academic Press, New York. pp. 799-815.

Bold, H.C. 1942. The cultivation of algae. *Bot. Rev.* **8**: 69-138.

Charters, A.C. and Neushul, M. 1979. A hydrodynamically defined culture system for benthic seaweeds. *Aquatic Bot.* **6**: 67-78.

Cheng, T.H. 1969. Production of kelp – a major aspect of China's exploitation of the sea. *Econ. Bot.* **23**: 215-236.

Doty, M.S. 1973a. Farming the red seaweed, *Eucheuma* for carrageenan. *Micronesica.* **9**: 59-73.

Doty, M.S. 1973b. *Eucheuma farming for carrageenan*. Univ. Hawaii Sea Grant Rep. pp. 21.

Drew, K.M. 1949. *Conchocelis*-phase in the life history of *Porphyra umbilicalis* (L.) Kuetz. .*Nature* **164**: 748-749.

Drew. K.M. 1954. Studies in the Bangioideae III. The life history of *Porphyra umbilicalis* (L.) *Kuetz.* var. *laciniata* (Lightf.) Ag. The *Conchocelis* phase in culture. *Ann. Bot.*. **18**: 183-211.

Drew, K.M. 1956. Reproduction in the Bangiophycideae. *Bot. Rev.* **30**: 343-390.

Droop, M.R. 1958. Optimum relative and actual ionic concentrations for growth of some euryhaline algae. *Verh. Int. Ver.Limnol.* **13**: 722-730.

Droop, M.R. 1961. Some chemical considerations in the design of synthetic culture media for marine algae. *Bot. Mar.* **2**: 231-246.

Duraiswamy, R. 1984. *Studies on the species of Caulerpa Lamonr. from Tamilnadu, India,* Ph.D Thesis, University of Madras. pp.186.

Fries, L. 1960. The influence of different B_{12} analogues on the growth of *Goniotrichum elegans*. *Physiol. Plant.* **13**: 264-275.

Fries, L. 1963. On the cultivation of axenic red algae. *Physiol. Plant.* **16**: 695-708.

Gerloff, G.C.,Fitzgerald, G.P. and Skoog, F. 1950. The isolation, purification and culture of blue-green algae. *Amer. J. Bot.* **37**: 216-218.

Goldstein, M.E. 1977. Artificial recruitment and mariculture of *Gracilaria* in P.E.I. Canada. *J. Phycol.* **13**: 24.

Guillard, R.R.L. and Ryther, J.H. 1962. Studies of marine planktonic diatoms, *Cyclotella nana* Hustedt and *Detonula confervacea* (Cleve) Gran. *Can. J. Microbiol.* **8**: 229-239.

Haines, K.C. 1975. Growth of the carrageenan producing tropical red seaweed *Hypnea musciformis* in surface water, 870m deep water, effluent from clam mariculture system and in deep water enriched with artificial fertilizer or domestic sewage. In: *Proc. 10th European Symp. Marine Biology*. Edited by Persoone, G. and Jaspers, E. **1**: 207-220.

Huguenin, J.E. 1975. Development of a marine aquaculture research complex. *Aquaculture* **5**: 135-150.

Huguenin, J.E. 1976. An examination of problems and potentials for future large scale intensive culture systems. *Aquaculture* **9**: 313-342.

Huguenin, J.E., 1977. A subjective analysis of alternate approaches to the mass culturing of seaweeds. In: *Proceedings of the Eighth Annual Meeting, World Mariculture Society, San Jose, Costa Rica*, Edited by Anault, J.W. Lousiana State Univ., Division of Continuing Education, Baton Rouge. pp. 387-400.

Humm, H.J. 1944. Agar resources of the South Atlantic and East Gulf Islands. *Science* **100**: 410-411.

Hutner,S.H. and Provasoli, L. 1964. Nutrition of algae. *Ann. Rev. Plant Physiol.* **15**: 37-56.

Isaac, W.E. 1956. *The ecology of Gracilaria confervoides (L.) Grev. in South Africa with special reference to its ecology in the Saldanha Langebaan lagoon*. Second International Seaweed Symposium, Trondheim. pp. 173-185.

Iwasaki, H. 1965. Nutritional studies of the edible seaweed *Porphyra tenera* I. The influence of different B_{12} analogues, plant hormones, purines and pyrimidines on the growth of *Conchocelis, Plant Cell Physiol.* **6**: 325-336.

Jones, R.F., Speer, H.L. and Kury, W. 1963. Studies on the growth of the red alga *Porphyridium cruentum*. *Physiol. Plant*. **16**: 636-643.

Krishnamurthy, V. 1954. Ecology and seasonal succession of the algae growing in a salt marsh at Madras. *J. Madras Univ*. **B 24**: 161-178.

Krishnamurthy, V. 1959. Cytological investigations on *Porphyra umbilicalis*(L.) Kuetz. var. *laciniata* (Lightf) J. Ag. *Ann. Bot*. **23**: 147-176.

Krishnamurthy, V. 1967. Marine algal cultivation, necessity, principles and problems. In: *Proceedigns of the Seminar on Sea, Salt and Plants*. Edited by Krishnamurthy, V. Bhavnagar. pp. 327-333.

Krishnamurthy, V. 1983. Seaweed cultivation in India. In: *Proceedings All India Applied Phycological Congress*. Edited by Shukhla, A.C., Kanpur. pp. 118-125.

Krishnamurthy, V. 1986. Seaweed cultivation , a review. In: *Marine Biology and Mariculture* . Edited by Krishnamurthy, V. pp. 117-128.

Krishnamurthy, V., Raju, P.V. and Thomas, P.C. 1977. On augmenting the seaweed resources of India. *Seaweed* Res. *Utlin*. **20**: 37-41.

Kurogi, M.1953a. Study of the life-history of *Porphyra*. 1. The germination and development of carpospores. *Bull. Tohoku Reg. Fish. Res. Lab*. **2**: 67-103.

Kurogi, M. 1953b. On the liberation of monospores from the filamentous thallus (Conchocelis stage) of *Porphyra tenera* Kjellm. *Bull. Tohoku Reg. Fish. Res. Lab*. **2**: 104 –108.

Kurogi, M. and Hirano, K. 1956. The influence of water temperature on the growth, formation of monosporangia and monospore liberation in the Conchocelis phase of *Porphyra tenera* Kjellm. *Bull. Tohoku. Reg. Fish. Res. Lab*. **8**: 45-61.

Kurogi, M. and Saito, S. 1962. Influence of light on the growth and maturation of Conchocelis thallus of *Porphyra* . II. Effect of different photo-periods. *Bull. Tohoku Reg. Fish. Res. Lab*. **20**: 127-137.

Kylin, H. 1927. Uber den Einfluss der Wasserstoffionen onzentration auf einige Meeresalgen. *Bot. Notiser*. **1927**: 243-254.

Lewin, R.A. 1959. The isolation of algae. *Rev.Algol*. **3**: 181-197.

Lin, M.N., Lin, G.Y. and Haung, T.L. 1979. Preliminary report on the improvement *of Gracilaria* farming. *Aquaculture* **2**: 22-31.

Mathieson, A.C. 1975. Seaweed aquaculture. *Mar. Fish. Rev*. **17**: 2-14.

Mathieson, A.C. 1977. Seaweed cultivation-a review. In: *Proceedings Sixth U.S. Japan Meeting on Aquaculture*. Edited by Sindermann, C.J., U.S. Rep.Commer. Rep. NMFS Cnc. 442. pp. 25-66.

Miura, A. 1974. Nori (seaweed) cultivation. In: *Japan's Food Industry*. Asahi Even. News Int. pp. 16-17.

Moeller, H.W., Churchill, A.C. and Giordano, S. 1973. Experimental farming of *Chondrus crispus* in temperate and tropical waters. *J. Phycol.* **9**: 11.

Mshigeni, K.F. 1976. Field cultivation of *Hypnea* (Rhodophyta) spores for carrageenan: prospects and problems. *Bot. Mar.* **19**: 227-230.

Mshigeni, K.F. 1978. Effects of nitrate fertilizer on the growth of the economic seaweed, *Hypnea* Lamouroux (Rhodophyta, Gigartinales). *Nova Hedw.* **19**: 231-236.

Myers, J. 1962. Laboratory cultures. In: *Physiology and Biochemistry of Algae.* Edited by Lewin, R.A. Academic Press, New York.. pp. 603-615.

Neish, A.C. and Fox, C.H. 1971. *Greenhouse Experiments on the Vegetative Propagation of Chondrus crispus (Irish moss).* NRC, Canada, Atlantic Regional Lab.**12**: 15.

Neushul, M., Fister, M.S., Coon, D.A., Woessner, J.W. and Harger, B.W.W. 1976. *An in situ* study of recruitment, growth and survival of subtidal marine algae, techniques and preliminary results. *J. Phycol.* **12**: 397-408.

Neushul, M., Charters, A.C., Coon, D.A., Wheeir, W., Harger, B.W.W. and Woessner, J.W. 1977. Studies on the growth and reproduction of the giant kelp, *Macrocystis. J.Phycol.* **13** : 77.

Ohno, M., Mairh, O.P., Chauhan, V.D., Tewari, A., Joshi, H.V., Pandey, R.S. and Rao, P.S. 1981. Mass cultivation of green alga *Enteromorpha* on the coast of Okha, India. *Rep. Mar. Biol. Inst.* **3**: 51-59.

Oza, R.M. 1971. Effect of IAA on the growth of fragments of *Gracilaria corticata* J.Ag. *Seaweed Res. Utiln.* **1**: 48-49.

Patel, N.B., Gopal, B.V., Nagulan, V.R., Subbaramaiah, K., Thomas, P.C.and Mehta, D.J. 1977. Experimental culture of *Gelidiella acerosa* in India (Abstract). *J. Phycol.* **13** (suppl.): 52.

Pringsheim, E.G. 1946. *Pure cultures of Algae.* Cambridge Univ. Press, London .

Provasoli, L. 1958a. Nutrition and ecology of protozoa and algae. *Ann. Rev. Microbiol.* **12**: 279-308.

Provasoli, L. 1958b. Effect of plant harmones on *Ulva. Biol. Bull.* 114: 375-384.

Provasoli, L. 1963. *Growing marine seaweeds.* Proc. 4th Int. Seaweed Symp. Pergamon Press, Oxford. pp. 9-17.

Provasoli, L. 1968. *Media and prospects for the cultivation of marine Algae*. Edited by Watanabe, H.A. and Hattori, A. Japanese Society of Plant Physiology. pp. 63-75.

Provasoli, L., McLachlan, J.J.A and Droop, M.R.1957. The development of artificial media for marine algae. *Arch. Microbiol.* **25**: 392-428.

Raju, P.V. and Thomas, P.C. 1971. Experimental field *cultivation of Gracilaria edulis* (Gmenl.)Silva. *Bot. Mar*. **14**: 71-75.

Rama Rao, K., Nair, M.R.P. and Subaramaiah, K. 1985. Experimental field cultivation of *Hypnea valentiae* (Turn.) Mont. at Krusadai island. In: *Marine Plants.* Edited by Krishnamurthy, V. and Untawale, A.G. pp. 205-208.

Saito, Y. 1976. Seaweed aquaculture in the north west Pacific. In: *Advances in aquaculture*. Edited by Pillai, T.V.R. and Dill, W.A., Kyoto, Japan. pp. 402-410.

Spencer, C.P., 1952. On the use of antibiotics for isolating bacteria-free cultures of marine phytoplankton organisms. *J. Mar. Biol. Ass. U.K.* **31**: 97-106.

Spencer, C.P. 1954. Studies on the culture of a marine diatom. *J. Mar. Biol. Ass. U.K.* **33**: 265-290.

Sreenivasa Rao, P., Parekh, K.S. and Parekh, H.H. 1979. Antibacterial activity of different fractions of seaweed extracts (Abstract). *International symposium on Marine Algae of the Indian Ocean region.*, Bhavnagar. pp. 47.

Subbaramaiah, K. and Krishnamurthy, V. 1967. Laboratory culture of seaweeds. In: *Proceedings of the Seminar on Sea, Salt and Plants.* Edited by Krishnamurthy, V., Bhavnagar. pp. 321-326.

Sweeney, B.M. 1951. Culture of the dinoflagellate *Gymnodinium* with soil extract. *Amer. J. Bot*. **38**: 669-677.

Tatewaki, M. and Provasoli, L. 1964. Vitamin requirements of three species of *Antithamnion*. *Bot. Mar*. **6**: 193-203.

Tewari, A., Rao, M.P. and Krishnamurthy, V. 1968. Chemical composition of a species of *Porphyra* from Visakhapatnam, South India. *Curr. Sci.* **37**: 138-139.

Tseng, C.K. 1981. Commercial cultivation. In: *Biology of Seaweeds.* Edited by Labban, C.S. and Wynne, M. J. pp. 680-725.

Umamaheswara Rao, M. and Sreeramulu, T. 1963. Vertical zonation and seasonal succession in the growth of *Porphyra* on Visakhapatnam coast. *Curr. Sci.* **32**: 173-174.

Waaland, R. 1977. Growth of Pacific northwest marine algae in semiclosed culture. In: *The marine plant biomasss of Pacific north west coast*. Edited by Krauss, R.W., Oregon State Univ. Press. Corvallis. pp. 117-137.

Wilcox, H.A. 1975a. The ocean food and energy farm project. *141st Annual Meeting of the American Association for the Advancement of Science*, New York. pp. 22.

Wilcox, H.A. 1975b. The ocean food and energy farm project. In: *Technology Assessment and the Ocean.* Edited by Willmot, P.D.and Slingerland, A., West View Press, Boulder, Colorado, U.S. pp. 67-76.

Wilcox, H.A., 1976a, Ocean farming. In: *Proceedings of the conference on capturing the sun through bioconversion.* Council on Solar Biofuels,Washington Centre for Metropolitan Studies. pp. 255-276.

Wilcox, H.A. 1976b. The ocean food and energy farm project. *Conference on Sharing the Sun*, International Solar Energy Society, Winnipeg., Canada. pp.29.

Wilcox, H.A. 1977. The ocean food and energy farm project. In: *Ocean Energy Resources.* Edited by Monney, N.T. The Energy Technology Conference, Houston, Texas, Amer. Soc. Mech. Eng., N.Y. **4**: 83-104.

Woessner, J.W. 1974. The measure and harvest of the California marine crop plant *Porphyra*. In: *Scientific Programme and Abstracts.* VIII International Seaweed Symposium, Bangor. pp.A61.

Woessner, J.W. 1979. Biological achievements for plant mariculture. *J. Phycol.* 15 (suppl.): 19.

Woessner, J.W., Sorenson, P. and Coon, D.A. 1977. The economic potential of harvesting and marketing an American *Porphyra. J. Phycol.* **13** (suppl.): 74.

Yendo, K. 1919. The germination and development of some marine algae. *Bot. Mag.Tokyo* **33**: 73-93.

Chapter 23

TRANSMISSION ELECTRON MICROSCOPY WITH ENERGY DISPERSIVE X-RAY MICRO ANALYSIS: PRINCIPLE AND TECHNIQUES ON MACROALGAE

K.Sivakumar

Abstract

Energy dispersive x-ray micro-analysis (EDAX) provides a unique approach for obtaining quantitative compositional analysis of an individual cell and intracellular compartments. Elemental quantification of ultrathin sections with electron probe x-ray micro analysis (EPMA) is generally based on the linear relationship between elemental concentration and the ratio of number of characteristics/continuum x-ray photons. The amount of different chemical elements present in the cell-wall, plastid and floridean starch of a carrageenophyte can be determined using EDAX. Complete methodology to use electron microscopy has been discussed.

Key words: Algae, micro analysis, technique, transmission electron microscope, ultramicrotomy

Introduction

Microscopy is that branch of science which covers methods and applications that use microscopes to distinguish material features too small to be seen by the unaided eye. The visible light microscope is one of the oldest and most useful scientific instruments. Electron microscope has made significant contributions in biological and biomedical sciences. It is used for ultrastructural studies, diagnostic purposes and also for quantitative ultrastructural studies.

Transmission electron microscope (TEM)

Major components

- Basic unit has microscope column, vacuum system and control panel.
- Power supply unit generates power supply to various parts of the basic unit including voltage to electron gun.
- Electromagnetic lenses and heaters generate heat. Water chiller (circulation of chilled water at -20^{0}C) is used to cool these down.
- The valves at various points in the column are to be opened and closed by an air compressor.

Microscope column – consists of an illumination system (electron gun and condenser lenses), imaging system (objective lens, projector lens) and image translating system (fluorescent screen, photographic recording).

Illumination system – contains two units i.e. the electron gun (the source of electrons), and the condenser which regulates the intensity of the beam and directs it on to the specimen. The electron gun consists of three components: filament, shield and anode. The filament (cathode) is simply a V-shaped piece of pure tungsten wire. As the voltage is applied it gets heated up and electron emission starts. The emitted electrons are accelerated in the space between cathode and anode by potential difference between this gap. Anode attracts electrons from the space surrounding the filament. After leaving the gun, the electrons pass down the column at a constant velocity. The electron beam emerging from the gun is focussed on to the specimen for optimal illumination by the condenser lens. The intensity of illumination can be increased or decreased by adjusting the focal length (lens current) of the condenser lens. It is also determined by the size of the aperture. Larger the aperture size more electrons will pass through the condenser and higher will be the intensity of illumination.

In imaging system – the objective lens forms the initial enlarged image of the illuminated portion of the specimen in a plane that is suitable for further enlargement by the projector lens. By varying the current in the objective lens, the image of the specimen is focussed. The projector lens projects the final magnified image on the screen or photographic emulsion. By varying the current in the projector lens, the size (magnification) of the projected can be varied.

Image translating system – gets the information provided by electron microscope contained in the final image, in the form of variations of electron intensity over its area. The information contained by the transmitted electrons must be converted into visible light. This is done by using fluorescent screen. Bombardment of fluorescent screen with electrons emits photons of visible light which can be easily seen with eyes. The image thus seen can be recorded permanently on photographic film coated with photosensitive silver halide. The electron beam liberates free silver from silver halide grains. This produces negatives after developing. These negatives can be further processed to get positives on the photographic paper. This image recorded on the photographic paper is known as an electron micrograph.

Specimen preparation for TEM

Fixation – is the first step in preparation of biological specimens for electron microscopy. It is a process to put the cell to death almost instantaneously using certain chemical compounds (fixatives) which prevent autolysis, change in volume and shape and preserve various chemical constituents of the cell.

Aims of fixation

- To preserve the structure of cells and tissues with minimum or least alteration from the living state.
- To protect them against alterations during embedding and sectioning
- To prepare them for subsequent treatments such as staining and exposure to the electron beam.

Transmission electronic microscopic studies (Hayat, 1986)

Fresh samples of 3.0-5.0mm length fixed in primary fixative (solution-B) on the spot were kept in an ice-bath.

Primary fixative

Solution A

Sodium cacodylate	–	2.14g
Sucrose	–	8.55 g
Distilled water	–	100ml
pH	–	6.8

Solution B

Solution A	–	80 ml
Glutaraldehyde (25%)	–	20ml

Samples were brought to the laboratory within 24 h and rinsed in the descending series of sucrose concentrations such as 0.25 M, 0.15 M and 0.05 M in cacodylate buffer at 4°C at an interval of 20 min. each. Final three washings of each 20 min were given only with cacodylate buffer (sodium cacodylate 2.14 g/ 100 ml distilled water, pH 6.8).

Primary fixation

The most commonly used primary fixative is 2.5% glutaraldehyde. For certain tissues, paraformaldehyde (4%) is used alongwith 1% glutaraldehyde (Karnovsky's fluid).

Fixation through immersion

The whole object or their parts are placed directly in the fixative. Since the ability of penetration of the fixative is poor, the size of the tissue has to be very small (1.2 mm thick). To facilitate penetration, additives like acroline (final concentration to 0.1%) is added to the primary fixative. Fixation is carried out at 4°C for 2-24 hours.

Washing

After fixation specimens should be rinsed thoroughly to wash off the excess fixative with buffer (same as used for preparing fixative). If traces of fixative are left in the specimen, they may react with the secondary fixative (osmium tetraoxide) thus producing precipitate (of reduced osmium) in the specimen.

Post fixation

OsO_4 (1% solution) is the most frequently used secondary fixative. It is generally used after glutaraldehyde fixation and it reacts principally with lipids. OsO_4 also acts as an electron dense stain.

Dehydration

The tissues contain about 75% water. It must be replaced with suitable organic solvent(s) before infiltration when water immiscible resins are used for embedding. The removal of water is unnecessary when tissues are embedded in water miscible resins.

The water is removed by passing the tissues through a series of solutions of ascending concentration of the dehydrating agent. Absolute ethanol or acetone is used as dehydrating agent. Acetone has an advantage over ethanol in being more readily miscible with clearing agents *viz.* xylene and toluene. Dry acetone should be used and it is prepared by adding sufficient amount of anhydrous copper sulphate to absolute acetone. Shake well, leave for some time and filter before use.

Clearing

Though acetone is easily miscible with resin, it is advantageous to clear it off with a clearing agent to facilitate penetration of resin. Xylene, toluene or epoxy propane is found to be good clearing agents and it has to be in pure form. Epoxy propane is known to be very effective as it can completely remove the traces of water in case of incomplete dehydration.

Infiltration

Very few tissues are rigid enough to be cut into thin sections without an additional support. Therefore, tissue must be attached to a block which is sufficiently strong to be handled safely during sectioning. This can be accomplished by infiltrating resin into the tissues and further hardening it, so that it will take the form of a block embedded with tissue. Infiltration is done by gradually decreasing the concentration of clearing agent and proportionately increasing the concentration of embedding medium. In the case of hard tissues, slow diffusion of the embedding medium is a problem. This can be overcome by carrying out the steps of infiltration under vacuum. Vacuum treatment is routinely employed for macroalgal tissues. This treatment also eliminates the air bubbles attached around the tissue. Infiltration is carried out with liquid resins. The resins are of different kinds and are used according to the requirement. Resins are toxic and allergic to skin, therefore, wear plastic or rubber gloves. All the operations must be carried out in a fume cupboard.

Kinds of resin (Water immiscible resins)

Highly viscous resins	1. Araldite CY212
	2. Epon812
	3. Methacrylate
Low viscous resins	1. Maraglas
	2. ERL 4206

Embedding

Embedding is done in the embedding medium using gelatin or beam capsule. After filling the capsule with the embedding medium, tissue is put and ensured that it settles at the tip (bottom when placed vertical) and filled with embedding medium if required. Paper label written with pencil is also put along the side of the wall of the capsule. While using capsules, ensure that they are properly dried. When a specific orientation is required, the use of flat embedding mould is recommended.

Polymerization

It is the process by which liquid resin is hardened. This is usually done by heat. The process should not be carried out too fast as this may cause blocks to polymerize non-homogenously and may lead to artifacts within the specimen. Another reason is that at 40-50°C, most embedding media become less viscous thus facilitating penetration into the tissue. So it is advisable to keep the specimen at 40-50°C for overnight. Then increase the temperature to 60°C so that resin gets hardened. It takes almost 24-48 hours for the blocks to get hardened. Use separate incubator with good temperature control for this purpose.

Removing the blocks from the mould

After polymerization, blocks can be removed as follows:

- Gelatine capsule - should be washed in water at 70°C
- Beem capsules - can be taken off easily by cutting them open with a razor blade.
- Flat embedding mould - blocks can be removed very easily by bending the mould sideways. After removal, blocks can be broken apart in case Reichert flat embedding trays are used.

Ultramicrotomy: Trimming of blocks

Blocks can be trimmed to produce a pyramid form at the tip of the block where the tissue is located. This can be done either manually or mechanically. Manually the block is fixed into the holder and mounted on the microtome with a clean razor blade trimming the four sides and the tip of the block.

Trimming can also be done with a pyramitome or block trimmer (Reichert TM 60) fitted with a rotating milling cutter. The block is fitted in the block holder of the ultramicrotome.

The following points must be taken care of during trimming : -

- Trim as small as possible
- Do not trim sides of the pyramid too steeply as this reduces the stability
- If possible trim away all free embedding medium
- Pyramid should be flat and free from plastic debris
- Trim the cutting surface of the block in the form of trapezium in order to get serial sections easily.

Making glass knives

Diamond knives produce very good sections but they are expensive. Therefore, in most of the laboratories glass knives are used. These are less expensive and easily available.

Belgium glass strip is usually used for making glass knives as other types of glass have different properties and hardness, and tend to break in a different manner. Glass knives can be made by glass knife maker by cutting the Belgium glass strips into squares and then cutting the squares diagonally to get triangles. The glass strips are cleaned with acetone before breaking to prevent any trace of grease. Knives with concave edge and irregular or partly broken edges are discarded. Glass knives should be used within days of their having been made. During this period they must be stored in a dust free container with their cutting edge upwards.

Filling the boat

To allow the sections to be cut without sticking to the knife edge, a trough (boat) is placed behind the edge. This is filled with water for collecting sections. The boat is either made from tape or from thin metal. It should be placed at the same height as the knife edge. The lower part of the boat is sealed with the knife to make it water tight seal using either wax or nail polish.

Preparation of semithin (thick) sections for optical microscopy (survey sections)

Before one proceeds with ultrathin sectioning, thick sections (0.5 to 2.0μm) are cut for scanning the tissue under an optical microscope. This enables the electron microscopist to determine the state of the embedded specimen, the fixation, selection of the area for ultrathin sectioning, the size and position of the cutting face for final trimming. The semithin sections are floated in water or 10% acetone. These are lifted with a glass-rod or a thin brush and placed on a clean glass slide. The slide is placed on a hot plate at about 80°C and dried. Methylene blue / azure II and / or toluidene blue are some of the stains commonly used for staining.

- Methylene blue / Azure II: Prepared by mixing equal parts of 1%
- Methylene blue in 1% Borax and 1% Azure II.
- Toluidene blue : 1% solution in 1% Borax
- Stain sections for 0.5 to 2.0 min (depending on the thickness). Wash thoroughly in running water and then in distilled water. Dry and mount cover-slip with DPX.
- Observe under the light microscope.

Preparation of ultrathin sections

After scanning the sections under the optical microscope, the area to be examined under the TEM is selected and the blocks are further trimmed (hand trimming is advised). Use fresh glass knife for cutting ultrathin sections. The ultrathin sections show interference colours on the liquid in the trough. This makes it possible to determine the thickness of the sections.

Gray: 60 nm (600λ): optimal for high resolution work.

Silver: 60 – 90 nm: ideal for most purposes.

Gold: 90 – 150 nm: useful for low magnification and autoradiography.

Purple, blue, green, and yellow: range from 150 – 320 nm: very thick sections and are suitable for scanning transmission electron microscopy.

The ultrathin sections floating on the water are stretched by exposing them to chloroform vapours. The sections are lifted from below on specially made metal grids. Copper grids are commonly used, however, grids made of nickel, stainless steel, gold and platinum are also available and are used for specific purposes.The grids are available as meshed or holed ones and grids of 100 to 300 mesh size are the most commonly used. One of the surfaces of the grid is shiny and the other matted. The sections are lifted using the matted surfaces of the grid which affords sticking of the sections firmly.

Staining ultrathin sections

To obtain a good contrast, a double staining method using uranyl acetate and lead citrate is routinely followed for histological studies.

- Uranyl acetate: Prepare a saturated solution by adding excess of uranyl acetate (Analar) to 10 ml of filtered 50% ethanol in a 15 ml centrifuge tube. Shake vigorously for 2 mins and spin down hard to allow the excess of uranyl acetate to settle down. The solution is ready for use. Stopper and store at 4°C.

- Lead citrate: Add one half pellet of sodium hydroxide to 12 ml of double distilled water in a 15 ml. centrifuge tube. Shake well to dissolve the NaOH. Add 50 mg of lead citrate (Analar) and shake well for 2 mins and centrifuge. The solution is ready for use. Stopper and store at 4°C.

Staining procedure

Filter or centrifuge the stains before use. Pipette out a small amount of uranyl acetate solution in a clean watch glass. Place the grid carrying the section down onto the stain. Place a wooden or card-board cover as the staining is effective when carried out in dark. Stain for 10-15 mins. Take each grid and wash in 2 lots of 50% ethanol and 2 lots of double distilled water with continuous agitation. Dry carefully on a filter paper (do not allow the section carrying surface of the grid to come in contact with the filter paper).

Place a few ml of lead citrate in a watch glass and place the grid onto the stain for 5 to 10 mins. Wash each grid briefly in 0.02 M sodium hydroxide and then in 2 lots of double distilled water. Dry the grids. Store the grids in grid boxes or a well covered petri-dish. The sections are ready for observation in the transmission electron microscope (TEM).

Photographic unit

A photographic unit is ancillary to an electron microscope laboratory and should be well equipped for production of high quality micrographs. This is very essential as all ultrastructural analyses and evaluations are carried out using the micrographs. Further, the acceptance of articles on ultrastructural studies in reputed scientific journals depends mainly on the quality of micrographs and hence production of electron micrographs of high quality are required.

Most of the equipment for photomicrography are often located in a 'Dark Room' in the vicinity of electron microscope laboratory and fitted with adequate safety-light arrangements. The loading of cameras with photographic films are carried out either under safety-light or in total darkness, to avoid photographic material getting exposed to light. Preparation of positive prints (using photographic paper) using the negatives (photographic film) is also carried out under safety-light. The 'Dark Room' should also be provided with good washing facility (with taps for running water), as the resultant photographic products need extensive washing after processing.

Basic equipment and other items

The most essential items are:

- Photographic enlarger for making prints.
- Equipment with camera for producing projection slides.
- Glazing machine to glaze and dry photographic prints
- Film developing tanks
- Trays for developing and fixing photographic prints.
- Film dryers.
- Trimmers, alarm stop-watches and adequate glass-ware.
- Thermo-regulated press for preparing plates.
- Adequate number of refrigerators for storing photographic material (films and paper) and chemicals.

Photomicrography in electron microscopy

The electron microscopes have in-built facilities for photomicrography such as cameras controlling the light setting and exposure time depending on the speed of the films (DIN / ASA) used. The condition for right light and exposure time is standardized and set for exposure.

TEMs often have inbuilt facility for using 35 mm plate or 70mm cameras. The films used are of very fine grain, slow speed and being so the cameras can be located with the film under safety light. The SEMs often have provision to use 35mm, 120mm or polaroid camera. The films used are often fast (coarse grains) films (125, 200 or 400 ASA) and their processing has to be carried out in total darkness. Following exposure, the films are developed, fixed and dried.

Routine method used for developing and fixing of exposed films and prints

35 mm film for TEM - Eastman Kodak fine grain release positive film non-perforated.

Cut film for plate camera for TEM - Agfa Geveart Scintia Film (6.5 x 9.0cm). Following exposure, the films are developed in D-19, high contrast developer, fixed in Agfa 301 fixer, washed in running water and dried.

Photographic prints - Various grades (soft, normal, hard and special) of papers are available and each of these are used depending on the kind of negative to get appropriate tone of the print. Exposed photographic papers are developed in Agfa 100 standard photopaper. Developer, fixed in Agfa 301 fixer washed in running water and glazed.

Preparation of developers and fixer

D-19 High contrast negative developer – Dissolved the substances in the order in which these are listed in 750 ml of tap water and make the final volume to 1000 ml. Store at 4°C in a brown bottle.

D-76 Fine-grain developer –

Metol	2 g
Sodium sulphite(anhydrous)	100g
Hydroquinone	5 g
Borax	2 g

Agfa 100 standard photopaper developer –

Metol	1 g
Sodium sulphite (anhydrous)	13 g
Hydroquinone	3 g
Sodium carbonate (anhydrous)	26 g
Potassium bromide	1 g

Agfa 301 fixer (acid fixing bath) –

Sodium thiosulphate	250 g
Sodium metabisulphite	13 g

Dissolve in 1000 ml tap water and stock at room temperature.

Procedure for developing and fixing

35 mm film for TEM – Develop in D-19 high-contrast negative developer at 20°C for 6 min. Wash (twice) in tap water and fix in Agfa 301 fixer (20°C - 24°C) for 15 mins. Wash in running tap water (20°C - 24°C) for 30 mins. Rinse in the 'Wetting Agent' (Agent A913) and dry.

Plate (cut) film for TEM – Develop in D-19 high-contrast negative developer at 18°C for 3.5 min. Wash (twice) in tap water and fix in Agfa 301 fixer (20°C - 24°C) for 30 mins. Wash in running tap water (20°C - 24°C) and rinse in the 'Wetting Agent' (Agent A913) and dry.

Energy dispersive x-ray micro analysis

Elemental quantification of ultrathin sections with electron probe x-ray micro analysis (EPMA) is generally based on the linear relationship between elemental concentration and the ratio of number of characteristics/continuum x-ray photons(Shuman *et. al.*, 1976; Silverberg, 1976; Kitazawa *et. al.*, 1983).

The EDAX system is under control of sophisticated software that, at a simple command from the operator, automatically collects an energy-dispersive X-ray spectrum, identifies the elements present, obtains X-ray count data, and presents all of the pertinent data in an appropriate display format (Fig. 23.1). The spectrum can be stored in disk memory and recalled as desired. Flexibility is central to the performance of the EDAX software. It easily adapts to fit virtually any user's needs. While peaks are identified and labelled automatically, a manual mode is available to allow the operator to

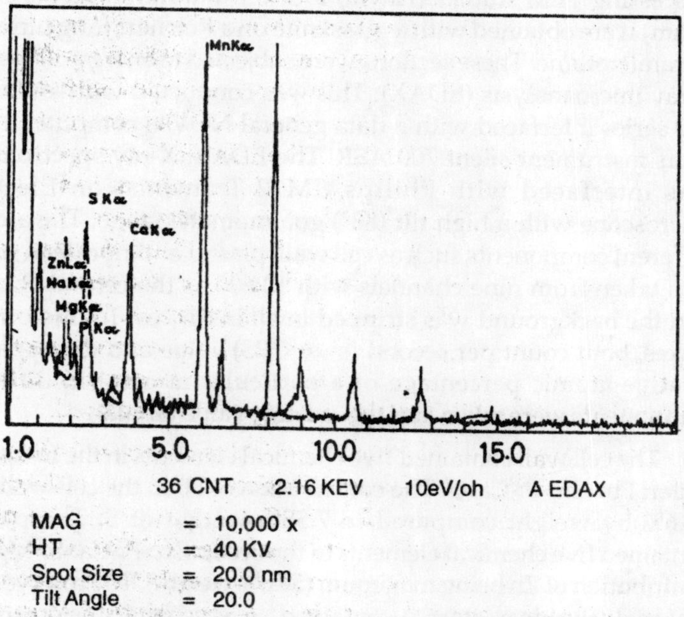

Fig 23.1: X-ray microanalysis of floridean starch of *H. musciformis*

identify peaks by elemental symbol, atomic number, or cursor position. This may be done exclusively by the operator or in conjunction with automatic mode. The operator may also modify search parameters to fine tune the search for a particular type of analysis. Elements that are not relevant to the analysis may be excluded from the search.

Energy Dispersive X-ray Micro-analysis system works on the principle of detection and characterization of characteristic X-rays generated in electron microscope. Whenever electron beam interacts with any elements or metals, X-rays are detected by a solid state detector. They are digitized, amplified and fed to a multichannel analyser, which sorts the X-ray signals depending upon their energy. The data generated were compared with existing data stored in computer programme and the computer provides the possibility of presence of various elements present in the samples.

Vegetative thalli of *Hypnea musciformis* (Wulf.) Lamour were cut into 2-3 mm length and fixed immediately in 3% glutaraldehyde for processing TEM studies (Hayat, 1986). The ulrathin sections (60-90nm) were obtained with a glassknife on a Reichert Jung ultracut-ultramicrotome. These sections were subjected to energy dispersive X-ray microanalysis (EDAX). This was conducted with an EDAX 700 series interfaced with a data general $NOVA_2$ computer and a texas instrument silent 700 ASR. The EDAX, X-ray spectrometer was interfaced with Philips CM12 Transmission Electron Microscope with a high tilt (80°) gonimometer stage. The area of different components such as cell wall, plastid and floridean starch was taken from nine channels with 12,600 as the central channel and the background was stripped in all instances. To find out the fluxes, both count per second (s^1 or CPS) value and the apparent relative atomic percentage of a particular mineral in different components were taken and the average value tabulated.

The cell wall contained five chemical elements in the following order: Pb>Zn>P>Ca>S. The contribution of Pb in the cell-wall was 34.6% by weight compared to 7.55% weight of S. The plastid contained five chemical elements in the order: Zn>P>Ca>S>Mg, the contribution of Zn being maximum (59.8% weight). The reserve food material (floridean starch) contained seven chemical elements in the order: Mn>Pb>Zn>Ca>P>Mg>S. Thus, the constituents of different chemical elements of cell-wall, plastid and floridean starch differed not only by quality but also in quantity.

The cell-wall

The pattern of chemical elements in the cell wall, plastid and floridean starch was found to be in the order such as Pb > Zn > P > Ca > S; Zn > P > Ca > S > Mg and Mn > Pb > Zn > Ca > P > Mg > S respectively (Table 23.1). The high values of Pb obtained on the cell-wall and floridean starch might be due to positive staining of their structures. The four chemical elements Zn, P, Ca and S were common in all the three components. Pb was present in both the cell-wall and floridean starch but absent in the plastid. Similarly the plastid and floridean starch contained Mg but it was absent in the cell-well. Among the chemical elements, Mn was specific to floridean starch.

Mn is found only in the floridean starch which is the characteristic photosynthetic reserve food substance of red algae. It's key role is nourishment of the nutritive material. Presence of Mn is needed for catalysing the photolysis of water and O_2 evolution associated with pigment system II.

Table 23.1: Compositional analysis of chemical elements present in the cell wall, plastid and floridean starch of *Hypnea musciformis*.

Chemical Elements	*Cell Wall (wt %)*	*Plastid (wt %)*	*Floridean Starch (wt %)*
Zn	26.150	59.839	20.659
P	20.637	26.996	2.202
S	7.549	4.804	0.957
Ca	11.061	7.454	10.694
Pb	34.603.	–	30.997
Mg	–	0.906	1.312
Mn	–	–	33.179

Mg is present in the plastid and floridean starch because Mg forms the nucleus for the porphyrin ring and hence its presence in the plastid is understandable. It's presence in floridean starch is also possible as it is the photosynthetic reserve food material in the red algae. Pb is present in the cell-wall and floridean starch only. For staining the sections uranyl acetate + lead citrate = positive staining (Reynolds, 1963) was used and hence there is every possibility for Pb to enter into the cell-wall and floridean starch. This study elucidates the various constituents within a cell which differed in their chemical composition, certain elements being specific to individual cell component. The above studies were made for the first time in the genus *Hypnea.*

Technique

Comparing specimens, or fields of view, it is obviously essential to operate under standard microscope conditions (beam current, spot size, intensity, etc.). It has been found to be advantageous, in addition, to use the Cu Kβ peak as an intensity standard, counting to the same peak height each time. This is obviously affected by the grid, but working close to the center of an opening gives reasonably reproducible results. Internal standards introduced in the embedding media may equally be used.

It is sometimes advantageous to check the calibration of the spectrometer before or after a vital experiment. This can be done using a molybdenum oxide specimen on a copper grid, which gives a series of peaks covering most of the spectrum. Sometimes, many superfluous peaks are present which can obscure useful information.

One source of these is, of course, the stains used in preparation and it is necessary to consider modifications to the technique carefully to avoid heavy metal stains.

Conclusion

The major concerns for elemental analysis with the energy-dispersive x-ray system are the limits of detectability and the area of analysis. The first depends largely on the energy-dispersive system and specimen preparation and the second also involves the microscope performance. In particular, a small, intense electron beam must be used to obtain meaningful results. The combination of a high performance transmission microscope with a scanning system and highly efficient energy-dispersive x-ray analysis system is the key to useful analysis of small areas of biological samples. This instrumentation is now available and its applications are being explored. Although further refinements are being made in instrumentation and techniques, x-ray microanalysis is now a reality for biological as well as material science applications.

References

Hayat, M.A. 1986. *Basic Techniques for Transmission Electron Microscopy*. Academic Press Inc., London. pp. 411.

Kitazawa, T. H., Somlyo, A.V. and Somlyo, A.P. 1983. Quantitative electron probe analysis; problems and solutions. *Ultramicroscopy* **11**: 251-262.

Reynolds, E.S. 1963. The use of lead citrate at high pH as an electron opaque stain in electron microscopy. *J. Cell Biol.* **17**: 208-212.

Shuman, H., Somlyo, A.V. and Somlyo, A.P. 1976. Quantitative electron probe microanalysis of biological thin sections: methods and validity. *Ultramicroscopy* **1**: 317-339.

Silverberg, B.A. 1976. Cadmium-induced ultrastructural changes in mitochondria of fresh water green algae. *Phycologia* **15**: 155-159.

Chapter 24

PHOTOBIOREACTORS FOR CULTURE OF *ARTHROSPIRA* AND OTHER MICROALGAE

N.Jeeji Bai

Abstract

Mass multiplication of useful algal organisms under controlled cultural conditions is required for their exploitation at a commercial stage. During the last about half a century, many types of photo-bioreactors have been developed for microalgal cultures. These photo-bioreactors have many advantages over routine culturing but for their cost and constraints in operation.

Key words: *Arthrospira*, mass culture, microalgae, photobioreactors.

Introduction

Photobioreactors were first developed for *Chlorella* cultures during 1951-1953. They were abandoned mainly due to the high cost involved for cooling. These were revived after 30 years, by research groups in France, Italy and United Kingdom followed by a few others. These have been developed for microalgal cultures because of several advantages such as:

- Being tubes or channels of relatively narrow gauge they are efficient collectors of solar radiation.
- Areal yields are high because of high culture densities achieved.

- Being closed systems, sterility can be maintained.
- Good mixing and CO_2 conversion efficiency ensure good photosynthetic production.
- Contamination is low.
- Monitoring and control of parameters is easy.

Tubular photobioreactors

The initial designs comprised tubular reactors kept horizontally over a flat surface. The French design (Gudin and Chaumont, 1991) developed for *Porphyridium* cultures consisted of 2 series of tubes, with those in upper level containing algal cultures and those at lower level serving as cooling systems. The upper tubes either float or are submerged in the lower ones. The evolution of different designs by the Italian group in Florence is given by Torzillo (1997). The first Italian design was made of polyethylene tubes with bends made of PVC pipes. At the bends, narrow tubes serve for release of excess O_2 developed due to the high rate of photosynthesis. A modified design involves the use of plexiglass tubes for giving mechanical strength. Inner diameters of 7.4 cm and 13 cm have been tried. The maximum length of loop used is 500 m on 80 m^2 area with 8m^3 volume. It consists of a feeding tank and a siphon system for a constant flow velocity of 0.26 m s-1. The yields of *Arthrospira* in this open-air system in summer is 33 t dry wt. ha^{-1} yr^{-1} (Torzillo *et. al.*, 1986). This enhancement in yield is attributed to a better temperature profile in cultures over the diurnal cycle and seasonal changes.

A second design of narrow (2.6 cm) interconnected tube system at two levels for efficient trapping of solar energy with separate air lifts for mixing and degassing was developed for *Arthrospira* cultures. This 145 litre capacity open reactor with a surface volume ratio of 49 m^{-1} yielded a net productivity of 28 g m^{-2} d^{-1} in summer. After several trials, they have arrived at an ideal system, consisting of 3.0 cm diameter and maximum length of 100 m. and flow speed of 0.3 m s^{-1} with a retention time cycle of 6 min. for *Arthrospira* cultures (Torzillo, 1997).

Tomaselli *et. al.* (1993) made short-term experiments in open-air conditions in their small tubular bioreactors, on biomass production, nitrogen uptake and lipid and carbohydrate biosynthesis, in a strain of *Arthrospira* grown on nitrate and ammonia and found slightly higher levels with ammonia. Net productivity in thermostated reactors over the summer months was much higher at 35°C as compared to 25°C.

Richmond *et. al.* (1993) have used transparent thin-walled extruded tubes made of polycarbonate, running parallel and connected by manifolds. An air-lift pump and a gas separator serve for mixing and degassing. As compared to loops, this system reduces minor heat losses and is more suitable for scaling up. A system consisting of a series of inclined tubes, interconnected with a solar heated circulation system for rural adoption was developed in France for *Arthrospira* cultures by Fox (1998).

Harker *et. al.* (1996) tried a laboratory scale sterile glass column bubble air-lift system of 30 litre capacity for growing *Haematococcus* with a temperature range of 14 – 27°C and an external illumination of 50 μ mol photon m^{-2} s^{-1} Filtered air was left in from a bottom sparger at 1.5 – 3.0 l min^{-1} . In this system, astaxanthin yield of 2.7% on dry wt. basis was obtained (Harker *et. al.*, 1996).

Vonshak (1988) reported about two solar reactors for culture of *Porphyridium*, the one consisting of an assembly of glass tubes of 3 cm diameter and 1 m length, and a larger system of thin polyethylene tubes of 6 cm diameter and 20 m length with a circulation pump yielding a linear flow speed of 20 cm s^{-1} . A carbonation column served to supply CO_2 and remove O_2. Their calculated biomass yield was as high as 42 – 76 tonnes ha^{-1} yr^{-1}.

Molina-Grima *et. al.* (1995) used photobioreactors made of transparent plexiglass tubes of 80.8 m length and 3 cm outer and 2.6 cm inner diameter for the culture of *Phaeodactylum tricornutum*. The tube loops were placed horizontally in open air with air-lift system for circulation and degassing. The 50 litre culture in 8.6 m^2 area was cooled thermostatically in a water bath to maintain the temperature at 21±1°C. They obtained a maximum biomass productivity of 0.78 gl^{-1} d^{-1} on 32nd day. The maximum EPA (eicosapentaenoic acid) concentration was 423 mgl^{-1} with a generation rate of 21 mg l^{-1} d^{-1} during linear growth phase.

The only large-scale tubular open-air system called Aquasearch Growth Module is reported by Olaizola (2000) for *Haematococcus* culture. It consists of parallel tubes of 0.18 – 0.41 m diameter laid on impermeable surface of about 100 m^2 area with a total capacity of

25,000 litre. The temperature is maintained at 15 – 25°C with deep sea water cooling. A biomass yield of 34 – 100 g m^{-2} with a productivity of 4 – 19 g m^{-2} d^{-1} with astaxanthin contents of 0.7-3.4% on a dry wt. basis was reported (Olaizola, 2000).

Yet another variation of tubular reactor is the vertical 'Biocoil" design. It consists of LDPE tube coiled around a circular framework. This was first developed for *Chlorella* with a yield of 10 g l^{-1} (Torzillo, 1997). A large version of 700 litre capacity with 2.4 cm diameter tubing has been developed and tested outdoors in Australia for *Isochrysis galbana* and *Tetraselmis chuii* with productivities of 1.2 g and 1.0 g l^{-1} d^{-1} respectively (Borowitzka,1997).

Flat plate reactors

A more sophisticated and efficient system consists of alveolar panels or plates. These flat reactors consist of transparent polyvinyl chloride, polycarbonate or polymethyl methacrylate sheets internally partitioned to form rectangular channels called alveoli. Commercial plates are of 4.5 to 40 mm thickness. As compared to tubular reactors these have an advantage of orientation according to the direction of solar radiation and their volumetric yields are much higher.

The French design of 6 m long, 0.25 m wide and 40 mm deep panel with triple wall design with two rows of alveoli with lower ones serving for cooling purpose was developed for *Chlorella* cultures in green house with 23-24 g m^{-2} d^{-1} yields in summer (Ramos de Ortega and Roux, 1986).

The Italian group also tested horizontal panels with pump circulation for *Arthrospira* cultures. They then patented a vertical alveolar panel with air bubbling from bottom for mixing and degassing (Tredici and Zittelli, 1997). Their device with a high surface to volume ratio of 80 m^{-1} yielded an areal productivity of 24.3 g m^{-2} d^{-1} at an air flow rate of 1.2 l/l of cell suspension/min. in summer. Although inclined plate yielded higher productivity, the vertical panel yielded higher areal biomass density.

A slightly modified system has been developed for *Arthrospira* culture by Pulz (1994) in Germany where several interconnected vertical panels with channels running horizontally at 20 cm intervals, gave a productivity of 1.3 g l^{-1} d^{-1} and an areal yield of 28 g m^{-2} d^{-1} on surface area basis and 120 g m^{-2} d^{-1} on occupied area basis.

Outdoor mass culture of *Porphyridium* in flat-plate glass reactors has been reported recently by the study group of Israel (Singh *et. al.*, 2000). The maximal volumetric concentration of soluble sulphated polysaccharide of 1.32 g l^{-1} was obtained in winter with 1 cm light path (LP) reactor and the highest areal concentration of 60 g m^{-2} was obtained with 20 cm LP reactor. Irrespective of the LP of the reactors, the sugar composition of the sulphated polyaccharides was the same. Cultivation through an entire year showed 20 cm LP to be optimal for maximal yields.

Conclusions

Both tubular as well as flat plate photobioreactors have a few disadvantages and constraints in operation.

1. The narrow gauge tubes and panels need frequent cleaning of algae sticking to the sides of tubes and channels.
2. In the hot tropics, efficient cooling systems are needed throughout the year and this escalates the working cost enormously.
3. The system needs degassing at frequent intervals to keep the oxygen level below 30 mg l^{-1}.
4. In these systems, the night biomass loss is high. In flat plate collectors, it is estimated to be as much as 30% of gross production.
5. The system collapses below optimal areal density, i.e., standing crop per unit surface.
6. The change in viscosity of culture with age and type of organism needs to be carefully studied for maintaining flow rate. For example, production of mucilage in *Arthrospira*, polysaccharide in *Porphyridium* and glycerol in *Dunaliella*.

Finally, it is to be noted that these systems are more expensive than the conventional raceway ponds and are suitable only for high value products. They are useful for temperate zones with long winters.

References

Borowitzka, M.A. 1997. Microalgae for aquaculture: Opportunities and constraints. *J. Appl. Phycol.* **9**: 393-401.

Fox, R.D. 1998. *Spirulina* farms- micro to macro. In: *Cyanobacterial Biotechnology*. Edited by Subramanian, G., Kaushik, B. D. and Venkataraman, G.S. Oxford & IBH Publ. Co. Pvt. Ltd., New Delhi, pp. 259-265.

Gudin, C. and Chaumont, D. 1991. The key problem of microalgal mass production in closed photobioreactors. *Biores. Technol.* **38**: 145.

Harker, M., Tsavalos, A.J. and Young, A.J. 1996. Autotrophic growth and carotenoid production in *Haematococcus pluvialis* in a 30 litre air-lift photobioreactor. *J. Ferment. Bioeng.* **82**: 713-718.

Molina-Grima, E., Sanchez-Perez, J.A., Garcia-Camacho, F., Fernandez-Savila, J.M., Acien-Fernandez, F.G. and Urda-Cardona, J. 1995. Biomass and eicosapentaenoic acid productivities from an outdoor batch culture of *Phaeodactylum tricornutum* UTEX 640 in an airlift tubular photobioreactor. *Appl. Microbiol. Biotechnol.* **42:** 658-663.

Olaizola, M. 2000. Commercial production of astaxanthin from *Haematococcus pluvialis* using 25,000 liter outdoor photobioreactors. *J. Appl. Phycol.* **12:** 499-506.

Pulz, O.1994. Open-air and semi-closed cultivation systems for the mass cultivation of microalgae. In: *Algal Biotechnology in Asia-Pacific Region*. Edited by Phang, S.M., Lee, K, Borowitzka, M.A. and Whitton, B. Institute of Advanced Studies, University Malaya, Kuala Lumpur, pp.113-117.

Ramos de Ortega, A. and Roux, J.C. 1986. Production of *Chlorella* biomass in different types of flat bioreactors in temperature zones. *Biomass* **10:** 141-156.

Richmond, A., Boussiba, S., Vonshak,A. and Kopel, R. 1993. A new tubular reactor for mass production of microalgae outdoors. *J. Appl. Phycol.* **5**: 327-332.

Singh, S., Arad, S. and Richmond, A. 2000. Extracellular polysaccharide production in outdoor mass cultures of *Porphyridium* sp. in flat plate glass reactors. *J. Appl. Phycol.* **12**: 269-275.

Tomaselli, L., Giovannetti, L. and Torzillo, G. 1993. Physiology of stress response in *Spirulina* sp. *Bull. Inst. Oceanogr. Monaco* **12**: 65-75.

Torzillo, G. 1997. Tubular bioreactors. In: *Spirulina platensis* (*Arthrospira*). Edited by Vonshak, A. Taylor & Francis, London. pp. 101-105.

Torzillo., G., Pushparaj, B., Bocci, F., Balloni, W., Materassi, R. and Florenzano, G. 1986. Production of *Spirulina* biomass in closed photobioreactors. *Biomass* **11**: 61-64.

Tredici, M.R. and Zittelli,G.C. 1997. Cultivation of Spirulina (*Arthrospira*) in flat plate reactors. In: *Spirulina platensis* (*Arthrospira*), Edited by Vonshak, A., Taylor & Francis, London. pp.117-130.

Vonshak, A. 1988. *Porphyridium*. In: *Micro-algal Biotechnology*. Edited by Borowitzka, M.A. and Borowitzka, L.J., Cambridge Univ. Press, N.Y. pp.122-134.

INDEX

A